Data Structures

and Algorithms

Data Structures

and Algorithms

ALFRED V. AHO

Bell Laboratories
Murray Hill, New Jersey

JOHN E. HOPCROFT

Cornell University
Ithaca, New York

JEFFREY D. ULLMAN

Stanford University
Stanford, California

▲▼ ADDISON-WESLEY PUBLISHING COMPANY
Reading, Massachusetts • Menlo Park, California
London • Amsterdam • Don Mills, Ontario • Sydney

This book is in the
ADDISON-WESLEY SERIES IN
COMPUTER SCIENCE AND INFORMATION PROCESSING

Michael A. Harrison
Consulting Editor

Library of Congress Cataloging in Publication Data

Aho, Alfred V.
    Data structures and algorithms.

    1. Data structures (Computer science)   2. Algorithms.
I. Hopcroft, John E., 1939–   . II. Ullman,
Jeffrey D., 1942–   . III. Title.
QA76.9.D35A38   1982   001.64   82-11596
ISBN 0-201-00023-7

Reproduced by Addison-Wesley from camera-ready copy supplied by the authors.

Reprinted with corrections, June 1983

ISBN 0-201-00023-7
FGHIJ-DO-8987654

Preface

This book presents the data structures and algorithms that underpin much of today's computer programming. The basis of this book is the material contained in the first six chapters of our earlier work, *The Design and Analysis of Computer Algorithms*. We have expanded that coverage and have added material on algorithms for external storage and memory management. As a consequence, this book should be suitable as a text for a first course on data structures and algorithms. The only prerequisite we assume is familiarity with some high-level programming language such as Pascal.

We have attempted to cover data structures and algorithms in the broader context of solving problems using computers. We use abstract data types informally in the description and implementation of algorithms. Although abstract data types are only starting to appear in widely available programming languages, we feel they are a useful tool in designing programs, no matter what the language.

We also introduce the ideas of step counting and time complexity as an integral part of the problem solving process. This decision reflects our long-held belief that programmers are going to continue to tackle problems of progressively larger size as machines get faster, and that consequently the time complexity of algorithms will become of even greater importance, rather than of less importance, as new generations of hardware become available.

The Presentation of Algorithms

We have used the conventions of Pascal to describe our algorithms and data structures primarily because Pascal is so widely known. Initially we present several of our algorithms both abstractly and as Pascal programs, because we feel it is important to run the gamut of the problem solving process from problem formulation to a running program. The algorithms we present, however, can be readily implemented in any high-level programming language.

Use of the Book

Chapter 1 contains introductory remarks, including an explanation of our view of the problem-to-program process and the role of abstract data types in that process. Also appearing is an introduction to step counting and "big-oh" and "big-omega" notation.

Chapter 2 introduces the traditional list, stack and queue structures, and the mapping, which is an abstract data type based on the mathematical notion of a function. The third chapter introduces trees and the basic data structures that can be used to support various operations on trees efficiently.

Chapters 4 and 5 introduce a number of important abstract data types that are based on the mathematical model of a set. Dictionaries and priority queues are covered in depth. Standard implementations for these concepts, including hash tables, binary search trees, partially ordered trees, tries, and 2-3 trees are covered, with the more advanced material clustered in Chapter 5.

Chapters 6 and 7 cover graphs, with directed graphs in Chapter 6 and undirected graphs in 7. These chapters begin a section of the book devoted more to issues of algorithms than data structures, although we do discuss the basics of data structures suitable for representing graphs. A number of important graph algorithms are presented, including depth-first search, finding minimal spanning trees, shortest paths, and maximal matchings.

Chapter 8 is devoted to the principal internal sorting algorithms: quicksort, heapsort, binsort, and the simpler, less efficient methods such as insertion sort. In this chapter we also cover the linear-time algorithms for finding medians and other order statistics.

Chapter 9 discusses the asymptotic analysis of recursive procedures, including, of course, recurrence relations and techniques for solving them.

Chapter 10 outlines the important techniques for designing algorithms, including divide-and-conquer, dynamic programming, local search algorithms, and various forms of organized tree searching.

The last two chapters are devoted to external storage organization and memory management. Chapter 11 covers external sorting and large-scale storage organization, including B-trees and index structures.

Chapter 12 contains material on memory management, divided into four subareas, depending on whether allocations involve fixed or varying sized blocks, and whether the freeing of blocks takes place by explicit program action or implicitly when garbage collection occurs.

Material from this book has been used by the authors in data structures and algorithms courses at Columbia, Cornell, and Stanford, at both undergraduate and graduate levels. For example, a preliminary version of this book was used at Stanford in a 10-week course on data structures, taught to a population consisting primarily of Juniors through first-year graduate students. The coverage was limited to Chapters 1-4, 9, 10, and 12, with parts of 5-7.

Exercises

A number of exercises of varying degrees of difficulty are found at the end of each chapter. Many of these are fairly straightforward tests of the mastery of the material of the chapter. Some exercises require more thought, and these have been singly starred. Doubly starred exercises are harder still, and are suitable for more advanced courses. The bibliographic notes at the end of each chapter provide references for additional reading.

Acknowledgments

We wish to acknowledge Bell Laboratories for the use of its excellent UNIX™-based text preparation and data communication facilities that significantly eased the preparation of a manuscript by geographically separated authors. Many of our colleagues have read various portions of the manuscript and have given us valuable comments and advice. In particular, we would like to thank Ed Beckham, Jon Bentley, Kenneth Chu, Janet Coursey, Hank Cox, Neil Immerman, Brian Kernighan, Steve Mahaney, Craig McMurray, Alberto Mendelzon, Alistair Moffat, Jeff Naughton, Kerry Nemovicher, Paul Niamkey, Rob Pike, Chris Rouen, Maurice Schlumberger, Stanley Selkow, Chengya Shih, Bob Tarjan, W. Van Snyder, Peter Weinberger, and Anthony Yeracaris for helpful suggestions. Finally, we would like to give our warmest thanks to Mrs. Claire Metzger for her expert assistance in helping prepare the manuscript for typesetting.

A.V.A.
J.E.H.
J.D.U.

Contents

Data Structures

and Algorithms

CHAPTER 1

Design and Analysis of Algorithms

There are many steps involved in writing a computer program to solve a given problem. The steps go from problem formulation and specification, to design of the solution, to implementation, testing and documentation, and finally to evaluation of the solution. This chapter outlines our approach to these steps. Subsequent chapters discuss the algorithms and data structures that are the building blocks of most computer programs.

1.1 From Problems to Programs

Half the battle is knowing what problem to solve. When initially approached, most problems have no simple, precise specification. In fact, certain problems, such as creating a "gourmet" recipe or preserving world peace, may be impossible to formulate in terms that admit of a computer solution. Even if we suspect our problem can be solved on a computer, there is usually considerable latitude in several problem parameters. Often it is only by experimentation that reasonable values for these parameters can be found.

If certain aspects of a problem can be expressed in terms of a formal model, it is usually beneficial to do so, for once a problem is formalized, we can look for solutions in terms of a precise model and determine whether a program already exists to solve that problem. Even if there is no existing program, at least we can discover what is known about this model and use the properties of the model to help construct a good solution.

Almost any branch of mathematics or science can be called into service to help model some problem domain. Problems essentially numerical in nature can be modeled by such common mathematical concepts as simultaneous linear equations (e.g., finding currents in electrical circuits, or finding stresses in frames made of connected beams) or differential equations (e.g., predicting population growth or the rate at which chemicals will react). Symbol and text processing problems can be modeled by character strings and formal grammars. Problems of this nature include compilation (the translation of programs written in a programming language into machine language) and information retrieval tasks such as recognizing particular words in lists of titles owned by a library.

Algorithms

Once we have a suitable mathematical model for our problem, we can attempt to find a solution in terms of that model. Our initial goal is to find a solution in the form of an *algorithm,* which is a finite sequence of instructions, each of which has a clear meaning and can be performed with a finite amount of effort in a finite length of time. An integer assignment statement such as $x := y + z$ is an example of an instruction that can be executed in a finite amount of effort. In an algorithm instructions can be executed any number of times, provided the instructions themselves indicate the repetition. However, we require that, no matter what the input values may be, an algorithm terminate after executing a finite number of instructions. Thus, a program is an algorithm as long as it never enters an infinite loop on any input.

There is one aspect of this definition of an algorithm that needs some clarification. We said each instruction of an algorithm must have a "clear meaning" and must be executable with a "finite amount of effort." Now what is clear to one person may not be clear to another, and it is often difficult to prove rigorously that an instruction can be carried out in a finite amount of time. It is often difficult as well to prove that on any input, a sequence of instructions terminates, even if we understand clearly what each instruction means. By argument and counterargument, however, agreement can usually be reached as to whether a sequence of instructions constitutes an algorithm. The burden of proof lies with the person claiming to have an algorithm. In Section 1.5 we discuss how to estimate the running time of common programming language constructs that can be shown to require a finite amount of time for their execution.

In addition to using Pascal programs as algorithms, we shall often present algorithms using a *pseudo-language* that is a combination of the constructs of a programming language together with informal English statements. We shall use Pascal as the programming language, but almost any common programming language could be used in place of Pascal for the algorithms we shall discuss. The following example illustrates many of the steps in our approach to writing a computer program.

Example 1.1. A mathematical model can be used to help design a traffic light for a complicated intersection of roads. To construct the pattern of lights, we shall create a program that takes as input a set of permitted turns at an intersection (continuing straight on a road is a "turn") and partitions this set into as few groups as possible such that all turns in a group are simultaneously permissible without collisions. We shall then associate a phase of the traffic light with each group in the partition. By finding a partition with the smallest number of groups, we can construct a traffic light with the smallest number of phases.

For example, the intersection shown in Fig. 1.1 occurs by a watering hole called JoJo's near Princeton University, and it has been known to cause some navigational difficulty, especially on the return trip. Roads C and E are one-way, the others two way. There are 13 turns one might make at this intersection. Some pairs of turns, like *AB* (from *A* to *B*) and *EC*, can be carried out simultaneously, while others, like *AD* and *EB*, cause lines of traffic to cross and therefore cannot be carried out simultaneously. The light at the intersection must permit turns in such an order that *AD* and *EB* are never permitted at the same time, while the light might permit *AB* and *EC* to be made simultaneously.

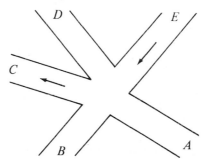

Fig. 1.1. An intersection.

We can model this problem with a mathematical structure known as a graph. A *graph* consists of a set of points called *vertices*, and lines connecting the points, called *edges*. For the traffic intersection problem we can draw a graph whose vertices represent turns and whose edges connect pairs of vertices whose turns cannot be performed simultaneously. For the intersection of Fig. 1.1, this graph is shown in Fig. 1.2, and in Fig. 1.3 we see another representation of this graph as a table with a 1 in row *i* and column *j* whenever there is an edge between vertices *i* and *j*.

The graph can aid us in solving the traffic light design problem. A *coloring* of a graph is an assignment of a color to each vertex of the graph so that no two vertices connected by an edge have the same color. It is not hard to see that our problem is one of coloring the graph of incompatible turns using as few colors as possible.

The problem of coloring graphs has been studied for many decades, and the theory of algorithms tells us a lot about this problem. Unfortunately, coloring an arbitrary graph with as few colors as possible is one of a large class of problems called "NP-complete problems," for which all known solutions are essentially of the type "try all possibilities." In the case of the coloring problem, "try all possibilities" means to try all assignments of colors to vertices using at first one color, then two colors, then three, and so on, until a legal coloring is found. With care, we can be a little speedier than this, but it is generally believed that no algorithm to solve this problem can be substantially more efficient than this most obvious approach.

We are now confronted with the possibility that finding an optimal solution for the problem at hand is computationally very expensive. We can adopt

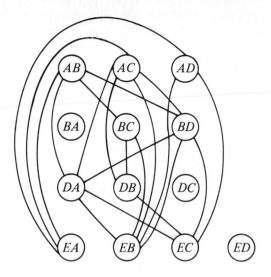

Fig. 1.2. Graph showing incompatible turns.

	AB	AC	AD	BA	BC	BD	DA	DB	DC	EA	EB	EC	ED
AB					1	1	1			1			
AC					1	1	1			1	1		
AD										1	1	1	
BA													
BC	1							1			1		
BD	1	1					1				1	1	
DA	1	1				1					1	1	
DB		1			1							1	
DC													
EA	1	1	1										
EB		1	1		1	1	1						
EC			1			1	1	1					
ED													

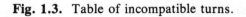

Fig. 1.3. Table of incompatible turns.

one of three approaches. If the graph is small, we might attempt to find an optimal solution exhaustively, trying all possibilities. This approach, however, becomes prohibitively expensive for large graphs, no matter how efficient we try to make the program. A second approach would be to look for additional information about the problem at hand. It may turn out that the graph has some special properties, which make it unnecessary to try all possibilities in finding an optimal solution. The third approach is to change the problem a little and look for a good but not necessarily optimal solution. We might be happy with a solution that gets close to the minimum number of colors on small graphs, and works quickly, since most intersections are not even as complex as Fig. 1.1. An algorithm that quickly produces good but not necessarily optimal solutions is called a *heuristic*.

One reasonable heuristic for graph coloring is the following "greedy" algorithm. Initially we try to color as many vertices as possible with the first color, then as many as possible of the uncolored vertices with the second color, and so on. To color vertices with a new color, we perform the following steps.

1.   Select some uncolored vertex and color it with the new color.
2.   Scan the list of uncolored vertices. For each uncolored vertex, determine whether it has an edge to any vertex already colored with the new color. If there is no such edge, color the present vertex with the new color.

This approach is called "greedy" because it colors a vertex whenever it can, without considering the potential drawbacks inherent in making such a move. There are situations where we could color more vertices with one color if we were less "greedy" and skipped some vertex we could legally color. For example, consider the graph of Fig. 1.4, where having colored vertex 1 red, we can color vertices 3 and 4 red also, provided we do not color 2 first. The greedy algorithm would tell us to color 1 and 2 red, assuming we considered vertices in numerical order.

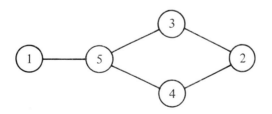

Fig. 1.4. A graph.

As an example of the greedy approach applied to Fig. 1.2, suppose we start by coloring *AB* blue. We can color *AC*, *AD*, and *BA* blue, because none of these four vertices has an edge in common. We cannot color *BC* blue because there is an edge between *AB* and *BC*. Similarly, we cannot color *BD*,

DA, or *DB* blue because each of these vertices is connected by an edge to one or more vertices already colored blue. However, we can color *DC* blue. Then *EA*, *EB*, and *EC* cannot be colored blue, but *ED* can.

Now we start a second color, say by coloring *BC* red. *BD* can be colored red, but *DA* cannot, because of the edge between *BD* and *DA*. Similarly, *DB* cannot be colored red, and *DC* is already blue, but *EA* can be colored red. Each other uncolored vertex has an edge to a red vertex, so no other vertex can be colored red.

The remaining uncolored vertices are *DA*, *DB*, *EB*, and *EC*. If we color *DA* green, then *DB* can be colored green, but *EB* and *EC* cannot. These two may be colored with a fourth color, say yellow. The colors are summarized in Fig. 1.5. The "extra" turns are determined by the greedy approach to be compatible with the turns already given that color, as well as with each other. When the traffic light allows turns of one color, it can also allow the extra turns safely.

color	turns	extras
blue	*AB*, *AC*, *AD*, *BA*, *DC*, *ED*	—
red	*BC*, *BD*, *EA*	*BA*, *DC*, *ED*
green	*DA*, *DB*	*AD*, *BA*, *DC*, *ED*
yellow	*EB*, *EC*	*BA*, *DC*, *EA*, *ED*

Fig. 1.5. A coloring of the graph of Fig. 1.2.

The greedy approach does not always use the minimum possible number of colors. We can use the theory of algorithms again to evaluate the goodness of the solution produced. In graph theory, a *k-clique* is a set of k vertices, every pair of which is connected by an edge. Obviously, k colors are needed to color a k-clique, since no two vertices in a clique may be given the same color.

In the graph of Fig. 1.2 the set of four vertices *AC*, *DA*, *BD*, *EB* is a 4-clique. Therefore, no coloring with three or fewer colors exists, and the solution of Fig. 1.5 is optimal in the sense that it uses the fewest colors possible. In terms of our original problem, no traffic light for the intersection of Fig. 1.1 can have fewer than four phases.

Therefore, consider a traffic light controller based on Fig. 1.5, where each phase of the controller corresponds to a color. At each phase the turns indicated by the row of the table corresponding to that color are permitted, and the other turns are forbidden. This pattern uses as few phases as possible. □

Pseudo-Language and Stepwise Refinement

Once we have an appropriate mathematical model for a problem, we can formulate an algorithm in terms of that model. The initial versions of the algorithm are often couched in general statements that will have to be refined subsequently into smaller, more definite instructions. For example, we described
the greedy graph coloring algorithm in terms such as "select some uncolored
vertex." These instructions are, we hope, sufficiently clear that the reader
grasps our intent. To convert such an informal algorithm to a program, however, we must go through several stages of formalization (called *stepwise
refinement*) until we arrive at a program the meaning of whose steps are formally defined by a language manual.

Example 1.2. Let us take the greedy algorithm for graph coloring part of the
way towards a Pascal program. In what follows, we assume there is a graph
G, some of whose vertices may be colored. The following program *greedy*
determines a set of vertices called *newclr*, all of which can be colored with a
new color. The program is called repeatedly, until all vertices are colored.
At a coarse level, we might specify *greedy* in pseudo-language as in Fig. 1.6.

```
        procedure greedy ( var G: GRAPH; var newclr: SET );
                { greedy assigns to newclr a set of vertices of G that may be
                    given the same color }
                begin
(1)                 newclr := ∅; †
(2)                 for each uncolored vertex v of G do
(3)                     if v is not adjacent to any vertex in newclr then begin
(4)                         mark v colored;
(5)                         add v to newclr
                    end
        end;  { greedy }
```

Fig. 1.6. First refinement of *greedy* algorithm.

We notice from Fig. 1.6 certain salient features of our pseudo-language.
First, we use boldface lower case keywords corresponding to Pascal reserved
words, with the same meaning as in standard Pascal. Upper case types such
as GRAPH and SET‡ are the names of "abstract data types." They will be
defined by Pascal type definitions and the operations associated with these
abstract data types will be defined by Pascal procedures when we create the
final program. We shall discuss abstract data types in more detail in the next
two sections.

The flow-of-control constructs of Pascal, like **if**, **for**, and **while**, are

---

† The symbol ∅ stands for the empty set.
‡ We distinguish the abstract data type SET from the built-in **set** type of Pascal.

available for pseudo-language statements, but conditionals, as in line (3), may be informal statements rather than Pascal conditional expressions. Note that the assignment at line (1) uses an informal expression on the right. Also, the **for**-loop at line (2) iterates over a set.

To be executed, the pseudo-language program of Fig. 1.6 must be refined into a conventional Pascal program. We shall not proceed all the way to such a program in this example, but let us give one example of refinement, transforming the **if**-statement in line (3) of Fig. 1.6 into more conventional code.

To test whether vertex v is adjacent to some vertex in *newclr*, we consider each member w of *newclr* and examine the graph G to see whether there is an edge between v and w. An organized way to make this test is to use *found*, a boolean variable to indicate whether an edge has been found. We can replace lines (3)−(5) of Fig. 1.6 by the code in Fig. 1.7.

```
        procedure greedy ( var G: GRAPH; var newclr: SET );
            begin
(1)             newclr := ∅;
(2)             for each uncolored vertex v of G do begin
(3.1)               found := false;
(3.2)               for each vertex w in newclr do
(3.3)                   if there is an edge between v and w in G then
(3.4)                       found := true;
(3.5)               if found = false then begin
                            { v is adjacent to no vertex in newclr }
(4)                         mark v colored;
(5)                         add v to newclr
                    end
            end
        end;  { greedy }
```

Fig. 1.7. Refinement of part of Fig. 1.6.

We have now reduced our algorithm to a collection of operations on two sets of vertices. The outer loop, lines (2)−(5), iterates over the set of uncolored vertices of G. The inner loop, lines (3.2)−(3.4), iterates over the vertices currently in the set *newclr*. Line (5) adds newly colored vertices to *newclr*.

There are a variety of ways to represent sets in a programming language like Pascal. In Chapters 4 and 5 we shall study several such representations. In this example we can simply represent each set of vertices by another abstract data type LIST, which here can be implemented by a list of integers terminated by a special value *null* (for which we might use the value 0). These integers might, for example, be stored in an array, but there are many other ways to represent LIST's, as we shall see in Chapter 2.

We can now replace the **for**-statement of line (3.2) in Fig. 1.7 by a loop, where *w* is initialized to be the first member of *newclr* and changed to be the next member, each time around the loop. We can also perform the same refinement for the **for**-loop of line (2) in Fig. 1.6. The revised procedure *greedy* is shown in Fig. 1.8. There is still more refinement to be done after Fig. 1.8, but we shall stop here to take stock of what we have done. □

```
procedure greedy ( var G: GRAPH; var newclr: LIST );
    { greedy assigns to newclr those vertices that may be
        given the same color }
var
    found: boolean;
    v, w: integer;
begin
    newclr := ∅;
    v := first uncolored vertex in G;
    while v <> null do begin
        found := false;
        w := first vertex in newclr;
        while w <> null do begin
            if there is an edge between v and w in G then
                found := true;
            w := next vertex in newclr;
        end;
        if found = false do begin
            mark v colored;
            add v to newclr;
        end;
        v := next uncolored vertex in G
    end
end; { greedy }
```

Fig. 1.8. Refined *greedy* procedure.

Summary

In Fig. 1.9 we see the programming process as it will be treated in this book. The first stage is modeling using an appropriate mathematical model such as a graph. At this stage, the solution to the problem is an algorithm expressed very informally.

At the next stage, the algorithm is written in pseudo-language, that is, a mixture of Pascal constructs and less formal English statements. To reach that stage, the informal English is replaced by progressively more detailed sequences of statements, in the process known as stepwise refinement. At some point the pseudo-language program is sufficiently detailed that the

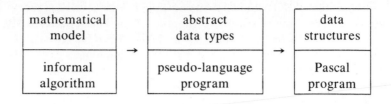

Fig. 1.9. The problem solving process.

operations to be performed on the various types of data become fixed. We then create abstract data types for each type of data (except for the elementary types such as integers, reals and character strings) by giving a procedure name for each operation and replacing uses of each operation by an invocation of the corresponding procedure.

In the third stage we choose an implementation for each abstract data type and write the procedures for the various operations on that type. We also replace any remaining informal statements in the pseudo-language algorithm by Pascal code. The result is a running program. After debugging it will be a working program, and we hope that by using the stepwise development approach outlined in Fig. 1.9, little debugging will be necessary.

1.2 Abstract Data Types

Most of the concepts introduced in the previous section should be familiar ideas from a beginning course in programming. The one possibly new notion is that of an abstract data type, and before proceeding it would be useful to discuss the role of abstract data types in the overall program design process. To begin, it is useful to compare an abstract data type with the more familiar notion of a procedure.

Procedures, an essential tool in programming, generalize the notion of an operator. Instead of being limited to the built-in operators of a programming language (addition, subtraction, etc.), by using procedures a programmer is free to define his own operators and apply them to operands that need not be basic types. An example of a procedure used in this way is a matrix multiplication routine.

Another advantage of procedures is that they can be used to *encapsulate* parts of an algorithm by localizing in one section of a program all the statements relevant to a certain aspect of a program. An example of encapsulation is the use of one procedure to read all input and to check for its validity. The advantage of encapsulation is that we know where to go to make changes to the encapsulated aspect of the problem. For example, if we decide to check that inputs are nonnegative, we need to change only a few lines of code, and we know just where those lines are.

Definition of Abstract Data Type

We can think of an *abstract data type* (ADT) as a mathematical model with a collection of operations defined on that model. Sets of integers, together with the operations of union, intersection, and set difference, form a simple example of an ADT. In an ADT, the operations can take as operands not only instances of the ADT being defined but other types of operands, e.g., integers or instances of another ADT, and the result of an operation can be other than an instance of that ADT. However, we assume that at least one operand, or the result, of any operation is of the ADT in question.

The two properties of procedures mentioned above — generalization and encapsulation — apply equally well to abstract data types. ADT's are generalizations of primitive data types (integer, real, and so on), just as procedures are generalizations of primitive operations (+, −, and so on). The ADT encapsulates a data type in the sense that the definition of the type and all operations on that type can be localized to one section of the program. If we wish to change the implementation of an ADT, we know where to look, and by revising one small section we can be sure that there is no subtlety elsewhere in the program that will cause errors concerning this data type. Moreover, outside the section in which the ADT's operations are defined, we can treat the ADT as a primitive type; we have no concern with the underlying implementation. One pitfall is that certain operations may involve more than one ADT, and references to these operations must appear in the sections for both ADT's.

To illustrate the basic ideas, consider the procedure *greedy* of the previous section which, in Fig. 1.8, was implemented using primitive operations on an abstract data type LIST (of integers). The operations performed on the LIST *newclr* were:

1. make a list empty,
2. get the first member of the list and return *null* if the list is empty,
3. get the next member of the list and return *null* if there is no next member, and
4. insert an integer into the list.

There are many data structures that can be used to implement such lists efficiently, and we shall consider the subject in depth in Chapter 2. In Fig. 1.8, if we replace these operations by the statements

1. MAKENULL(*newclr*);
2. w := FIRST(*newclr*);
3. w := NEXT(*newclr*);
4. INSERT(v, *newclr*);

then we see an important aspect of abstract data types. We can implement a type any way we like, and the programs, such as Fig. 1.8, that use objects of that type do not change; only the procedures implementing the operations on the type need to change.

Turning to the abstract data type GRAPH we see need for the following operations:

1. get the first uncolored vertex,
2. test whether there is an edge between two vertices,
3. mark a vertex colored, and
4. get the next uncolored vertex.

There are clearly other operations needed outside the procedure *greedy*, such as inserting vertices and edges into the graph and making all vertices uncolored. There are many data structures that can be used to support graphs with these operations, and we shall study the subject of graphs in Chapters 6 and 7.

It should be emphasized that there is no limit to the number of operations that can be applied to instances of a given mathematical model. Each set of operations defines a distinct ADT. Some examples of operations that might be defined on an abstract data type SET are:

1. MAKENULL(A). This procedure makes the null set be the value for set A.
2. UNION(A, B, C). This procedure takes two set-valued arguments A and B, and assigns the union of A and B to be the value of set C.
3. SIZE(A). This function takes a set-valued argument A and returns an object of type integer whose value is the number of elements in the set A.

An *implementation* of an ADT is a translation, into statements of a programming language, of the declaration that defines a variable to be of that abstract data type, plus a procedure in that language for each operation of the ADT. An implementation chooses a *data structure* to represent the ADT; each data structure is built up from the basic data types of the underlying programming language using the available data structuring facilities. Arrays and record structures are two important data structuring facilities that are available in Pascal. For example, one possible implementation for variable S of type SET would be an array that contained the members of S.

One important reason for defining two ADT's to be different if they have the same underlying model but different operations is that the appropriateness of an implementation depends very much on the operations to be performed. Much of this book is devoted to examining some basic mathematical models such as sets and graphs, and developing the preferred implementations for various collections of operations.

Ideally, we would like to write our programs in languages whose primitive data types and operations are much closer to the models and operations of our ADT's. In many ways Pascal is not well suited to the implementation of various common ADT's but none of the programming languages in which ADT's can be declared more directly is as well known. See the bibliographic notes for information about some of these languages.

1.3 Data Types, Data Structures and Abstract Data Types

Although the terms "data type" (or just "type"), "data structure" and "abstract data type" sound alike, they have different meanings. In a programming language, the *data type* of a variable is the set of values that the variable may assume. For example, a variable of type boolean can assume either the value true or the value false, but no other value. The basic data types vary from language to language; in Pascal they are integer, real, boolean, and character. The rules for constructing composite data types out of basic ones also vary from language to language; we shall mention how Pascal builds such types momentarily.

An abstract data type is a mathematical model, together with various operations defined on the model. As we have indicated, we shall design algorithms in terms of ADT's, but to implement an algorithm in a given programming language we must find some way of representing the ADT's in terms of the data types and operators supported by the programming language itself. To represent the mathematical model underlying an ADT we use *data structures,* which are collections of variables, possibly of several different data types, connected in various ways.

The *cell* is the basic building block of data structures. We can picture a cell as a box that is capable of holding a value drawn from some basic or composite data type. Data structures are created by giving names to aggregates of cells and (optionally) interpreting the values of some cells as representing connections (e.g., pointers) among cells.

The simplest aggregating mechanism in Pascal and most other programming languages is the (one-dimensional) *array*, which is a sequence of cells of a given type, which we shall often refer to as the celltype. We can think of an array as a mapping from an index set (such as the integers 1, 2, . . . , n) into the celltype. A cell within an array can be referenced by giving the array name together with a value from the index set of the array. In Pascal the index set may be an enumerated type, such as (north, east, south, west), or a subrange type, such as 1..10. The values in the cells of an array can be of any one type. Thus, the declaration

> *name*: **array**[indextype] **of** celltype;

declares *name* to be a sequence of cells, one for each value of type indextype; the contents of the cells can be any member of type celltype.

Incidentally, Pascal is somewhat unusual in its richness of index types. Many languages allow only subrange types (finite sets of consecutive integers) as index types. For example, to index an array by letters in Fortran, one must simulate the effect by using integer indices, such as by using index 1 to stand for 'A', 2 to stand for 'B', and so on.

Another common mechanism for grouping cells in programming languages is the *record structure*. A *record* is a cell that is made up of a collection of cells, called *fields*, of possibly dissimilar types. Records are often grouped into arrays; the type defined by the aggregation of the fields of a record becomes the "celltype" of the array. For example, the Pascal declaration

var
    reclist: **array**[1..4] **of record**
        data: real;
        next: integer
    end

declares *reclist* to be a four-element array, whose cells are records with two fields, *data* and *next*.

    A third grouping method found in Pascal and some other languages is the *file*. The file, like the one-dimensional array, is a sequence of values of some particular type. However, a file has no index type; elements can be accessed only in the order of their appearance in the file. In contrast, both the array and the record are "random-access" structures, meaning that the time needed to access a component of an array or record is independent of the value of the array index or field selector. The compensating benefit of grouping by file, rather than by array, is that the number of elements in a file can be time-varying and unlimited.

Pointers and Cursors

In addition to the cell-grouping features of a programming language, we can represent relationships between cells using pointers and cursors. A *pointer* is a cell whose value indicates another cell. When we draw pictures of data structures, we indicate the fact that cell A is a pointer to cell B by drawing an arrow from A to B.

    In Pascal, we can create a pointer variable *ptr* that will point to cells of a given type, say celltype, by the declaration

var
    ptr: ↑ celltype

A postfix up-arrow is used in Pascal as the dereferencing operator, so the expression *ptr*↑ denotes the value (of type celltype) in the cell pointed to by *ptr*.

    A *cursor* is an integer-valued cell, used as a pointer to an array. As a method of connection, the cursor is essentially the same as a pointer, but a cursor can be used in languages like Fortran that do not have explicit pointer types as Pascal does. By treating a cell of type integer as an index value for some array, we effectively make that cell point to one cell of the array. This technique, unfortunately, works only when cells of arrays are pointed to; there is no reasonable way to interpret an integer as a "pointer" to a cell that is not part of an array.

    We shall draw an arrow from a cursor cell to the cell it "points to." Sometimes, we shall also show the integer in the cursor cell, to remind us that it is not a true pointer. The reader should observe that the Pascal pointer mechanism is such that cells in arrays can only be "pointed to" by cursors, never by true pointers. Other languages, like PL/I or C, allow components of arrays to be pointed to by either cursors or true pointers, while in Fortran or Algol, there being no pointer type, only cursors can be used.

Example 1.3. In Fig. 1.10 we see a two-part data structure that consists of a chain of cells containing cursors to the array *reclist* defined above. The purpose of the field *next* in *reclist* is to point to another record in the array. For example, *reclist*[4].*next* is 1, so record 4 is followed by record 1. Assuming record 4 is first, the *next* field of *reclist* orders the records 4, 1, 3, 2. Note that the *next* field is 0 in record 2, indicating that there is no following record. It is a useful convention, one we shall adopt in this book, to use 0 as a "NIL pointer," when cursors are being used. This idea is sound only if we also make the convention that arrays to which cursors "point" must be indexed starting at 1, never at 0.

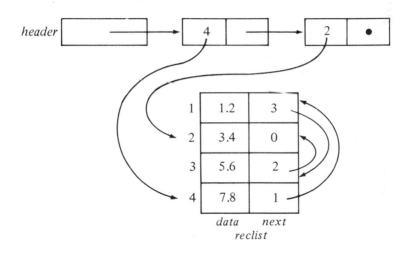

Fig. 1.10. Example of a data structure.

The cells in the chain of records in Fig. 1.10 are of the type

```
type
    recordtype = record
        cursor: integer;
        ptr: ↑ recordtype
    end
```

The chain is pointed to by a variable named *header*, which is of type ↑ record-type; *header* points to an anonymous record of type recordtype.† That record

---
† The record has no known name because it was created by a call *new(header)*, which

has a value 4 in its *cursor* field; we regard this 4 as an index into the array *reclist*. The record has a true pointer in field *ptr* to another anonymous record. The record pointed to has an index in its *cursor* field indicating position 2 of *reclist*; it also has a **nil** pointer in its *ptr* field. □

1.4 The Running Time of a Program

When solving a problem we are faced frequently with a choice among algorithms. On what basis should we choose? There are two often contradictory goals.

1.   We would like an algorithm that is easy to understand, code, and debug.
2.   We would like an algorithm that makes efficient use of the computer's resources, especially, one that runs as fast as possible.

When we are writing a program to be used once or a few times, goal (1) is most important. The cost of the programmer's time will most likely exceed by far the cost of running the program, so the cost to optimize is the cost of writing the program. When presented with a problem whose solution is to be used many times, the cost of running the program may far exceed the cost of writing it, especially, if many of the program runs are given large amounts of input. Then it is financially sound to implement a fairly complicated algorithm, provided that the resulting program will run significantly faster than a more obvious program. Even in these situations it may be wise first to implement a simple algorithm, to determine the actual benefit to be had by writing a more complicated program. In building a complex system it is often desirable to implement a simple prototype on which measurements and simulations can be performed, before committing oneself to the final design. It follows that programmers must not only be aware of ways of making programs run fast, but must know when to apply these techniques and when not to bother.

Measuring the Running Time of a Program

The running time of a program depends on factors such as:

1.   the input to the program,
2.   the quality of code generated by the compiler used to create the object program,
3.   the nature and speed of the instructions on the machine used to execute the program, and
4.   the time complexity of the algorithm underlying the program.

The fact that running time depends on the input tells us that the running time of a program should be defined as a function of the input. Often, the running time depends not on the exact input but only on the "size" of the

---

made *header* point to this newly-created record. Internal to the machine, however, there is a memory address that can be used to locate the cell.

input. A good example is the process known as *sorting,* which we shall discuss in Chapter 8. In a sorting problem, we are given as input a list of items to be sorted, and we are to produce as output the same items, but smallest (or largest) first. For example, given 2, 1, 3, 1, 5, 8 as input we might wish to produce 1, 1, 2, 3, 5, 8 as output. The latter list is said to be *sorted smallest first.* The natural size measure for inputs to a sorting program is the number of items to be sorted, or in other words, the length of the input list. In general, the length of the input is an appropriate size measure, and we shall assume that measure of size unless we specifically state otherwise.

It is customary, then, to talk of $T(n)$, the running time of a program on inputs of size n. For example, some program may have a running time $T(n) = cn^2$, where c is a constant. The units of $T(n)$ will be left unspecified, but we can think of $T(n)$ as being the number of instructions executed on an idealized computer.

For many programs, the running time is really a function of the particular input, and not just of the input size. In that case we define $T(n)$ to be the *worst case* running time, that is, the maximum, over all inputs of size n, of the running time on that input. We also consider $T_{avg}(n)$, the average, over all inputs of size n, of the running time on that input. While $T_{avg}(n)$ appears a fairer measure, it is often fallacious to assume that all inputs are equally likely. In practice, the average running time is often much harder to determine than the worst-case running time, both because the analysis becomes mathematically intractable and because the notion of "average" input frequently has no obvious meaning. Thus, we shall use worst-case running time as the principal measure of time complexity, although we shall mention average-case complexity wherever we can do so meaningfully.

Now let us consider remarks (2) and (3) above: that the running time of a program depends on the compiler used to compile the program and the machine used to execute it. These facts imply that we cannot express the running time $T(n)$ in standard time units such as seconds. Rather, we can only make remarks like "the running time of such-and-such an algorithm is proportional to n^2." The constant of proportionality will remain unspecified since it depends so heavily on the compiler, the machine, and other factors.

Big-Oh and Big-Omega Notation

To talk about growth rates of functions we use what is known as *"big-oh"* notation. For example, when we say the running time $T(n)$ of some program is $O(n^2)$, read "big oh of n squared" or just "oh of n squared," we mean that there are positive constants c and n_0 such that for n equal to or greater than n_0, we have $T(n) \le cn^2$.

Example 1.4. Suppose $T(0) = 1$, $T(1) = 4$, and in general $T(n) = (n+1)^2$. Then we see that $T(n)$ is $O(n^2)$, as we may let $n_0 = 1$ and $c = 4$. That is, for $n \ge 1$, we have $(n+1)^2 \le 4n^2$, as the reader may prove easily. Note that we cannot let $n_0 = 0$, because $T(0) = 1$ is not less than $c0^2 = 0$ for any constant c. □

In what follows, we assume all running-time functions are defined on the nonnegative integers, and their values are always nonnegative, although not necessarily integers. We say that $T(n)$ is $O(f(n))$ if there are positive constants c and n_0 such that $T(n) \le cf(n)$ whenever $n \ge n_0$. A program whose running time is $O(f(n))$ is said to have *growth rate* $f(n)$.

Example 1.5. The function $T(n) = 3n^3 + 2n^2$ is $O(n^3)$. To see this, let $n_0 = 0$ and $c = 5$. Then, the reader may show that for $n \ge 0$, $3n^3 + 2n^2 \le 5n^3$. We could also say that this $T(n)$ is $O(n^4)$, but this would be a weaker statement than saying it is $O(n^3)$.

As another example, let us prove that the function 3^n is not $O(2^n)$. Suppose that there were constants n_0 and c such that for all $n \ge n_0$, we had $3^n \le c2^n$. Then $c \ge (3/2)^n$ for any $n \ge n_0$. But $(3/2)^n$ gets arbitrarily large as n gets large, so no constant c can exceed $(3/2)^n$ for all n. □

When we say $T(n)$ is $O(f(n))$, we know that $f(n)$ is an upper bound on the growth rate of $T(n)$. To specify a lower bound on the growth rate of $T(n)$ we can use the notation $T(n)$ is $\Omega(g(n))$, read "big omega of $g(n)$" or just "omega of $g(n)$," to mean that there exists a constant c such that $T(n) \ge cg(n)$ infinitely often (for an infinite number of values of n).[†]

Example 1.6. To verify that the function $T(n) = n^3 + 2n^2$ is $\Omega(n^3)$, let $c = 1$. Then $T(n) \ge cn^3$ for $n = 0, 1, \ldots$.

For another example, let $T(n) = n$ for odd $n \ge 1$ and $T(n) = n^2/100$ for even $n \ge 0$. To verify that $T(n)$ is $\Omega(n^2)$, let $c = 1/100$ and consider the infinite set of n's: $n = 0, 2, 4, 6, \ldots$. □

The Tyranny of Growth Rate

We shall assume that programs can be evaluated by comparing their running-time functions, with constants of proportionality neglected. Under this assumption a program with running time $O(n^2)$ is better than one with running time $O(n^3)$, for example. Besides constant factors due to the compiler and machine, however, there is a constant factor due to the nature of the program itself. It is possible, for example, that with a particular compiler-machine combination, the first program takes $100n^2$ milliseconds, while the second takes $5n^3$ milliseconds. Might not the $5n^3$ program be better than the $100n^2$ program?

The answer to this question depends on the sizes of inputs the programs are expected to process. For inputs of size $n < 20$, the program with running time $5n^3$ will be faster than the one with running time $100n^2$. Therefore, if the program is to be run mainly on inputs of small size, we would indeed

---

† Note the asymmetry between big-oh and big-omega notation. The reason such asymmetry is often useful is that there are many times when an algorithm is fast on many but not all inputs. For example, there are algorithms to test whether their input is of prime length that run very fast whenever that length is even, so we could not get a good lower bound on running time that held for all $n \ge n_0$.

prefer the program whose running time was $O(n^3)$. However, as n gets large, the ratio of the running times, which is $5n^3/100n^2 = n/20$, gets arbitrarily large. Thus, as the size of the input increases, the $O(n^3)$ program will take significantly more time than the $O(n^2)$ program. If there are even a few large inputs in the mix of problems these two programs are designed to solve, we can be much better off with the program whose running time has the lower growth rate.

Another reason for at least considering programs whose growth rates are as low as possible is that the growth rate ultimately determines how big a problem we can solve on a computer. Put another way, as computers get faster, our desire to solve larger problems on them continues to increase. However, unless a program has a low growth rate such as $O(n)$ or $O(n\log n)$, a modest increase in computer speed makes very little difference in the size of the largest problem we can solve in a fixed amount of time.

Example 1.7. In Fig. 1.11 we see the running times of four programs with different time complexities, measured in seconds, for a particular compiler-machine combination. Suppose we can afford 1000 seconds, or about 17 minutes, to solve a given problem. How large a problem can we solve? In 10^3 seconds, each of the four algorithms can solve roughly the same size problem, as shown in the second column of Fig. 1.12.

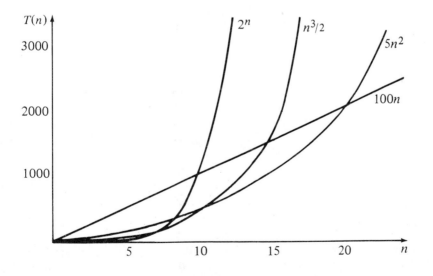

Fig. 1.11. Running times of four programs.

Suppose that we now buy a machine that runs ten times faster at no additional cost. Then for the same cost we can spend 10^4 seconds on a problem where we spent 10^3 seconds before. The maximum size problem we can now solve using each of the four programs is shown in the third column of Fig. 1.12, and the ratio of the third and second columns is shown in the fourth column. We observe that a 1000% improvement in computer speed yields only a 30% increase in the size of problem we can solve if we use the $O(2^n)$ program. Additional factors of ten speedup in the computer yield an even smaller percentage increase in problem size. In effect, the $O(2^n)$ program can solve only small problems no matter how fast the underlying computer.

Running Time $T(n)$	Max. Problem Size for 10^3 sec.	Max. Problem Size for 10^4 sec.	Increase in Max. Problem Size
$100n$	10	100	10.0
$5n^2$	14	45	3.2
$n^3/2$	12	27	2.3
2^n	10	13	1.3

Fig. 1.12. Effect of a ten-fold speedup in computation time.

In the third column of Fig. 1.12 we see the clear superiority of the $O(n)$ program; it returns a 1000% increase in problem size for a 1000% increase in computer speed. We see that the $O(n^3)$ and $O(n^2)$ programs return, respectively, 230% and 320% increases in problem size for 1000% increases in speed. These ratios will be maintained for additional increases in speed. □

As long as the need for solving progressively larger problems exists, we are led to an almost paradoxical conclusion. As computation becomes cheaper and machines become faster, as will most surely continue to happen, our desire to solve larger and more complex problems will continue to increase. Thus, the discovery and use of efficient algorithms, those whose growth rates are low, becomes more rather than less important.

A Few Grains of Salt

We wish to re-emphasize that the growth rate of the worst case running time is not the sole, or necessarily even the most important, criterion for evaluating an algorithm or program. Let us review some conditions under which the running time of a program can be overlooked in favor of other issues.

1. If a program is to be used only a few times, then the cost of writing and debugging dominate the overall cost, so the actual running time rarely affects the total cost. In this case, choose the algorithm that is easiest to implement correctly.

2. If a program is to be run only on "small" inputs, the growth rate of the running time may be less important than the constant factor in the formula for running time. What is a "small" input depends on the exact running times of the competing algorithms. There are some algorithms, such as the integer multiplication algorithm due to Schonhage and Strassen [1971], that are asymptotically the most efficient known for their problem, but have never been used in practice even on the largest problems,

because the constant of proportionality is so large in comparison to other simpler, less "efficient" algorithms.

3. A complicated but efficient algorithm may not be desirable because a person other than the writer may have to maintain the program later. It is hoped that by making the principal techniques of efficient algorithm design widely known, more complex algorithms may be used freely, but we must consider the possibility of an entire program becoming useless because no one can understand its subtle but efficient algorithms.

4. There are a few examples where efficient algorithms use too much space to be implemented without using slow secondary storage, which may more than negate the efficiency.

5. In numerical algorithms, accuracy and stability are just as important as efficiency.

1.5 Calculating the Running Time of a Program

Determining, even to within a constant factor, the running time of an arbitrary program can be a complex mathematical problem. In practice, however, determining the running time of a program to within a constant factor is usually not that difficult; a few basic principles suffice. Before presenting these principles, it is important that we learn how to add and multiply in "big oh" notation.

Suppose that $T_1(n)$ and $T_2(n)$ are the running times of two program fragments P_1 and P_2, and that $T_1(n)$ is $O(f(n))$ and $T_2(n)$ is $O(g(n))$. Then $T_1(n) + T_2(n)$, the running time of P_1 followed by P_2, is $O(\max(f(n), g(n)))$. To see why, observe that for some constants c_1, c_2, n_1, and n_2, if $n \geq n_1$ then $T_1(n) \leq c_1 f(n)$, and if $n \geq n_2$ then $T_2(n) \leq c_2 g(n)$. Let $n_0 = \max(n_1, n_2)$. If $n \geq n_0$, then $T_1(n) + T_2(n) \leq c_1 f(n) + c_2 g(n)$. From this we conclude that if $n \geq n_0$, then $T_1(n) + T_2(n) \leq (c_1 + c_2)\max(f(n), g(n))$. Therefore $T_1(n) + T_2(n)$ is $O(\max(f(n), g(n)))$.

Example 1.8. The *rule for sums* given above can be used to calculate the running time of a sequence of program steps, where each step may be an arbitrary program fragment with loops and branches. Suppose that we have three steps whose running times are, respectively, $O(n^2)$, $O(n^3)$ and $O(n\log n)$. Then the running time of the first two steps executed sequentially is $O(\max(n^2, n^3))$ which is $O(n^3)$. The running time of all three together is $O(\max(n^3, n\log n))$ which is $O(n^3)$. □

In general, the running time of a fixed sequence of steps is, to within a constant factor, the running time of the step with the largest running time. In rare circumstances there will be two or more steps whose running times are *incommensurate* (neither is larger than the other, nor are they equal). For example, we could have steps of running times $O(f(n))$ and $O(g(n))$, where

$$f(n) = \begin{cases} n^4 & \text{if } n \text{ is even} \\ n^2 & \text{if } n \text{ is odd} \end{cases} \qquad g(n) = \begin{cases} n^2 & \text{if } n \text{ is even} \\ n^3 & \text{if } n \text{ is odd} \end{cases}$$

In such cases the sum rule must be applied directly; the running time is $O(\max(f(n), g(n)))$, that is, n^4 if n is even and n^3 if n is odd.

Another useful observation about the sum rule is that if $g(n) \leq f(n)$ for all n above some constant n_0, then $O(f(n)+g(n))$ is the same as $O(f(n))$. For example, $O(n^2+n)$ is the same as $O(n^2)$.

The *rule for products* is the following. If $T_1(n)$ and $T_2(n)$ are $O(f(n))$ and $O(g(n))$, respectively, then $T_1(n)T_2(n)$ is $O(f(n)g(n))$. The reader should prove this fact using the same ideas as in the proof of the sum rule. It follows from the product rule that $O(cf(n))$ means the same thing as $O(f(n))$ if c is any positive constant. For example, $O(n^2/2)$ is the same as $O(n^2)$.

Before proceeding to the general rules for analyzing the running times of programs, let us take a simple example to get an overview of the process.

Example 1.9. Consider the sorting program *bubble* of Fig. 1.13, which sorts an array of integers into increasing order. The net effect of each pass of the inner loop of statements $(3)-(6)$ is to "bubble" the smallest element toward the front of the array.

```
            procedure bubble ( var A: array [1..n] of integer );
                { bubble sorts array A into increasing order }
            var
                i, j, temp: integer;
                begin
(1)                 for i := 1 to n−1 do
(2)                     for j := n downto i+1 do
(3)                         if A[j−1] > A[j] then begin
                                { swap A[j−1] and A[j] }
(4)                             temp := A[j−1];
(5)                             A[j−1] := A[j];
(6)                             A[j] := temp
                            end
            end;  { bubble }
```

Fig. 1.13. Bubble sort.

The number n of elements to be sorted is the appropriate measure of input size. The first observation we make is that each assignment statement takes some constant amount of time, independent of the input size. That is to say, statements (4), (5) and (6) each take $O(1)$ time. Note that $O(1)$ is "big oh" notation for "some constant amount." By the sum rule, the combined running time of this group of statements is $O(\max(1, 1, 1)) = O(1)$.

Now we must take into account the conditional and looping statements. The if- and for-statements are nested within one another, so we may work from the inside out to get the running time of the conditional group and each loop. For the if-statement, testing the condition requires $O(1)$ time. We don't know whether the body of the if-statement (lines $(4)-(6)$) will be executed. Since we are looking for the worst-case running time, we assume the worst and suppose that it will. Thus, the if-group of statements $(3)-(6)$ takes $O(1)$ time.

Proceeding outward, we come to the for-loop of lines $(2)-(6)$. The general rule for a loop is that the running time is the sum, over each iteration of the loop, of the time spent executing the loop body for that iteration. We must, however, charge at least $O(1)$ for each iteration to account for incrementing the index, for testing to see whether the limit has been reached, and for jumping back to the beginning of the loop. For lines $(2)-(6)$ the loop body takes $O(1)$ time for each iteration. The number of iterations of the loop is $n-i$, so by the product rule, the time spent in the loop of lines $(2)-(6)$ is $O((n-i)\times 1)$ which is $O(n-i)$.

Now let us progress to the outer loop, which contains all the executable statements of the program. Statement (1) is executed $n-1$ times, so the total running time of the program is bounded above by some constant times

$$\sum_{i=1}^{n-1}(n-i) = n(n-1)/2 = n^2/2 - n/2$$

which is $O(n^2)$. The program of Fig. 1.13, therefore, takes time proportional to the square of the number of items to be sorted. In Chapter 8, we shall give sorting programs whose running time is $O(n\log n)$, which is considerably smaller, since for large n, $\log n$† is very much smaller than n. □

Before proceeding to some general analysis rules, let us remember that determining a precise upper bound on the running time of programs is sometimes simple, but at other times it can be a deep intellectual challenge. There are no complete sets of rules for analyzing programs. We can only give the reader some hints and illustrate some of the subtler points by examples throughout this book.

Now let us enumerate some general rules for the analysis of programs. In general, the running time of a statement or group of statements may be parameterized by the input size and/or by one or more variables. The only permissible parameter for the running time of the whole program is n, the input size.

1. The running time of each assignment, read, and write statement can usually be taken to be $O(1)$. There are a few exceptions, such as in PL/I, where assignments can involve arbitrarily large arrays, and in any language that allows function calls in assignment statements.

2. The running time of a sequence of statements is determined by the sum rule. That is, the running time of the sequence is, to within a constant factor, the largest running time of any statement in the sequence.

3. The running time of an if-statement is the cost of the conditionally

---

† Unless otherwise specified all logarithms are to the base 2. Note that $O(\log n)$ does not depend on the base of the logarithm since $\log_a n = c\log_b n$, where $c = \log_a b$.

executed statements, plus the time for evaluating the condition. The time to evaluate the condition is normally $O(1)$. The time for an if-then-else construct is the time to evaluate the condition plus the larger of the time needed for the statements executed when the condition is true and the time for the statements executed when the condition is false.

4. The time to execute a loop is the sum, over all times around the loop, of the time to execute the body and the time to evaluate the condition for termination (usually the latter is $O(1)$). Often this time is, neglecting constant factors, the product of the number of times around the loop and the largest possible time for one execution of the body, but we must consider each loop separately to make sure. The number of iterations around a loop is usually clear, but there are times when the number of iterations cannot be computed precisely. It could even be that the program is not an algorithm, and there is no limit to the number of times we go around certain loops.

Procedure Calls

If we have a program with procedures, none of which is recursive, then we can compute the running time of the various procedures one at a time, starting with those procedures that make no calls on other procedures. (Remember to count a function invocation as a "call.") There must be at least one such procedure, else at least one procedure is recursive. We can then evaluate the running time of procedures that call only procedures that make no calls, using the already-evaluated running times of the called procedures. We continue this process, evaluating the running time of each procedure after the running times of all procedures it calls have been evaluated.

If there are recursive procedures, then we cannot find an ordering of all the procedures so that each calls only previously evaluated procedures. What we must now do is associate with each recursive procedure an unknown time function $T(n)$, where n measures the size of the arguments to the procedure. We can then get a *recurrence* for $T(n)$, that is, an equation for $T(n)$ in terms of $T(k)$ for various values of k.

Techniques for solving many different kinds of recurrences exist; we shall present some of these in Chapter 9. Here we shall show how to analyze a simple recursive program.

Example 1.10. Figure 1.14 gives a recursive program to compute $n!$, the product of all the integers from 1 to n inclusive.

An appropriate size measure for this function is the value of n. Let $T(n)$ be the running time for *fact*(n). The running time for lines (1) and (2) is $O(1)$, and for line (3) it is $O(1) + T(n-1)$. Thus, for some constants c and d,

$$T(n) = \begin{cases} c + T(n-1) & \text{if } n > 1 \\ d & \text{if } n \leq 1 \end{cases} \qquad (1.1)$$

```
             function fact ( n: integer ) : integer;
                { fact(n) computes n! }
                begin
(1)                 if n <= 1 then
(2)                     fact := 1
                    else
(3)                     fact := n * fact(n−1)
                end; { fact }
```

Fig. 1.14. Recursive program to compute factorials.

Assuming $n > 2$, we can expand $T(n-1)$ in (1.1) to obtain

$$T(n) = 2c + T(n-2) \quad \text{if } n > 2$$

That is, $T(n-1) = c + T(n-2)$, as can be seen by substituting $n-1$ for n in (1.1). Thus, we may substitute $c + T(n-2)$ for $T(n-1)$ in the equation $T(n) = c + T(n-1)$. We can then use (1.1) to expand $T(n-2)$ to obtain

$$T(n) = 3c + T(n-3) \quad \text{if } n > 3$$

and so on. In general,

$$T(n) = ic + T(n-i) \quad \text{if } n > i$$

Finally, when $i = n-1$ we get

$$T(n) = c(n-1) + T(1) = c(n-1) + d \qquad (1.2)$$

From (1.2) we can conclude that $T(n)$ is $O(n)$. We should note that in this analysis we have assumed that the multiplication of two integers is an $O(1)$ operation. In practice, however, we cannot use the program in Fig. 1.14 to compute $n!$ for large values of n, because the size of the integers being computed will exceed the word length of the underlying machine. □

The general method for solving recurrence equations, as typified by Example 1.10, is repeatedly to replace terms $T(k)$ on the right side of the equation by the entire right side with k substituted for n, until we obtain a formula in which T does not appear on the right as in (1.2). Often we must then sum a series or, if we cannot sum it exactly, get a close upper bound on the sum to obtain an upper bound on $T(n)$.

Programs with GOTO's

In analyzing the running time of a program we have tacitly assumed that all flow of control within a procedure was determined by branching and looping constructs. We relied on this fact as we determined the running time of progressively larger groups of statements by assuming that we needed only the sum rule to group sequences of statements together. Goto statments, however, make the logical grouping of statements more complex. For this reason, goto statements should be avoided, but Pascal lacks break- and continue-statements to jump out of loops. The goto-statement is often used as a substitute for statements of this nature in Pascal.

We suggest the following approach to handling goto's that jump from a loop to code that is guaranteed to follow the loop, which is generally the only kind of goto that is justified. As the goto is presumably executed conditionally within the loop, we may pretend that it is never taken. Because the goto takes us to a statement that will be executed after the loop completes, this assumption is conservative; we can never underestimate the worst case running time of the program if we assume the loop runs to completion. However, it is a rare program in which ignoring the goto is so conservative that it causes us to overestimate the growth rate of the worst case running time for the program. Notice that if we were faced with a goto that jumped back to previously executed code we could not ignore it safely, since that goto may create a loop that accounts for the bulk of the running time.

We should not leave the impression that the use of backwards goto's by themselves make running times unanalyzable. As long as the loops of a program have a reasonable structure, that is, each pair of loops are either disjoint or nested one within the other, then the approach to running time analysis described in this section will work. (However, it becomes the responsibility of the analyzer to ascertain what the loop structure is.) Thus, we should not hesitate to apply these methods of program analysis to a language like Fortran, where goto's are essential, but where programs written in the language tend to have a reasonable loop structure.

Analyzing a Pseudo-Program

If we know the growth rate of the time needed to execute informal English statements, we can analyze pseudo-programs just as we do real ones. Often, however, we do not know the time to be spent on not-fully-implemented parts of a pseudo-program. For example, if we have a pseudo-program in which the only unimplemented parts are operations on ADT's, one of several implementations for an ADT may be chosen, and the overall running time may depend heavily on the implementation. Indeed, one of the reasons for writing programs in terms of ADT's is so we can consider the trade-offs among the running times of the various operations that we obtain by different implementations.

To analyze pseudo-programs consisting of programming language statements and calls to unimplemented procedures, such as operations on ADT's, we compute the running time as a function of unspecified running times for each procedure. The running time for a procedure will be parameterized by the "size" of the argument or arguments for that procedure. Just as for "input size," the appropriate measure of size for an argument is a matter for the analyzer to decide. If the procedure is an operation on an ADT, then the underlying mathematical model for the ADT often indicates the logical notion of size. For example, if the ADT is based on sets, the number of elements in a set is often the right notion of size. In the remaining chapters we shall see many examples of analyzing the running time of pseudo-programs.

1.6 Good Programming Practice

There are a substantial number of ideas we should bear in mind when designing an algorithm and implementing it as a program. These ideas often appear platitudinous, because by-and-large they can be appreciated only through their successful use in real problems, rather than by development of a theory. They are sufficiently important, however, that they are worth repeating here. The reader should watch for the application of these ideas in the programs designed in this book, as well as looking for opportunities to put them into practice in his own programs.

1. *Plan the design of a program.* We mentioned in Section 1.1 how a program can be designed by first sketching the algorithm informally, then as a pseudo-program, and gradually refining the pseudo-program until it becomes executable code. This strategy of sketch-then-detail tends to produce a more organized final program that is easier to debug and maintain.

2. *Encapsulate.* Use procedures and ADT's to place the code for each principal operation and type of data in one place in the program listing. Then, if changes become necessary, the section of code requiring change will be localized.

3. *Use or modify an existing program.* One of the chief inefficiencies in the programming process is that usually a project is tackled as if it were the first program ever written. One should first look for an existing program that does all or a part of the task. Conversely, when writing a program, one should consider making it available to others for possibly unanticipated uses.

4. *Be a toolsmith.* In programming parlance, a *tool* is a program with a variety of uses. When writing a program, consider whether it could be written in a somewhat more general way with little extra effort. For example, suppose one is assigned the task of writing a program to schedule final examinations. Instead, write a tool that takes an arbitrary graph and colors the vertices with as few colors as possible, so that no two vertices connected by an edge have the same color. In the context of examination scheduling, the vertices are classes, the colors are examination periods, and an edge between two classes means that the classes have a student in common. The coloring program, together with routines that translate class lists into graphs and colors into specific times and days, is the examination scheduler. However, the coloring program can be used for problems totally unrelated to examination scheduling, such as the traffic light problem of Section 1.1.

5. *Program at the command level.* Often we cannot find in a library the one program needed to do a job, nor can we adapt one tool to do the job. A well-designed operating system will allow us to connect a network of

available programs together without writing any programs at all, except
for one list of operating system commands. To make commands compos-
able, it is generally necessary that each behave as a *filter,* a program with
one input file and one output file. Notice that any number of filters can
be composed, and if the command language of the operating system is
intelligently designed, merely listing the commands in the order in which
they are to be performed will suffice as a program.

Example 1.11. As an example, let us consider the program *spell*, as it was
originally written by S. C. Johnson from UNIX† commands. This program
takes as input a file f_1 consisting of English text and produces as output all
those words in f_1 that are not found in a small dictionary.‡ *spell* tends to list
proper names as misspellings and may also list real words not in its dictionary,
but the typical output of *spell* is short enough that it can be scanned by eye,
and human intelligence can be used to determine whether a word in the output
of *spell* is a misspelling. (This book was checked using *spell*.)

The first filter used by *spell* is a command called *translate* that, given
appropriate parameters, replaces capital letters by lower case letters and
blanks by newlines, leaving other characters unchanged. The output of
translate consists of a file f_2 that has the words of f_1, uncapitalized, one to a
line. Next comes a command *sort* that sorts the lines of its input file into lexicographic (alphabetical) order. The output of *sort* is a file f_3 that has all the
words of f_2 in alphabetical order, with repetitions. Then a command *unique*
removes duplicate lines from its input file, producing an output file f_4 that has
the words of the original file, without capitalization or duplicates, in alphabetical order. Finally, a command *diff*, with a parameter indicating a second file
f_5 that holds the alphabetized list of words in the dictionary, one to a line, is
applied to f_4. The result is all words in f_4 (and hence f_1) but not in f_5, i.e.,
all words in the original input that are not in the dictionary. The complete
program *spell* is just the following sequence of commands.

> *spell*: *translate* [A-Z] → [a-z], blank → newline
> *sort*
> *unique*
> *diff* dictionary

Command level programming requires discipline from a community of
programmers; they must write programs as filters wherever possible, and they
must write tools instead of special purpose programs wherever possible. Yet
the reward, in terms of the overall ratio of work to results, is substantial. □

---

† UNIX is a Trademark of Bell Laboratories.
‡ We could use an unabridged dictionary, but many misspellings are real words one has
never heard of.

1.7 Super Pascal

Most of the programs written in this book are in Pascal. To make programs more readable, however, we occasionally use three constructs not found in standard Pascal, each of which can be mechanically translated into pure Pascal. One such construct is the nonnumeric label. The few times we need labels, we shall use nonnumeric labels since they make programs easier to understand. For example, "**goto** *output*" is invariably more meaningful than "**goto** 561." To convert a program containing nonnumeric labels into pure Pascal, we must replace each nonnumeric label by a distinct numeric label and we must then declare those labels with a label declaration at the beginning of the program. This process can be done mechanically.

The second nonstandard construct is the return statement, which we use because it allows us to write more understandable programs without using goto statements to interrupt the flow of control. The return statement we use has the form

>   **return** (*expression*)

where the (*expression*) is optional. We can convert a procedure containing return statements into a standard Pascal program quite simply. First, we declare a new label, say 999, and let it label the last end statement of the procedure. If the statement **return** (*x*) appears in a function *zap*, say, we replace this statement with the block

```
begin
     zap := x;
     goto 999
end
```

In a procedure, the statement **return**, which can have no argument, is simply replaced by **goto** 999.

Example 1.12. Figure 1.15 shows the factorial program written using return statements. Figure 1.16 shows the resulting Pascal program if we apply this transformation systematically to Fig. 1.15. □

```
function fact ( n: integer ) : integer;
    begin
        if n <= 1 then
            return (1)
        else
            return (n * fact(n−1))
    end; { fact }
```

Fig. 1.15. Factorial program with return statements.

The third extension is that we use expressions as names of types

```
function fact ( n: integer ) : integer;
    label
        999;
    begin
        if n <= 1 then
            begin
                fact := 1;
                goto 999
            end
        else
            begin
                fact := n * fact(n−1);
                goto 999
            end
    999:
    end;  { fact }
```

Fig. 1.16. Resulting Pascal program.

uniformly throughout a program. For example, an expression like ↑ celltype, while permissible everywhere else, is not permitted as the type of a parameter of a procedure or the type of the value returned by a function. Technically, Pascal requires that we invent a name for the type expression, say ptrtocell. In this book, we shall allow such expressions, expecting that the reader could invent such a type name and mechanically replace type expressions by the type name. Thus, we shall write statements like

function *zap* (*A*: **array**[1..10] **of** integer) : ↑ celltype

to stand for

function *zap* (*A*: arrayoftenints) : ptrtocell

Finally, a note on our typesetting conventions for programs. Pascal reserved words are in boldface, types are in roman, and procedure, function, and variable names are in italic. We distinguish between upper and lower case letters.

Exercises

1.1 There are six teams in the football league: the Vultures, the Lions, the Eagles, the Beavers, the Tigers, and the Skunks. The Vultures have already played the Lions and the Eagles; the Lions have also played the Beavers and Skunks. The Tigers have played the Eagles and Skunks. Each team plays one game per week. Find a schedule so that all teams will have played each other in the fewest number of weeks. *Hint.* Create a graph whose vertices are the pairs of teams that have not yet played each other. What should the edges be so that in a legal coloring of the graph, each color can represent the games played in one week?

*1.2 Consider a robot arm that is fixed at one end. The arm contains two elbows at each of which it is possible to rotate the arm 90 degrees up and down in a vertical plane. How would you mathematically model the possible movements of the end of the arm? Describe an algorithm to move the end of the robot arm from one permissible position to another.

*1.3 Suppose we wish to multiply four matrices of real numbers $M_1 \times M_2 \times M_3 \times M_4$ where M_1 is 10 by 20, M_2 is 20 by 50, M_3 is 50 by 1, and M_4 is 1 by 100. Assume that the multiplication of a $p \times q$ matrix by a $q \times r$ matrix requires pqr scalar operations, as it does in the usual matrix multiplication algorithm. Find the optimal order in which to multiply the matrices so as to minimize the total number of scalar operations. How would you find this optimal ordering if there are an arbitrary number of matrices?

**1.4 Suppose we wish to partition the square roots of the integers from 1 to 100 into two piles of fifty numbers each, such that the sum of the numbers in the first pile is as close as possible to the sum of the numbers in the second pile. If we could use two minutes of computer time to help answer this question, what computations would you perform in those two minutes?

1.5 Describe a greedy algorithm for playing chess. Would you expect it to perform very well?

1.6 In Section 1.2 we considered an ADT SET, with operations MAKE-NULL, UNION, and SIZE. Suppose for convenience that we assume all sets are subsets of {0, 1, . . . ,31} and let the ADT SET be interpreted as the Pascal data type **set of** 0..31. Write Pascal procedures for these operations using this implementation of SET.

1.7 The *greatest common divisor* of two integers p and q is the largest integer d that divides both p and q evenly. We wish to develop a program for computing the greatest common divisor of two integers p and q using the following algorithm. Let r be the remainder of p

divided by q. If r is 0, then q is the greatest common divisor. Otherwise, set p equal to q, then q equal to r, and repeat the process.

a) Show that this process does find the correct greatest common divisor.

b) Refine this algorithm into a pseudo-language program.

c) Convert your pseudo-language program into a Pascal program.

1.8 We want to develop a program for a text formatter that will place words on lines that are both left and right justified. The program will have a word buffer and a line buffer. Initially both are empty. A word is read into the word buffer. If there is sufficient room in the line buffer, the word is transferred to the line buffer. Otherwise, additional spaces are inserted between words in the line buffer to fill out the line, and then the line buffer is emptied by printing the line.

a) Refine this algorithm into a pseudo-language program.

b) Convert your pseudo-language program to a Pascal program.

1.9 Consider a set of n cities and a table of distances between pairs of cities. Write a pseudo-language program for finding a short path that goes through each city exactly once and returns to the city from which it started. There is no known method for obtaining the shortest such tour except by exhaustive searching. Thus try to find an efficient algorithm for this problem using some reasonable heuristic.

1.10 Consider the following functions of n:

$$f_1(n) = n^2$$

$$f_2(n) = n^2 + 1000n$$

$$f_3(n) = \begin{cases} n & \text{if } n \text{ is odd} \\ n^3 & \text{if } n \text{ is even} \end{cases}$$

$$f_4(n) = \begin{cases} n & \text{if } n \leq 100 \\ n^3 & \text{if } n > 100 \end{cases}$$

Indicate for each distinct pair i and j whether $f_i(n)$ is $O(f_j(n))$ and whether $f_i(n)$ is $\Omega(f_j(n))$.

1.11 Consider the following functions of n:

$$g_1(n) = \begin{cases} n^2 & \text{for even } n \geq 0 \\ n^3 & \text{for odd } n \geq 1 \end{cases}$$

$$g_2(n) = \begin{cases} n & \text{for } 0 \leq n \leq 100 \\ n^3 & \text{for } n > 100 \end{cases}$$

$$g_3(n) = n^{2.5}$$

Indicate for each distinct pair i and j whether $g_i(n)$ is $O(g_j(n))$ and whether $g_i(n)$ is $\Omega(g_j(n))$.

1.12 Give, using "big oh" notation, the worst case running times of the following procedures as a function of n.

a)
```
procedure matmpy ( n: integer);
    var
        i, j, k: integer;
    begin
        for i := 1 to n do
            for j := 1 to n do begin
                C[i, j] := 0;
                for k := 1 to n do
                    C[i, j] := C[i, j] + A[i, k] * B[k, j]
            end
    end
```

b)
```
procedure mystery ( n: integer );
    var
        i, j, k: integer;
    begin
        for i := 1 to n−1 do
        for j := i + 1 to n do
            for k := 1 to j do
                { some statement requiring O(1) time }
    end
```

c)
```
procedure veryodd ( n: integer );
    var
        i, j, x, y: integer;
    begin
        for i := 1 to n do
            if odd(i) then begin
                for j := i to n do
                    x := x + 1;
                for j := 1 to i do
                    y := y + 1
            end
    end
```

*d)
```
function recursive (n: integer ) : integer;
    begin
        if n <= 1 then
            return (1)
        else
            return (recursive(n−1) + recursive(n−1))
    end
```

1.13 Show that the following statements are true.

    a) 17 is $O(1)$.

    b) $n(n-1)/2$ is $O(n^2)$.

    c) $\max(n^3, 10n^2)$ is $O(n^3)$.

    d) $\sum_{i=1}^{n} i^k$ is $O(n^{k+1})$ and $\Omega(n^{k+1})$ for integer k.

    e) If $p(x)$ is any k^{th} degree polynomial with a positive leading coefficient, then $p(n)$ is $O(n^k)$ and $\Omega(n^k)$.

*1.14 Suppose $T_1(n)$ is $\Omega(f(n))$ and $T_2(n)$ is $\Omega(g(n))$. Which of the following statements are true?

    a) $T_1(n) + T_2(n)$ is $\Omega(\max(f(n), g(n)))$.

    b) $T_1(n)T_2(n)$ is $\Omega(f(n)g(n))$.

*1.15 Some authors define big omega by saying $f(n)$ is $\Omega(g(n))$ if there is some n_0 and $c > 0$ such that for all $n \geq n_0$ we have $f(n) \geq cg(n)$.

    a) Is it true for this definition that $f(n)$ is $\Omega(g(n))$ if and only if $g(n)$ is $O(f(n))$?

    b) Is (a) true for the definition of big omega in Section 1.4?

    c) Does Exercise 1.14(a) or (b) hold for this definition of big omega?

1.16 Order the following functions by growth rate: (a) n, (b) $\sqrt{n}$, (c) $\log n$, (d) $\log\log n$, (e) $\log^2 n$, (f) $n/\log n$, (g) $\sqrt{n}\log^2 n$, (h) $(1/3)^n$, (i) $(3/2)^n$, (j) 17.

1.17 Assume the parameter n in the procedure below is a positive power of 2, i.e., $n = 2, 4, 8, 16 , \ldots$. Give the formula that expresses the value of the variable *count* in terms of the value of n when the procedure terminates.

```
procedure mystery ( n: integer );
    var
        x, count: integer;
    begin
        count := 0;
        x := 2;
        while x < n do begin
            x := 2 * x;
            count := count + 1
        end;
        writeln(count)
    end
```

1.18 Here is a function *max(i, n)* that returns the largest element in positions i through $i+n-1$ of an integer array A. You may assume for convenience that n is a power of 2.

```
function max ( i, n: integer ) : integer;
    var
        m1, m2: integer;
    begin
        if n = 1 then
            return (A[i])
        else begin
            m1 := max(i, n div 2);
            m2 := max(i+n div 2, n div 2);
            if m1 < m2 then
                return (m2)
            else
                return (m1)
        end
    end
```

a) Let $T(n)$ be the worst-case time taken by *max* with second argument n. That is, n is the number of elements of which the largest is found. Write an equation expressing $T(n)$ in terms of $T(j)$ for one or more values of j less than n and a constant or constants that represent the times taken by individual statements of the *max* program.

b) Give a tight big oh upper bound on $T(n)$. Your answer should be equal to the big omega lower bound, and be as simple as possible.

Bibliographic Notes

The concept of an abstract data type can be traced to the *class* type in the language SIMULA 67 (Birtwistle et al. [1973]). Since that time, a variety of other languages that support abstract data types have been developed including Alphard (Shaw, Wulf, and London [1977]), C with classes (Stroustrup [1982]), CLU (Liskov, et al. [1977]), MESA (Geschke, Morris, and Satterthwaite [1977]), and Russell (Demers and Donahue [1979]). The ADT concept is further discussed in works such as Gotlieb and Gotlieb [1978] and Wulf et al. [1981].

Knuth [1968] was the first major work to advocate the systematic study of the running time of programs. Aho, Hopcroft, and Ullman [1974] relate the time and space complexity of algorithms to various models of computation, such as Turing machines and random-access machines. See also the bibliographic notes to Chapter 9 for more references to the subject of analysis of algorithms and programs.

For additional material on structured programming see Hoare, Dahl, and Dijkstra [1972], Wirth [1973], Kernighan and Plauger [1974], and Yourdon and Constantine [1975]. Organizational and psychological problems arising in the development of large software projects are discussed in Brooks [1974] and

Weinberg [1971]. Kernighan and Plauger [1981] show how to build useful software tools for a programming environment.

CHAPTER 2

Basic
Abstract
Data
Types

In this chapter we shall study some of the most fundamental abstract data types. We consider lists, which are sequences of elements, and two special cases of lists: stacks, where elements are inserted and deleted at one end only, and queues, where elements are inserted at one end and deleted at the other. We then briefly study the mapping or associative store, an ADT that behaves as a function. For each of these ADT's we consider several implementations and compare their relative merits.

2.1 The Abstract Data Type "List"

Lists are a particularly flexible structure because they can grow and shrink on demand, and elements can be accessed, inserted, or deleted at any position within a list. Lists can also be concatenated together or split into sublists. Lists arise routinely in applications such as information retrieval, programming language translation, and simulation. Storage management techniques of the kind we discuss in Chapter 12 use list-processing techniques extensively. In this section we shall introduce a number of basic list operations, and in the remainder of this chapter present data structures for lists that support various subsets of these operations efficiently.

Mathematically, a *list* is a sequence of zero or more elements of a given type (which we generally call the elementtype). We often represent such a list by a comma-separated sequence of elements

$$a_1, a_2, \ldots , a_n$$

where $n \geq 0$, and each a_i is of type elementtype. The number n of elements is said to be the *length* of the list. Assuming $n \geq 1$, we say that a_1 is the *first* element and a_n is the *last* element. If $n = 0$, we have an *empty list*, one which has no elements.

An important property of a list is that its elements can be linearly ordered according to their position on the list. We say a_i *precedes* a_{i+1} for $i = 1, 2, \ldots , n-1$, and a_i *follows* a_{i-1} for $i = 2, 3, \ldots , n$. We say that the element a_i is at *position i*. It is also convenient to postulate the existence of a position following the last element on a list. The function END(L) will return the position following position n in an n-element list L. Note that position END(L) has a distance from the beginning of the list that varies as the list grows or shrinks, while all other positions have a fixed distance from the beginning of the list.

To form an abstract data type from the mathematical notion of a list we must define a set of operations on objects of type LIST.† As with many other ADT's we discuss in this book, no one set of operations is suitable for all applications. Here, we shall give one representative set of operations. In the next section we shall offer several data structures to represent lists and we shall write procedures for the typical list operations in terms of these data structures.

To illustrate some common operations on lists, let us consider a typical application in which we have a mailing list from which we wish to purge duplicate entries. Conceptually, this problem can be solved quite simply: for each item on the list, remove all equivalent following items. To present this algorithm, however, we need to define operations that find the first element on a list, step through all successive elements, and retrieve and delete elements.

We shall now present a representative set of list operations. In what follows, L is a list of objects of type elementtype, x is an object of that type, and p is of type position. Note that "position" is another data type whose implementation will vary for different list implementations. Even though we informally think of positions as integers, in practice, they may have another representation.

1. INSERT(x, p, L). Insert x at position p in list L, moving elements at p and following positions to the next higher position. That is, if L is $a_1, a_2, \ldots, a_n$, then L becomes $a_1, a_2, \ldots, a_{p-1}, x, a_p, \ldots, a_n$. If p is END(L), then L becomes $a_1, a_2, \ldots, a_n, x$. If list L has no position p, the result is undefined.

2. LOCATE(x, L). This function returns the position of x on list L. If x appears more than once, then the position of the first occurrence is returned. If x does not appear at all, then END(L) is returned.

3. RETRIEVE(p, L). This function returns the element at position p on list L. The result is undefined if p = END(L) or if L has no position p. Note that the elements must be of a type that can be returned by a function if RETRIEVE is used. In practice, however, we can always modify RETRIEVE to return a pointer to an object of type elementtype.

---

† Strictly speaking, the type is "LIST of elementtype." However, the implementations of lists we propose do not depend on what elementtype is; indeed, it is that independence that justifies the importance we place on the list concept. We shall use "LIST" rather than "LIST of elementtype," and similarly treat other ADT's that depend on the types of elements.

4. DELETE(p, L). Delete the element at position p of list L. If L is $a_1, a_2, \ldots , a_n$, then L becomes $a_1, a_2, \ldots , a_{p-1}, a_{p+1}, \ldots , a_n$. The result is undefined if L has no position p or if p = END(L).

5. NEXT(p, L) and PREVIOUS(p, L) return the positions following and preceding position p on list L. If p is the last position on L, then NEXT(p, L) = END(L). NEXT is undefined if p is END(L). PREVIOUS is undefined if p is 1. Both functions are undefined if L has no position p.

6. MAKENULL(L). This function causes L to become an empty list and returns position END(L).

7. FIRST(L). This function returns the first position on list L. If L is empty, the position returned is END(L).

8. PRINTLIST(L). Print the elements of L in the order of occurrence.

Example 2.1. Let us write, using these operators, a procedure PURGE that takes a list as argument and eliminates duplicates from the list. The elements of the list are of type elementtype, and a list of such elements has type LIST, a convention we shall follow throughout this chapter. There is a function $same(x,y)$, where x and y are of elementtype, that is true if x and y are "the same" and false if not. The notion of sameness is purposely left vague. If elementtype is real, for example, we might want $same(x,y)$ true if and only if $x = y$. However, if elementtype is a record containing the account number, name, and address of a subscriber as in

> **type**
>> elementtype = **record**
>>> *acctno*: integer;
>>> *name*: **packed array** [1..20] **of** char;
>>> *address*: **packed array** [1..50] **of** char
>> **end**

then we might want $same(x, y)$ to be true whenever $x.acctno = y.acctno$.[†]

Figure 2.1 shows the code for PURGE. The variables p and q are used to represent two positions in the list. As the program proceeds, duplicate copies of any elements to the left of position p have been deleted from the list. In one iteration of the loop (2)−(8), q is used to scan the list following position p to delete any duplicates of the element at position p. Then p is moved to the next position and the process is repeated.

---

† In this case, if we eliminate records that are "the same" we might wish to check that the names and addresses are also equal; if the account numbers are equal but the other fields are not, two people may have inadvertently gotten the same account number. More likely, however, is that the same subscriber appears on the list more than once with distinct account numbers and slightly different names and/or addresses. In such cases, it is difficult to eliminate all duplicates.

In the next section we shall provide appropriate declarations for LIST and position, and implementations for the operations so that PURGE becomes a working program. As written, the program is independent of the manner in which lists are represented so we are free to experiment with various list implementations.

```
              procedure PURGE ( var L: LIST ) ;
                  { PURGE removes duplicate elements from list L }
                  var
                      p, q: position;  { p will be the "current" position
                              in L, and q will move ahead to find equal elements }
                  begin
(1)                   p := FIRST(L);
(2)                   while p <> END(L) do begin
(3)                       q := NEXT(p, L);
(4)                       while q <> END(L) do
(5)                           if same(RETRIEVE(p, L), RETRIEVE(q, L)) then
(6)                               DELETE(q, L)
                          else
(7)                               q := NEXT(q, L);
(8)                       p := NEXT(p, L)
                  end
              end;  { PURGE }
```

Fig. 2.1. Program to remove duplicates.

A point worth observing concerns the body of the inner loop, lines (4)−(7) of Fig. 2.1. When we delete the element at position q at line (6), the elements that were at positions $q+1$, $q+2$, ..., and so on, move up one position in the list. In particular, should q happen to be the last position on L, the value of q would become END(L). If we then executed line (7), NEXT(END(L), L) would produce an undefined result. Thus, it is essential that either (6) or (7), but never both, is executed between the tests for $q =$ END(L) at line (4). □

2.2 Implementation of Lists

In this section we shall describe some data structures that can be used to represent lists. We shall consider array, pointer, and cursor implementations of lists. Each of these implementations permits certain list operations to be done more efficiently than others.

Array Implementation of Lists

In an array implementation of a list, the elements are stored in contiguous cells of an array. With this representation a list is easily traversed and new elements can be appended readily to the tail of the list. Inserting an element into the middle of the list, however, requires shifting all following elements one place over in the array to make room for the new element. Similarly, deleting any element except the last also requires shifting elements to close up the gap.

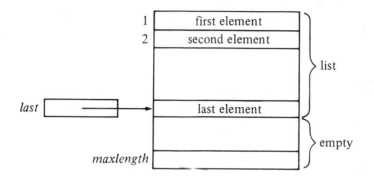

Fig. 2.2. Array implementation of a list.

In the array implementation we define the type LIST to be a record having two fields. The first field is an array of elements whose length is sufficient to hold the maximum size list that will be encountered. The second field is an integer *last* indicating the position of the last list element in the array. The ith element of the list is in the ith cell of the array, for $1 \le i \le last$, as shown in Fig. 2.2. Positions in the list are represented by integers, the ith position by the integer i. The function END(L) has only to return $last + 1$. The important declarations are:

```
const
    maxlength = 100  { some suitable constant };
type
    LIST = record
        elements: array[1..maxlength] of elementtype;
        last: integer
    end;
    position: integer;
function END ( var L: LIST ) : position;†
    begin
        return (L.last + 1)
    end;  { END }
```

Figure 2.3 shows how we might implement the operations INSERT, DELETE, and LOCATE using this array-based implementation. INSERT moves the elements at locations $p, p+1, \ldots, last$ into locations $p+1, p+2, \ldots, last+1$ and then inserts the new element at location p. If there is no room in the array for an additional element, the routine *error* is invoked, causing its argument to be printed, followed by termination of execution of the program. DELETE removes the element at position p by moving the elements at positions $p+1, p+2, \ldots, last$ into positions $p, p+1, \ldots, last-1$. LOCATE sequentially scans the array to look for a given element. If the element is not found, LOCATE returns $last+1$.

It should be clear how to encode the other list operations using this implementation of lists. For example, FIRST always returns 1; NEXT returns one more than its argument and PREVIOUS returns one less, each first checking that the result is in range; MAKENULL(L) sets $L.last$ to 0.

If procedure PURGE of Fig. 2.1 is preceded by

1. the definitions of elementtype and the function *same*,
2. the definitions of LIST, position and END(L) as above,
3. the definition of DELETE from Fig. 2.3, and
4. suitable definitions for the trivial procedures FIRST, NEXT, and RETRIEVE,

then a working procedure PURGE results.

At first, it may seem tedious writing procedures to govern all accesses to the underlying structures. However, if we discipline ourselves to writing programs in terms of the operations for manipulating abstract data types rather than making use of particular implementation details, then we can modify programs more readily by reimplementing the operations rather than searching all programs for places where we have made accesses to the underlying data structures. This flexibility can be particularly important in large software efforts, and the reader should not judge the concept by the necessarily tiny examples found in this book.

Pointer Implementation of Lists

Our second implementation of lists, singly-linked cells, uses pointers to link successive list elements. This implementation frees us from using contiguous memory for storing a list and hence from shifting elements to make room for new elements or to close up gaps created by deleted elements. However, one price we pay is extra space for pointers.

In this representation, a list is made up of cells, each cell consisting of an element of the list and a pointer to the next cell on the list. If the list is $a_1, a_2, \ldots, a_n$, the cell holding a_i has a pointer to the cell holding a_{i+1}, for

---

† Even though L is not modified, we pass L by reference because frequently it will be a big structure and we don't want to waste time copying it.

```
procedure INSERT ( x: elementtype; p: position; var L: LIST );
    { INSERT places x at position p in list L }
    var
        q: position;
    begin
        if L.last >= maxlength then
            error('list is full')
        else if (p > L.last + 1) or (p < 1) then
            error('position does not exist')
        else begin
            for q := L.last downto p do
                { shift elements at p, p+1, . . . down one position }
                L.elements[q+1] := L.elements[q];
            L.last := L.last + 1;
            L.elements[p] := x
        end
    end; { INSERT }


procedure DELETE ( p: position; var L: LIST );
    { DELETE removes the element at position p of list L }
    var
        q: position;
    begin
        if (p > L.last) or (p < 1) then
            error('position does not exist')
        else begin
            L.last := L.last − 1;
            for q := p to L.last do
                { shift elements at p+1, p+2, . . . up one position }
                L.elements[q] := L.elements[q+1]
        end
    end; { DELETE }


function LOCATE ( x: elementtype; var L: LIST ) : position;
    { LOCATE returns the position of x on list L }
    var
        q: position;
    begin
        for q := 1 to L.last do
            if L.elements[q] = x then
                return (q);
        return (L.last + 1) { if not found }
    end; { LOCATE }
```

Fig. 2.3. Array-based implementation of some list operations.

$i = 1, 2, \ldots, n-1$. The cell holding a_n has a **nil** pointer. There is also a *header* cell that points to the cell holding a_1; the header holds no element.† In the case of an empty list, the header's pointer is **nil**, and there are no other cells. Figure 2.4 shows a linked list of this form.

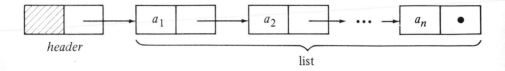

header

list

Fig. 2.4. A linked list.

For singly-linked lists, it is convenient to use a definition of position that is somewhat different than the definition of position in an array implementation. Here, position i will be a pointer to the cell holding the pointer to a_i for $i = 2, 3, \ldots, n$. Position 1 is a pointer to the header, and position END(L) is a pointer to the last cell of L.

The type of a list happens to be the same as that of a position — it is a pointer to a cell, the header in particular. We can formally define the essential parts of a linked list data structure as follows.

```
type
    celltype = record
        element: elementtype;
        next: ↑ celltype
    end;
    LIST = ↑ celltype;
    position = ↑ celltype;
```

The function END(L) is shown in Fig. 2.5. It works by moving pointer q down the list from the header, until it reaches the end, which is detected by the fact that q points to a cell with a **nil** pointer. Note that this implementation of END is inefficient, as it requires us to scan the entire list every time we need to compute END(L). If we need to do so frequently, as in the PURGE program of Fig. 2.1, we could either

1. Use a representation of lists that includes a pointer to the last cell, or

2. Replace uses of END(L) where possible. For example, the condition

_____

† Making the header a complete cell simplifies the implementation of list operations in Pascal. We can use pointers for headers if we are willing to implement our operations so they do insertions and deletions at the beginning of a list in a special way. See the discussion under cursor-based implementation of lists in this section.

$p <>$ END(L) in line (2) of Fig. 2.1 could be replaced by $p\uparrow.next <>$ **nil**, at a cost of making that program dependent on one particular implementation of lists.

```
        function END ( L: LIST ) : position;
            { END returns a pointer to the last cell of L }
            var
                q: position;
            begin
(1)             q := L;
(2)             while q↑.next <> nil do
(3)                 q := q↑.next;
(4)             return (q)
        end; { END }
```

Fig. 2.5. The function END.

Figure 2.6 contains routines for the four operations INSERT, DELETE, LOCATE, and MAKENULL using this pointer implementation of lists. The other operations can be implemented as one-step routines, with the exception of PREVIOUS, which requires a scan of the list from the beginning. We leave these other routines as exercises. Note that many of the commands do not use parameter L, the list, and we omit it from those that do not.

The mechanics of the pointer manipulations of the INSERT procedure in Fig. 2.6 are shown in Fig. 2.7. Figure 2.7(a) shows the situation before executing INSERT. We wish to insert a new element in front of the cell containing b, so p is a pointer to the list cell that contains the pointer to b. At line (1), *temp* is set to point to the cell containing b. At line (2) a new list cell is created and the *next* field of the cell containing a is made to point to this cell. At line (3) the *element* field of the newly-created cell is made to hold x, and at line (4) the *next* field is given the value of *temp*, thus making it point to the cell containing b. Figure 2.7(b) shows the result of executing INSERT. The new pointers are shown dashed, and marked with the step at which they were created.

The DELETE procedure is simpler. Figure 2.8 shows the pointer manipulations of the DELETE procedure in Fig. 2.6. Old pointers are solid and the new pointers dashed.

We should note that a position in a linked-list implementation of a list behaves differently from a position in an array implementation. Suppose we have a list with three elements a, b, c and a variable p, of type position, which currently has position 3 as its value; i.e., it points to the cell holding b, and thus represents the position of c. If we execute a command to insert x at position 2, the list becomes a, x, b, c, and element b now occupies position 3. If we use the array implementation of lists described earlier, b and c would be moved down the array, so b would indeed occupy the third position.

```
        procedure INSERT ( x: elementtype; p: position );
            var
                temp: position;
            begin
(1)             temp := p↑.next;
(2)             new(p↑.next);
(3)             p↑.next↑.element := x;
(4)             p↑.next↑.next := temp
            end;  { INSERT }


        procedure DELETE ( p: position );
            begin
                p↑.next := p↑.next↑.next
            end;  { DELETE }


        function LOCATE ( x: elementtype; L: LIST ) : position;
            var
                p: position;
            begin
                p := L;
                while p↑.next <> nil do
                    if p↑.next↑.element = x then
                        return (p)
                    else
                        p := p↑.next;
                return (p)  { if not found }
            end;  { LOCATE }


        function  MAKENULL ( var L: LIST )  : position;
            begin
                new(L);
                return (L)
            end;  { MAKENULL }
```

Fig. 2.6. Some operations using the linked-list implementation.

However, if the linked-list implementtation is used, the value of p, which is a pointer to the cell containing b, does not change because of the insertion, so afterward, the value of p is "position 4," not 3. This position variable must be updated, if it is to be used subsequently as the position of b.†

---

† Of course, there are many situations where we would like p to continue to represent the position of c.

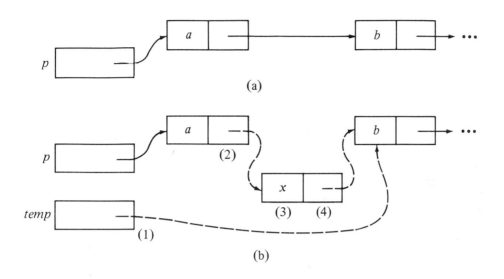

Fig. 2.7. Diagram of INSERT.

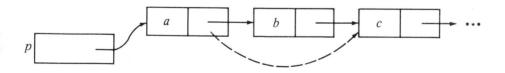

Fig. 2.8. Diagram of DELETE.

Comparison of Methods

We might wonder whether it is best to use a pointer-based or array-based implementation of lists in a given circumstance. Often the answer depends on which operations we intend to perform, or on which are performed most frequently. Other times, the decision rests on how long the list is likely to get. The principal issues to consider are the following.

1. The array implementation requires us to specify the maximum size of a list at compile time. If we cannot put a bound on the length to which the list will grow, we should probably choose a pointer-based implementation.

2. Certain operations take longer in one implementation than the other. For

example, INSERT and DELETE take a constant number of steps for a linked list, but require time proportional to the number of following elements when the array implementation is used. Conversely, executing PREVIOUS and END require constant time with the array implementation, but time proportional to the length of the list if pointers are used.

3.   If a program calls for insertions or deletions that affect the element at the position denoted by some position variable, and the value of that variable will be used later on, then the pointer representation cannot be used as we have described it here. As a general principle, pointers should be used with great care and restraint.

4.   The array implementation may waste space, since it uses the maximum amount of space independent of the number of elements actually on the list at any time. The pointer implementation uses only as much space as is needed for the elements currently on the list, but requires space for the pointer in each cell. Thus, either method could wind up using more space than the other in differing circumstances.

Cursor-Based Implementation of Lists

Some languages, such as Fortran and Algol, do not have pointers. If we are working with such a language, we can simulate pointers with cursors, that is, with integers that indicate positions in arrays. For all the lists of elements whose type is elementtype, we create one array of records; each record consists of an element and an integer that is used as a cursor. That is, we define

```
        var
             SPACE: array [1..maxlength] of record
                    element: elementtype;
                    next: integer
             end
```

If L is a list of elements, we declare an integer variable say $Lhead$, as a header for L. We can treat $Lhead$ as a cursor to a header cell in $SPACE$ with an empty element field. The list operations can then be implemented as in the pointer-based implementation just described.

Here, we shall describe an alternative implementation that avoids the use of header cells by making special cases of insertions and deletions at position 1. This same technique can also be used with pointer-based linked-lists to avoid the use of header cells. For a list L, the value of $SPACE[Lhead].element$ is the first element of L. The value of $SPACE[Lhead].next$ is the index of the cell containing the second element, and so on. A value of 0 in either $Lhead$ or the field $next$ indicates a "nil pointer"; that is, there is no element following.

A list will have type integer, since the header is an integer variable that represents the list as a whole. Positions will also be of type integer. We adopt the convention that position i of list L is the index of the cell of $SPACE$ holding element $i-1$ of L, since the $next$ field of that cell will hold the cursor to element i. As a special case, position 1 of any list is represented by 0. Since the name of the list is always a parameter of operations that use positions, we can distinguish among the first positions of different lists. The position END(L) is the index of the last element on L.

Figure 2.9 shows two lists, $L = a, b, c$ and $M = d, e$, sharing the array SPACE, with maxlength = 10. Notice that all the cells of the array that are not on either list are linked on another list called available. This list is necessary so we can obtain an empty cell when we want to insert into some list, and so we can have a place to put deleted cells for later reuse.

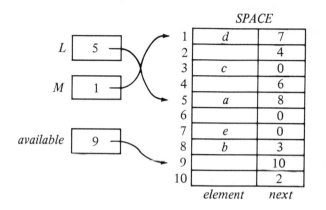

Fig. 2.9. A cursor implementation of linked lists.

To insert an element x into a list L we take the first cell in the available list and place it in the correct position of list L. Element x is then placed in the element field of this cell. To delete an element x from list L we remove the cell containing x from L and return it to the beginning of the available list. These two actions can be viewed as special cases of the act of taking a cell C pointed to by one cursor p and causing another cursor q to point to C, while making p point where C had pointed and making C point where q had pointed. Effectively, C is inserted between q and whatever q pointed to. For example, if we delete b from list L in Fig. 2.9, C is row 8 of SPACE, p is SPACE[5].next, and q is available. The cursors before (solid) and after (dashed) this action are shown in Fig. 2.10, and the code is embodied in the function move of Fig. 2.11, which performs the move if C exists and returns false if C does not exist.

Figure 2.12 shows the procedures INSERT and DELETE, and a procedure initialize that links the cells of the array SPACE into an available space list. These procedures omit checks for errors; the reader may insert them as an exercise. Other operations are left as exercises and are similar to those for pointer-based linked lists.

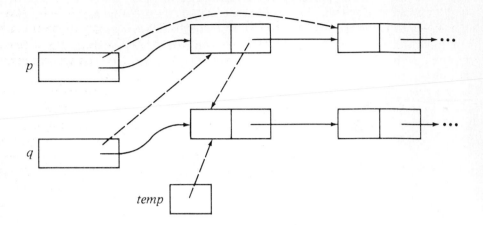

Fig. 2.10. Moving a cell C from one list to another.

```
function move ( var p, q: integer ) : boolean;
    { move puts cell pointed to by p ahead of cell pointed to by q }
    var
        temp: integer;
    begin
        if p = 0 then begin  { cell nonexistent }
            writeln('cell does not exist');
            return (false)
        end
        else begin
            temp := q;
            q := p;
            p := SPACE[q].next;
            SPACE[q].next := temp;
            return (true)
        end
    end;  { move }
```

Fig. 2.11. Code to move a cell.

Doubly-Linked Lists

In a number of applications we may wish to traverse a list both forwards and
backwards efficiently. Or, given an element, we may wish to determine the
preceding and following elements quickly. In such situations we might wish

```
procedure INSERT ( x: elementtype; p: position; var L: LIST );
    begin
        if p = 0 then begin
            { insert at first position }
            if move(available, L) then
                SPACE[L].element := x
        end
        else  { insert at position other than first }
            if move(available, SPACE[p].next) then
                { cell for x now pointed to by SPACE[p].next }
                SPACE[SPACE[p].next].element := x
    end;  { INSERT }

procedure DELETE ( p: position; var L: LIST );
    begin
        if p = 0 then
            move(L, available)
        else
            move(SPACE[p].next, available)
    end;  { DELETE }

procedure initialize;
    { initialize links SPACE into one available list }
    var
        i: integer;
    begin
        for i := maxsize − 1 downto 1 do
            SPACE[i].next := i + 1;
        available := 1;
        SPACE[maxsize].next := 0 { mark end of available list }
    end;  { initialize }
```

Fig. 2.12. Some procedures for cursor-based linked lists.

to give each cell on a list a pointer to both the next and previous cells on the list, as suggested by the doubly-linked list in Fig. 2.13. Chapter 12 mentions some specific situations where doubly-linked lists are essential for efficiency.

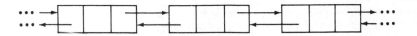

Fig. 2.13. A doubly linked list.

Another important advantage of doubly linked lists is that we can use a pointer to the cell containing the ith element of a list to represent position i, rather than using the less natural pointer to the previous cell. The only price we pay for these features is the presence of an additional pointer in each cell, and somewhat lengthier procedures for some of the basic list operations. If we use pointers (rather than cursors), we may declare cells consisting of an element and two pointers by

```
type
    celltype = record
        element: elementtype;
        next, previous: ↑ celltype
    end;
    position = ↑ celltype;
```

A procedure for deleting an element at position p in a doubly-linked list is given in Fig. 2.14. Figure 2.15 shows the changes in pointers caused by Fig. 2.14, with old pointers drawn solid and new pointers drawn dashed, on the assumption that the deleted cell is neither first nor last.† We first locate the preceding cell using the *previous* field. We make the *next* field of this cell point to the cell following the one in position p. Then we make the *previous* field of this following cell point to the cell preceding the one in position p. The cell pointed to by p becomes useless and should be reused automatically by the Pascal run-time system if needed.

```
procedure  DELETE (var p: position );
    begin
        if p↑.previous <> nil then
            { deleted cell not the first }
            p↑.previous↑.next := p↑.next;
        if p↑.next <> nil then
            { deleted cell not the last }
            p↑.next↑.previous := p↑.previous
    end; { DELETE }
```

Fig. 2.14. Deletion from a doubly linked list.

---

† Incidentally, it is common practice to make the header of a doubly linked list be a cell that effectively "completes the circle." That is, the header's *previous* field points to the last cell and *next* points to the first cell. In this manner, we need not check for **nil** pointers in Fig. 2.14.

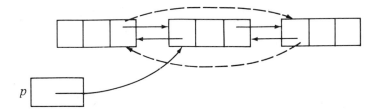

Fig. 2.15. Pointer changes for implementation of a deletion.

2.3 Stacks

A *stack* is a special kind of list in which all insertions and deletions take place at one end, called the *top*. Other names for a stack are "pushdown list," and "LIFO" or "last-in-first-out" list. The intuitive model of a stack is a pile of poker chips on a table, books on a floor, or dishes on a shelf, where it is only convenient to remove the top object on the pile or add a new one above the top. An abstract data type in the STACK family often includes the following five operations.

1. MAKENULL(S). Make stack S be an empty stack. This operation is exactly the same as for general lists.

2. TOP(S). Return the element at the top of stack S. If, as is normal, we identify the top of a stack with position 1, then TOP(S) can be written in terms of list operations as RETRIEVE(FIRST(S), S).

3. POP(S). Delete the top element of the stack, that is, DELETE(FIRST(S), S). Sometimes it is convenient to implement POP as a function that returns the element it has just popped, although we shall not do so here.

4. PUSH(x, S). Insert the element x at the top of stack S. The old top element becomes next-to-top, and so on. In terms of list primitives this operation is INSERT(x, FIRST(S), S).

5. EMPTY(S). Return true if S is an empty stack; return false otherwise.

Example 2.2. Text editors always allow some character (for example, "backspace") to serve as an *erase character*, which has the effect of canceling the previous uncanceled character. For example, if '#' is the erase character, then the string *abc#d##e* is really the string *ae*. The first '#' cancels *c*, the second *d*, and the third *b*.

Text editors also have a *kill character*, whose effect is to cancel all previous characters on the current line. For the purposes of this example, we shall use '@' as the kill character.

A text editor can process a line of text using a stack. The editor reads
one character at a time. If the character read is neither the kill nor the erase
character, it pushes the character onto the stack. If the character read is the
erase character, the editor pops the stack, and if it is the kill character, the
editor makes the stack empty. A program that executes these actions is shown
in Fig. 2.16.

```
procedure EDIT;
    var
        S: STACK;
        c: char;
    begin
        MAKENULL(S);
        while not eoln do begin
            read(c);
            if c = '#' then
                POP(S)
            else if c = '@' then
                MAKENULL(S)
            else { c is an ordinary character }
                PUSH(c, S)
        end;
        print S in reverse order
    end; { EDIT }
```

Fig. 2.16. Program to carry out effects of erase and kill characters.

In this program, the type of STACK must be declared as a list of characters. The process of writing the stack in reverse order in the last line of the
program is a bit tricky. Popping the stack one character at a time gets the line
in reverse order. Some stack implementations, such as the array-based implementation to be discussed next, allow us to write a simple procedure to print
the stack from the bottom. In general, however, to reverse a stack, we must
pop each element and push it onto another stack; the characters can then be
popped from the second stack and printed in the order they are popped. □

An Array Implementation of Stacks

Every implementation of lists we have described works for stacks, since a
stack with its operations is a special case of a list with its operations. The
linked-list representation of a stack is easy, as PUSH and POP operate only on
the header cell and the first cell on the list. In fact, headers can be pointers
or cursors rather than complete cells, since there is no notion of "position" for
stacks, and thus no need to represent position 1 in a way analogous to other
positions.

However, the array-based implementation of lists we gave in Section 2.3 is not a particularly good one for stacks, as every PUSH or POP requires moving the entire list up or down, thus taking time proportional to the number of elements on the stack. A better arrangement for using an array takes account of the fact that insertions and deletions occur only at the top. We can anchor the bottom of the stack at the bottom (high-indexed end) of the array, and let the stack grow towards the top (low-indexed end) of the array. A cursor called *top* indicates the current position of the first stack element. This idea is shown in Fig. 2.17.

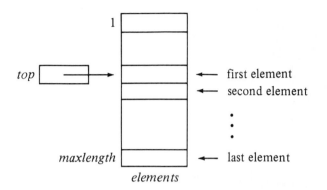

Fig. 2.17. An array implementation of a stack.

For this array-based implementation of stacks we define the abstract data type STACK by

```
type
    STACK = record
        top: integer;
        elements: array[1..maxlength] of elementtype
    end;
```

An instance of the stack consists of the sequence *elements*[*top*], *elements*[*top* + 1], . . . , *elements*[*maxlength*]. Note that if *top* = *maxlength* + 1, then the stack is empty.

The five typical stack operations are implemented in Fig. 2.18. Note that for TOP to return an elementtype, that type must be legal as the result of a function. If not, TOP must be a procedure that modifies its second argument by assigning it the value TOP(*S*), or a function that returns a pointer to elementtype.

```
procedure MAKENULL ( var S: STACK );
    begin
        S.top := maxlength + 1
    end; { MAKENULL }
```

```
function EMPTY ( S: STACK ) : boolean;
    begin
        if S.top > maxlength then
                return (true)
        else
                return (false)
    end; { EMPTY }
```

```
function TOP ( var S: STACK ) : elementtype;
    begin
        if EMPTY(S) then
            error('stack is empty')
        else
            TOP := S.elements[S.top]
    end; { TOP }
```

```
procedure POP ( var S: STACK );
    begin
        if EMPTY(S) then
            error('stack is empty')
        else
            S.top := S.top + 1
    end; { POP }
```

```
procedure PUSH ( x: elementtype; S: STACK );
    begin
        if S.top := 1 then
            error('stack is full')
        else begin
            S.top := S.top - 1;
            S.elements[S.top] := x
        end
    end; { PUSH }
```

Fig. 2.18. Operations on stacks.

2.4 Queues

A *queue* is another special kind of list, where items are inserted at one end (the *rear*) and deleted at the other end (the *front*). Another name for a queue is a "FIFO" or "first-in first-out" list. The operations for a queue are analogous to those for a stack, the substantial differences being that insertions go at the end of the list, rather than the beginning, and that traditional terminology for stacks and queues is different. We shall use the following operations on queues.

1. MAKENULL(Q) makes queue Q an empty list.

2. FRONT(Q) is a function that returns the first element on queue Q. FRONT(Q) can be written in terms of list operations as RETRIEVE(FIRST(Q), Q).

3. ENQUEUE(x, Q) inserts element x at the end of queue Q. In terms of list operations, ENQUEUE(x, Q) is INSERT(x, END(Q), Q).

4. DEQUEUE(Q) deletes the first element of Q; that is, DEQUEUE(Q) is DELETE(FIRST(Q), Q).

5. EMPTY(Q) returns true if and only if Q is an empty queue.

A Pointer Implementation of Queues

As for stacks, any list implementation is legal for queues. However, we can take advantage of the fact that insertions are only done at the rear to make ENQUEUE efficient. Instead of running down the list from beginning to end each time we wish to enqueue something, we can keep a pointer (or cursor) to the last element. As for all kinds of lists, we also keep a pointer to the front of the list; for queues that pointer is useful for executing FRONT and DEQUEUE commands. In Pascal, we shall use a dummy cell as a header and have the front pointer point to it. This convention allows us to handle an empty queue conveniently.

Here we shall develop an implementation for queues that is based on Pascal pointers. The reader may develop a cursor-based implementation that is analogous, but we have available, in the case of queues, a better array-oriented representation than would be achieved by mimicking pointers with cursors directly. We shall discuss this so-called "circular array" implementation at the end of this section. To proceed with the pointer-based implementation, let us define cells as before:

```
type
    celltype = record
        element: elementtype;
        next: ↑ celltype
    end;
```

Then we can define a list to consist of pointers to the front and rear of the queue. The first cell on a queue is a header cell in which the element field is ignored. This convention, as mentioned above, allows a simple representation for an empty queue. We define:

```
type
    QUEUE = record
        front, rear: ↑ celltype
    end;
```

Figure 2.19 shows programs for the five queue operations. In MAKENULL the first statement *new(Q.front)* allocates a variable of type celltype and assigns its address to *Q.front*. The second statement puts **nil** into the *next* field of that cell. The third statement makes the header both the first and last cell of the queue.

The procedure DEQUEUE(Q) deletes the first element of Q by disconnecting the old header from the queue. The first element on the list becomes the new dummy header cell.

Figure 2.20 shows the results created by the sequence of commands MAKENULL(Q), ENQUEUE(x, Q), ENQUEUE(y, Q), DEQUEUE(Q). Note that after dequeueing, the element x being in the element field of the header cell, is no longer considered part of the queue.

A Circular Array Implementation of Queues

The array representation of lists discussed in Section 2.2 can be used for queues, but it is not very efficient. True, with a pointer to the last element, we can execute ENQUEUE in a fixed number of steps, but DEQUEUE, which removes the first element, requires that the entire queue be moved up one position in the array. Thus DEQUEUE takes $\Omega(n)$ time if the queue has length n.

To avoid this expense, we must take a different viewpoint. Think of an array as a circle, where the first position follows the last, as suggested in Fig. 2.21. The queue is found somewhere around the circle in consecutive positions,† with the rear of the queue somewhere clockwise from the front. To enqueue an element, we move the *Q.rear* pointer one position clockwise and write the element in that position. To dequeue, we simply move *Q.front* one position clockwise. Thus, the queue migrates in a clockwise direction as we enqueue and dequeue elements. Observe that the procedures ENQUEUE and DEQUEUE can be written to take some constant number of steps if the circular array model is used.

There is one subtlety that comes up in the representation of Fig. 2.21 and in any minor variation of that strategy (e.g., if *Q.rear* points one position clockwise of the last element, rather than to that element itself). The problem is that there is no way to tell an empty queue from one that occupies the entire circle, short of maintaining a bit that is true if and only if the queue is empty. If we are not willing to maintain this bit, we must prevent the queue from ever filling the entire array.

To see why, suppose the queue of Fig. 2.21 had *maxlength* elements. Then *Q.rear* would point one position counterclockwise of *Q.front*. What if the queue were empty? To see how an empty queue is represented, let us first consider a queue of one element. Then *Q.front* and *Q.rear* point to the same position. If we dequeue the one element, *Q.front* moves one position

---

† Note that "consecutive" must be taken in a circular sense. That is, a queue of length four could occupy the last two and first two positions of the array, for example.

```
procedure MAKENULL ( var Q: QUEUE );
    begin
        new(Q.front);    { create header cell }
        Q.front↑.next := nil;
        Q.rear := Q.front    { header is both first and last cell }
    end; { MAKENULL }

function EMPTY ( Q: QUEUE) : boolean;
    begin
        if Q.front = Q.rear then
            return (true)
        else
            return (false)
    end;  { EMPTY }

function  FRONT ( Q: QUEUE ) : elementtype;
    begin
        if EMPTY(Q) then
            error('queue is empty')
        else
            return (Q.front↑.next↑.element)
    end;  { FRONT }

procedure  ENQUEUE ( x: elementtype; var Q: QUEUE );
    begin
        new(Q.rear↑.next); { add new cell to rear of queue }
        Q.rear := Q.rear↑.next;
        Q.rear↑.element := x;
        Q.rear↑.next := nil
    end; { ENQUEUE }

procedure  DEQUEUE ( var Q: QUEUE );
    begin
        if EMPTY(Q) then
            error('queue is empty')
        else
            Q.front := Q.front↑.next
    end;  { DEQUEUE }
```

Fig. 2.19. Implementation of queue commands.

clockwise; forming an empty queue. Thus an empty queue has $Q.rear$ one position counterclockwise of $Q.front$, which is exactly the same relative position as when the queue had *maxlength* elements. We thus see that even though the array has *maxlength* places, we cannot let the queue grow longer than *maxlength*-1, unless we introduce another mechanism to distinguish empty queues.

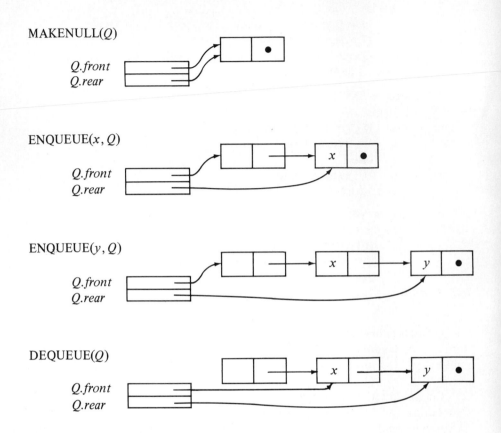

MAKENULL(Q)

ENQUEUE(x, Q)

ENQUEUE(y, Q)

DEQUEUE(Q)

Fig. 2.20. A sequence of queue operations.

Let us now write the five queue commands using this representation for a queue. Formally, queues are defined by:

type
    QUEUE = **record**
        elements: **array**[1..*maxlength*] **of** elementtype;
        front, rear: integer
    end;

The commands appear in Fig. 2.22. The function *addone*(i) adds one to position i in the circular sense.

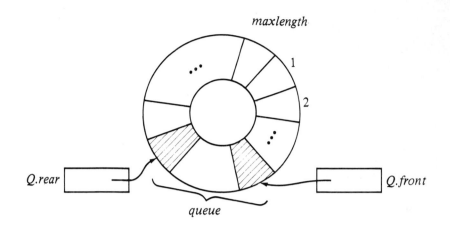

Fig. 2.21. A circular implementation of queues.

2.5 Mappings

A *mapping* or *associative store* is a function from elements of one type, called the *domain type* to elements of another (possibly the same) type, called the *range type*. We express the fact that the mapping M associates element r of range type rangetype with element d of domain type domaintype by $M(d) = r$.

Certain mappings such as $square(i) = i^2$ can be implemented easily as a Pascal function by giving an arithmetic expression or other simple means for calculating $M(d)$ from d. However, for many mappings there is no apparent way to describe $M(d)$ other than to store for each d the value of $M(d)$. For example, to implement a payroll function that associates with each employee a weekly salary seems to require that we store the current salary for each employee. In the remainder of this section we describe a method of implementing functions such as the "payroll" function.

Let us consider what operations we might wish to perform on a mapping M. Given an element d of some domain type, we may wish to obtain $M(d)$ or know whether $M(d)$ is defined (i.e., whether d is currently in the domain of M). Or we may wish to enter new elements into the current domain of M and state their associated range values. Alternatively, we might wish to change the value of $M(d)$. We also need a way to initialize a mapping to the *null mapping*, the mapping whose domain is empty. These operations are summarized by the following three commands.

1. MAKENULL(M). Make M be the null mapping.
2. ASSIGN(M, d, r). Define $M(d)$ to be r, whether or not $M(d)$ was defined previously.

```
function addone ( i : integer ) : integer;
    begin
        return (i mod maxlength + 1)
    end;  { addone }

procedure  MAKENULL ( var Q: QUEUE );
    begin
        Q.front := 1;
        Q.rear := maxlength
    end;  { MAKENULL }

function  EMPTY ( var Q: QUEUE ) : boolean;
    begin
        if addone(Q.rear) = Q.front then
            return (true)
        else
            return (false)
    end;  { EMPTY }

function  FRONT ( var Q: QUEUE ) : elementtype;
    begin
        if EMPTY(Q) then
            error('queue is empty')
        else
            return (Q.elements[Q.front])
    end;  { FRONT }

procedure ENQUEUE ( x: elementtype; var Q: QUEUE );
    begin
        if addone(addone(Q.rear)) = Q.front then
            error('queue is full')
        else begin
            Q.rear := addone(Q.rear);
            Q.elements[Q.rear] := x
        end
    end;  { ENQUEUE }

procedure DEQUEUE ( var Q: QUEUE );
    begin
        if EMPTY(Q) then
            error('queue is empty')
        else
            Q.front := addone(Q.front)
    end;  { DEQUEUE }
```

Fig. 2.22. Circular queue implementation.

3. COMPUTE(M, d, r). Return true and set variable r to $M(d)$ if the latter is defined; return false otherwise.

Array Implementation of Mappings

Many times, the domain type of a mapping will be an elementary type that can be used as an index type of an array. In Pascal, the index types include all the finite subranges of integers, like 1..100 or 17..23, the type char and subranges of char like $'A'..'Z'$, and enumerated types like (north, east, south, west). For example, a cipher-breaking program might keep a mapping *crypt*, with $'A'..'Z'$ as both its domain type and its range type, such that *crypt*(*plaintext*) is the letter currently guessed to stand for the letter *plaintext*.

Such mappings can be implemented simply by arrays, provided there is some range type value that can stand for "undefined." For example, the above mapping *crypt* might be defined to have range type char, rather than $'A'..'Z'$, and $'?'$ could be used to denote "undefined."

Suppose the domain and range types are domaintype and rangetype, and domaintype is a basic Pascal type. Then we can define the type MAPPING (strictly speaking, mapping from domaintype to rangetype) by the declaration

> **type**
> MAPPING = **array**[domaintype] **of** rangetype;

On the assumption that "undefined" is a constant of rangetype, and that first-value and lastvalue are the first and last values in domaintype,[†] we can implement the three commands on mappings as in Fig. 2.23.

List Implementations of Mappings

There are many possible implementations of mappings with finite domains. For example, hash tables are an excellent choice in many situations, but one whose discussion we shall defer to Chapter 4. Any mapping with a finite domain can be represented by the list of pairs $(d_1, r_1), (d_2, r_2), \ldots, (d_k, r_k)$, where $d_1, d_2, \ldots, d_k$ are all the current members of the domain, and r_i is the value that the mapping associates with d_i, for $i = 1, 2, \ldots, k$. We can then use any implementation of lists we choose for this list of pairs.

To be precise, the abstract data type MAPPING can be implemented by lists of elementtype, if we define

> **type**
> elementtype = **record**
> *domain*: domaintype;
> *range*: rangetype
> **end;**

and then define MAPPING as we would define type LIST (of elementtype) in

---

† For example, firstvalue = $'A'$ and lastvalue = $'Z'$ if domaintype is $'A'..'Z'$.

```
procedure MAKENULL ( var M: MAPPING );
    var
        i: domaintype;
    begin
        for i := firstvalue to lastvalue do
            M[i] := undefined
    end;  { MAKENULL }


procedure ASSIGN ( var M: MAPPING;
        d: domaintype; r: rangetype );
    begin
        M[d] := r
    end;  { ASSIGN }


function COMPUTE ( var M: MAPPING;
        d: domaintype; var r: rangetype ) : boolean;
    begin
        if M[d] = undefined then
            return (false)
        else begin
            r := M[d];
            return (true)
        end
    end;  { COMPUTE }
```

Fig 2.23. Array implementation of mappings.

whatever implementation of lists we choose. The three mapping commands
are defined in terms of commands on type LIST in Fig. 2.24.

2.6 Stacks and Recursive Procedures

One important application of stacks is in the implementation of recursive procedures in programming languages. The *run-time organization* for a programming language is the set of data structures used to represent the values of the
program variables during program execution. Every language that, like Pascal, allows recursive procedures, uses a stack of *activation records* to record
the values for all the variables belonging to each active procedure of a program. When a procedure P is called, a new activation record for P is placed
on the stack, regardless of whether there is already another activation record
for P on the stack. When P returns, its activation record must be on top of
the stack, since P cannot return until all procedures it has called have
returned to P. Thus, we may pop the activation record for this call of P to
cause control to return to the point at which P was called (that point, known
as the *return address,* was placed in P's activation record when the call to P

```
procedure MAKENULL ( var M: MAPPING );
    { same as for list }

procedure ASSIGN ( var M: MAPPING;
        d: domaintype; r: rangetype );
    var
        x: elementtype; { the pair (d, r) }
        p: position; { used to go from first to last position on list M }
    begin
        x.domain := d;
        x.range := r;
        p := FIRST(M);
        while p <> END(M) do
            if RETRIEVE(p, M).domain = d then
                DELETE(p, M)   { remove element with domain d }
            else
                p := NEXT(p, M);
        INSERT(x, FIRST(M), M)   { put (d, r) at front of list }
    end; { ASSIGN }

function COMPUTE ( var M: MAPPING;
        d: domaintype; var r: rangetype ) : boolean;
    var
        p: position;
    begin
        p := FIRST(M);
        while p <> END(M) do begin
            if RETRIEVE(p, M).domain = d then begin
                r := RETRIEVE(p, M).range;
                return (true)
            end;
            p := NEXT(p, M)
        end;
        return (false) { if d is not in domain }
    end; { COMPUTE }
```

Fig. 2.24. Mapping implementation in terms of lists.

was made).

Recursion simplifies the structure of many programs. In some languages, however, procedure calls are much more costly than assignment statements, so a program may run faster by a large constant factor if we eliminate recursive procedure calls from it. We do not advocate that recursion or other procedure calls be eliminated habitually; most often the structural simplicity is well worth the running time. However, in the most frequently executed portions of programs, we may wish to eliminate recursion, and it is the purpose of this discussion to illustrate how recursive procedures can be converted to nonrecursive ones by the introduction of a user-defined stack.

Example 2.3. Let us consider recursive and nonrecursive solutions to a simplified version of the classic *knapsack problem* in which we are given *target t* and a collection of positive integer *weights* $w_1, w_2, \ldots, w_n$. We are asked to determine whether there is some selection from among the weights that totals exactly t. For example, if $t = 10$, and the weights are 7, 5, 4, 4, and 1, we could select the second, third, and fifth weights, since $5+4+1 = 10$.

The image that justifies the name "knapsack problem" is that we wish to carry on our back no more than t pounds, and we have a choice of items with given weights to carry. We presumably find the items' utility to be proportional to their weight,† so we wish to pack our knapsack as closely to the target weight as we can.

In Fig. 2.25 we see a function *knapsack* that operates on an array

$$weights: \textbf{array } [1..n] \textbf{ of integer.}$$

A call to *knapsack*(s, i) determines whether there is a collection of the elements in *weight*[i] through *weight*[n] that sums to exactly s, and prints these weights if so. The first thing *knapsack* does is determine if it can respond immediately. Specifically, if $s = 0$, then the empty set of weights is a solution. If $s < 0$, there can be no solution, and if $s > 0$ and $i > n$, then we are out of weights to consider and therefore cannot find a sum equal to s.

If none of these cases applies, then we simply call *knapsack*($s-w_i$, $i+1$) to see if there is a solution that includes w_i. If there is such a solution, then the total problem is solved, and the solution includes w_i, so we print it. If there is no solution, then we call *knapsack*(s, $i+1$) to see if there is a solution that does not use w_i. □

Elimination of Tail Recursion

Often, we can eliminate mechanically the last call a procedure makes to itself. If a procedure $P(x)$ has, as its last step, a call to $P(y)$, then we can replace the call to $P(y)$ by an assignment $x := y$, followed by a jump to the beginning of the code for P. Here, y could be an expression, but x must be a parameter passed by value, so its value is stored in a location private to this call to P.‡ P could have more than one parameter, of course, and if so, they are each treated exactly as x and y above.

This change works because rerunning P with the new value of x has exactly the same effect as calling $P(y)$ and then returning from that call.

---

† In the "real" knapsack problem, we are given utility values as well as weights and are asked to maximize the utility of the items carried, subject to a weight constraint.

‡ Alternatively, x could be passed by reference if y is x.

```
        function knapsack ( target: integer; candidate: integer ) : boolean;
            begin
(1)             if target = 0 then
(2)                 return (true)
(3)             else if (target < 0) or (candidate > n) then
(4)                 return (false)
            else { consider solutions with and without candidate }
(5)             if knapsack(target−weights[candidate], candidate+1) then
                    begin
(6)                     writeln(weights[candidate]);
(7)                         return (true)
                    end
            else { only possible solution is without candidate }
(8)                     return (knapsack(target, candidate+1))
        end;  { knapsack }
```

Fig. 2.25. Recursive knapsack solution.

Notice that the fact that some of P's local variables have values the second time around is of no consequence. P could not use any of those values, or had we called $P(y)$ as originally intended, the value used would not have been defined.

Another variant of tail recursion is illustrated by Fig. 2.25, where the last step of the function *knapsack* just returns the result of calling itself with other parameters. In such a situation, again provided the parameters are passed by value (or by reference if the same parameter is passed to the call), we can replace the call by assignments to the parameters and a jump to the beginning of the function. In the case of Fig. 2.25, we can replace line (8) by

$$candidate := candidate + 1;$$
$$\textbf{goto } beginning$$

where *beginning* stands for a label to be assigned to statement (1). Note no change to *target* is needed, since it is passed intact as the first parameter. In fact, we could observe that since *target* has not changed, the tests of statements (1) and (3) involving *target* are destined to fail, and we could instead skip statements (1) and (2), test only for *candidate* $> n$ at line (3), and proceed directly to line (5).

Complete Recursion Elimination

The tail recursion elimination procedure removes recursion completely only when the recursive call is at the end of the procedure, and the call has the correct form. There is a more general approach that converts any recursive procedure (or function) into a nonrecursive one, but this approach introduces a user-defined stack. In general, a cell of this stack will hold:

1. The current values of the parameters of the procedure;

2. The current values of any local variables of the procedure; and

3. An indication of the return address, that is, the place to which control returns when the current invocation of the procedure is done.

In the case of the function *knapsack,* we can do something simpler. First, observe that whenever we make a call (push a record onto the stack), *candidate* increases by 1. Thus, we can keep *candidate* as a global variable, incrementing it by one every time we push the stack and decreasing it by one when we pop.

A second simplification we can make is to keep a modified "return address" on the stack. Strictly speaking, the return address for this function is either a place in some other procedure that calls *knapsack*, or the call at line (5), or the call at line (8). We can represent these three conditions by a "status," which has one of three values:

1. *none*, indicating that the call is from outside the function *knapsack*,

2. *included*, indicating the call at line (5), which includes *weights[candidate]* in the solution, or

3. *excluded*, indicating the call at line (8), which excludes *weights[candidate]*.

If we store this status symbol as the return address, then we can treat *target* as a global variable. When changing from status *none* to *included,* we subtract *weights[candidate]* from *target,* and we add it back in when changing from status *included* to *excluded.* To help represent the effect of the *knapsack*'s return indicating whether a solution has been found, we use a global *winflag*. Once set to true, *winflag* remains true and causes the stack to be popped and those weights with status *included* to be printed. With these modifications, we can declare our stack to be a list of statuses, by

> **type**
>   statuses = (*none, included, excluded*);
>   STACK = { suitable declaration for stack of statuses }

Figure 2.26 shows the resulting nonrecursive procedure *knapsack* operating on an array *weights* as before. Although this procedure may be faster than the original function *knapsack*, it is clearly longer and more difficult to understand. For this reason, recursion elimination should be used only when speed is very important.

```
procedure knapsack ( target: integer ) ;
    var
        candidate: integer;
        winflag: boolean;
        S: STACK;
    begin
        candidate := 1;
        winflag := false;
        MAKENULL(S);
        PUSH(none, S); { initialize stack to consider weights[1] }
        repeat
            if winflag then begin
                { pop stack, printing weights included in solution }
                if TOP(S) = included then
                    writeln(weights[candidate]);
                candidate := candidate - 1;
                POP(S)
            end
            else if target = 0 then begin  { solution found }
                winflag := true;
                candidate := candidate - 1;
                POP(S)
            end
            else if (((target < 0) and (TOP(S) = none))
                    or (candidate > n)) then begin
                { no solution possible with choices made }
                candidate := candidate - 1;
                POP(S)
            end
            else { no resolution yet; consider status of current candidate }
            if TOP(S) = none then begin  { first try including candidate }
                target := target − weights[candidate];
                candidate := candidate + 1;
                POP(S); PUSH(included, S); PUSH(none, S)
            end
            else if TOP(S) = included then begin  { try excluding candidate }
                target := target + weights[candidate];
                candidate := candidate + 1;
                POP(S); PUSH(excluded, S); PUSH(none, S)
            end
            else begin  { TOP(S) = excluded; give up on current choice }
                POP(S);
                candidate := candidate − 1
            end
        until EMPTY(S)
    end; { knapsack }
```

Fig. 2.26. Nonrecursive knapsack procedure.

Exercises

2.1 Write a program to print the elements on a list. Throughout these exercises use list operations to implement your programs.

2.2 Write programs to insert, delete, and locate an element on a sorted list using

a) array,

b) pointer, and

c) cursor implementations of lists.

What is the running time of each of your programs?

2.3 Write a program to merge

a) two sorted lists,

b) n sorted lists.

2.4 Write a program to concatenate a list of lists.

2.5 Suppose we wish to manipulate polynomials of the form $p(x) = c_1 x^{e_1} + c_2 x^{e_2} + \cdots + c_n x^{e_n}$, where $e_1 > e_2 > \cdots > e_n \geq 0$. Such a polynomial can be represented by a linked list in which each cell has three fields: one for the coefficient c_i, one for the exponent e_i, and one for the pointer to the next cell. Write a program to differentiate polynomials represented in this manner.

2.6 Write programs to add and multiply polynomials of the form in Exercise 2.5. What is the running time of your programs as a function of the number of terms?

*2.7 Suppose we declare cells by

```
type
    celltype = record
        bit: 0..1;
        next: ↑ celltype
    end;
```

A binary number $b_1 b_2 \cdots b_n$, where each b_i is 0 or 1, has numerical value $\sum_{i=1}^{n} b_i 2^{n-i}$. This number can be represented by the list $b_1, b_2, \ldots, b_n$. That list, in turn, can be represented as a linked list of cells of type celltype. Write a procedure *increment(bnumber)* that adds one to the binary number pointed to by *bnumber*. *Hint*: Make *increment* recursive.

2.8 Write a procedure to interchange the elements at positions p and NEXT(p) in a singly linked list.

*2.9 The following procedure was intended to remove all occurrences of element x from list L. Explain why it doesn't always work and suggest a way to repair the procedure so it performs its intended task.

```
procedure delete ( x: elementtype; var L: LIST );
    var
        p: position;
    begin
        p := FIRST(L);
        while p <> END(L) do begin
            if RETRIEVE(p, L) = x then
                DELETE(p, L);
            p := NEXT(p, L)
        end
    end; { delete }
```

2.10 We wish to store a list in an array A whose cells consist of two fields, *data* to store an element and *position* to give the (integer) position of the element. An integer *last* indicates that $A[1]$ through $A[last]$ are used to hold the list. The type LIST can be defined by

```
type
    LIST = record
        last: integer;
        elements: array[1..maxlength] of record
            data: elementtype;
            position: integer
        end
    end;
```

Write a procedure DELETE(p, L) to remove the element at position p. Include all necessary error checks.

2.11 Suppose L is a LIST and p, q, and r are positions. As a function of n, the length of list L, determine how many times the functions FIRST, END, and NEXT are executed by the following program.

```
p := FIRST(L);
while p <> END(L) do begin
    q := p;
    while q <> END(L) do begin
        q := NEXT(q, L);
        r := FIRST(L);
        while r <> q do
            r := NEXT(r, L)
    end;
    p := NEXT(p, L)
end;
```

2.12 Rewrite the code for the LIST operations assuming a linked list representation, but without a header cell. Assume true pointers are used and position 1 is represented by **nil**.

2.13 Add the necessary error checks in the procedure of Fig. 2.12.

2.14 Another array representation of lists is to insert as in Section 2.2, but when deleting, simply replace the deleted element by a special value "deleted," which we assume does not appear on lists otherwise. Rewrite the list operations to implement this strategy. What are the advantages and disadvantages of the approach compared with our original array representation of lists?

2.15 Suppose we wish to use an extra bit in queue records to indicate whether a queue is empty. Modify the declarations and operations for a circular queue to accommodate this feature. Would you expect the change to be worthwhile?

2.16 A *dequeue* (double-ended queue) is a list from which elements can be inserted or deleted at either end. Develop array, pointer, and cursor implementations for a dequeue.

2.17 Define an ADT to support the operations ENQUEUE, DEQUEUE, and ONQUEUE. ONQUEUE(x) is a function returning true or false depending on whether x is or is not on the queue.

2.18 How would one implement a queue if the elements that are to be placed on the queue are arbitrary length strings? How long does it take to enqueue a string?

2.19 Another possible linked-list implementation of queues is to use no header cell, and let *front* point directly to the first cell. If the queue is empty, let *front* = *rear* = **nil**. Implement the queue operations for this representation. How does this implementation compare with the list implementation given for queues in Section 2.4 in terms of speed, space utilization, and conciseness of the code?

2.20 A variant of the circular queue records the position of the front element and the length of the queue.
   a) Is it necessary in this implementation to limit the length of a queue to *maxlength* − 1?
   b) Write the five queue operations for this implementation.
   c) Compare this implementation with the circular queue implementation of Section 2.4.

2.21 It is possible to keep two stacks in a single array, if one grows from position 1 of the array, and the other grows from the last position. Write a procedure PUSH(x, S) that pushes element x onto stack S, where S is one or the other of these two stacks. Include all necessary error checks in your procedure.

2.22 We can store k stacks in a single array if we use the data structure suggested in Fig. 2.27, for the case $k = 3$. We push and pop from each stack as suggested in connection with Fig. 2.17 in Section 2.3. However, if pushing onto stack i causes TOP(i) to equal BOTTOM($i-1$), we first move all the stacks so that there is an appropriate size gap between each adjacent pair of stacks. For example, we might make the gaps above all stacks equal, or we might make the gap above stack i proportional to the current size of stack i (on the theory that larger stacks are likely to grow sooner, and we want to postpone as long as possible the next reorganization).

a) On the assumption that there is a procedure *reorganize* to call when stacks collide, write code for the five stack operations.

b) On the assumption that there is a procedure *makenewtops* that computes *newtop*[i], the "appropriate" position for the top of stack i, for $1 \le i \le k$, write the procedure *reorganize*. *Hint.* Note that stack i could move up or down, and it is necessary to move stack i before stack j if the new position of stack j overlaps the old position of stack i. Consider stacks $1, 2, \ldots, k$ in order, but keep a stack of "goals," each goal being to move a particular stack. If on considering stack i, we can move it safely, do so, and then reconsider the stack whose number is on top of the goal stack. If we cannot safely move stack i, push i onto the goal stack.

c) What is an appropriate implementation for the goal stack in (b)? Do we really need to keep it as a list of integers, or will a more succinct representation do?

d) Implement *makenewtops* in such a way that space above each stack is proportional to the current size of that stack.

e) What modifications of Fig. 2.27 are needed to make the implementation work for queues? For general lists?

2.23 Modify the implementations of POP and ENQUEUE in Sections 2.3 and 2.4 to return the element removed from the stack or queue. What modifications must be made if the element type is not a type that can be returned by a function?

2.24 Use a stack to eliminate recursion from the following procedures.

a)
```
        function comb ( n, m: integer ) : integer;
```
$$\{ \text{computes } \binom{n}{m} \text{ assuming } 0 \le m \le n \text{ and } n \ge 1 \}$$
```
            begin
                if (n = 1) or (m = 0) or (m = n) then
                    return (1)
                else
                    return (comb(n-1, m) + comb(n-1, m-1))
            end; { comb }
```

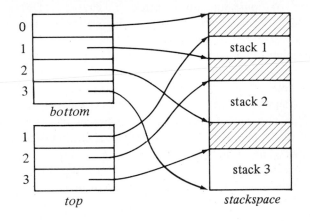

Fig. 2.27. Many stacks in one array.

b)   **procedure** *reverse* (**var** *L*: LIST);
            { reverse list *L* }
            var
                  x: elementtype;
            begin
                  if not EMPTY(*L*) **then begin**
                        x := RETRIEVE(FIRST(*L*), *L*);
                        DELETE(FIRST(*L*), *L*);
                        reverse(*L*);
                        INSERT(*x*, END(*L*), *L*)
                  end
            end; { *reverse* }

*2.25 Can we eliminate the tail recursion from the programs in Exercise 2.24? If so, do it.

Bibliographic Notes

Knuth [1968] contains additional information on the implementation of lists, stacks, and queues. A number of programming languages, such as LISP and SNOBOL, support lists and strings in a convenient manner. See Sammet [1969], Nicholls [1975], Pratt [1975], or Wexelblat [1981] for a history and description of many of these languages.

CHAPTER 3

Trees

A tree imposes a hierarchical structure on a collection of items. Familiar examples of trees are genealogies and organization charts. Trees are used to help analyze electrical circuits and to represent the structure of mathematical formulas. Trees also arise naturally in many different areas of computer science. For example, trees are used to organize information in database systems and to represent the syntactic structure of source programs in compilers. Chapter 5 describes applications of trees in the representation of data. Throughout this book, we shall use many different variants of trees. In this chapter we introduce the basic definitions and present some of the more common tree operations. We then describe some of the more frequently used data structures for trees that can be used to support these operations efficiently.

3.1 Basic Terminology

A tree is a collection of elements called *nodes*, one of which is distinguished as a *root*, along with a relation ("parenthood") that places a hierarchical structure on the nodes. A node, like an element of a list, can be of whatever type we wish. We often depict a node as a letter, a string, or a number with a circle around it. Formally, a *tree* can be defined recursively in the following manner.

1. A single node by itself is a tree. This node is also the root of the tree.
2. Suppose n is a node and $T_1, T_2, \ldots, T_k$ are trees with roots $n_1, n_2, \ldots, n_k$, respectively. We can construct a new tree by making n be the parent of nodes $n_1, n_2, \ldots, n_k$. In this tree n is the root and $T_1, T_2, \ldots, T_k$ are the *subtrees* of the root. Nodes $n_1, n_2, \ldots, n_k$ are called the *children* of node n.

Sometimes, it is convenient to include among trees the *null tree*, a "tree" with no nodes, which we shall represent by Λ.

Example 3.1. Consider the table of contents of a book, as suggested by Fig. 3.1(a). This table of contents is a tree. We can redraw it in the manner shown in Fig. 3.1(b). The parent-child relationship is depicted by a line. Trees are normally drawn top-down as in Fig. 3.1(b), with the parent above the child.

   The root, the node called "Book," has three subtrees with roots corresponding to the chapters C1, C2, and C3. This relationship is represented by the lines downward from Book to C1, C2, and C3. Book is the parent of C1, C2, and C3, and these three nodes are the children of Book.

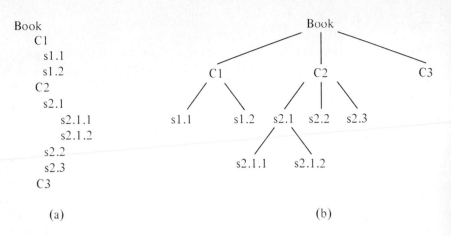

(a)   (b)

Fig. 3.1. A table of contents and its tree representation.

The third subtree, with root C3, is a tree of a single node, while the other two subtrees have a nontrivial structure. For example, the subtree with root C2 has three subtrees, corresponding to the sections s2.1, s2.2, and s2.3; the last two are one-node trees, while the first has two subtrees corresponding to the subsections s2.1.1 and s2.1.2. □

Example 3.1 is typical of one kind of data that is best represented as a tree. In this example, the parenthood relationship stands for containment; a parent node is comprised of its children, as Book is comprised of C1, C2, and C3. Throughout this book we shall encounter a variety of other relationships that can be represented by parenthood in trees.

If $n_1, n_2, \ldots, n_k$ is a sequence of nodes in a tree such that n_i is the parent of n_{i+1} for $1 \le i < k$, then this sequence is called a *path* from node n_1 to node n_k. The *length* of a path is one less than the number of nodes in the path. Thus there is a path of length zero from every node to itself. For example, in Fig. 3.1 there is a path of length two, namely (C2, s2.1, s2.1.2) from C2 to s2.1.2.

If there is a path from node a to node b, then a is an *ancestor* of b, and b is a *descendant* of a. For example, in Fig. 3.1, the ancestors of s2.1, are itself, C2, and Book, while its descendants are itself, s2.1.1, and s2.1.2. Notice that any node is both an ancestor and a descendant of itself.

An ancestor or descendant of a node, other than the node itself, is called a *proper* ancestor or *proper* descendant, respectively. In a tree, the root is the only node with no proper ancestors. A node with no proper descendants is called a *leaf*. A subtree of a tree is a node, together with all its descendants.

The *height* of a node in a tree is the length of a longest path from the node to a leaf. In Fig. 3.1 node C1 has height 1, node C2 height 2, and node C3 height 0. The *height of a tree* is the height of the root. The *depth* of a node is the length of the unique path from the root to that node.

The Order of Nodes

The children of a node are usually ordered from left-to-right. Thus the two trees of Fig. 3.2 are different because the two children of node *a* appear in a different order in the two trees. If we wish explicitly to ignore the order of children, we shall refer to a tree as an *unordered* tree.

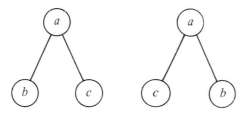

Fig. 3.2. Two distinct (ordered) trees.

The "left-to-right" ordering of *siblings* (children of the same node) can be extended to compare any two nodes that are not related by the ancestor-descendant relationship. The relevant rule is that if *a* and *b* are siblings, and *a* is to the left of *b*, then all the descendants of *a* are to the left of all the descendants of *b*.

Example 3.2. Consider the tree in Fig. 3.3. Node 8 is to the right of node 2, to the left of nodes 9, 6, 10, 4, and 7, and neither left nor right of its ancestors 1, 3, and 5.

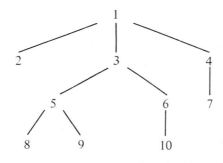

Fig. 3.3. A tree.

A simple rule, given a node *n*, for finding those nodes to its left and those to its right, is to draw the path from the root to *n*. All nodes branching off to the left of this path, and all descendants of such nodes, are to the left of *n*. All nodes and descendants of nodes branching off to the right are to the right of *n*. □

Preorder, Postorder, and Inorder

There are several useful ways in which we can systematically order all nodes
of a tree. The three most important orderings are called preorder, inorder
and postorder; these orderings are defined recursively as follows.

- If a tree T is null, then the empty list is the preorder, inorder and postorder listing of T.
- If T consists a single node, then that node by itself is the preorder,
  inorder, and postorder listing of T.

Otherwise, let T be a tree with root n and subtrees $T_1, T_2, \ldots, T_k$, as suggested in Fig. 3.4.

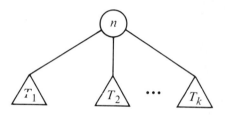

Fig. 3.4. Tree T.

1. The *preorder listing* (or *preorder traversal*) of the nodes of T is the root n
   of T followed by the nodes of T_1 in preorder, then the nodes of T_2 in
   preorder, and so on, up to the nodes of T_k in preorder.
2. The *inorder listing* of the nodes of T is the nodes of T_1 in inorder, followed by node n, followed by the nodes of $T_2, \ldots, T_k$, each group of
   nodes in inorder.
3. The *postorder listing* of the nodes of T is the nodes of T_1 in postorder,
   then the nodes of T_2 in postorder, and so on, up to T_k, all followed by
   node n.

Figure 3.5(a) shows a sketch of a procedure to list the nodes of a tree in
preorder. To make it a postorder procedure, we simply reverse the order of
steps (1) and (2). Figure 3.5(b) is a sketch of an inorder procedure. In each
case, we produce the desired ordering of the tree by calling the appropriate
procedure on the root of the tree.

Example 3.3. Let us list the tree of Fig. 3.3 in preorder. We first list 1 and
then call PREORDER on the first subtree of 1, the subtree with root 2. This
subtree is a single node, so we simply list it. Then we proceed to the second
subtree of 1, the tree rooted at 3. We list 3, and then call PREORDER on
the first subtree of 3. That call results in listing 5, 8, and 9, in that order.

```
        procedure PREORDER ( n: node );
          begin
(1)           list n;
(2)           for each child c of n, if any, in order from the left do
                  PREORDER(c)
          end;  { PREORDER }
```

(a) PREORDER procedure.

```
        procedure INORDER ( n: node );
          begin
            if n is a leaf then
              list n
            else begin
              INORDER(leftmost child of n);
              list n;
              for each child c of n, except for the leftmost,
                  in order from the left do
                    INORDER(c)
            end
          end;  { INORDER }
```

(b) INORDER procedure.

Fig. 3.5. Recursive ordering procedures.

Continuing in this manner, we obtain the complete preorder traversal of Fig. 3.3: 1, 2, 3, 5, 8, 9, 6, 10, 4, 7.

Similarly, by simulating Fig. 3.5(a) with the steps reversed, we can discover that the postorder of Fig. 3.3 is 2, 8, 9, 5, 10, 6, 3, 7, 4, 1. By simulating Fig. 3.5(b), we find that the inorder listing of Fig. 3.3 is 2, 1, 8, 5, 9, 3, 10, 6, 7, 4. □

A useful trick for producing the three node orderings is the following. Imagine we walk around the outside of the tree, starting at the root, moving counterclockwise, and staying as close to the tree as possible; the path we have in mind for Fig. 3.3 is shown in Fig. 3.6.

For preorder, we list a node the first time we pass it. For postorder, we list a node the last time we pass it, as we move up to its parent. For inorder, we list a leaf the first time we pass it, but list an interior node the second time we pass it. For example, node 1 in Fig. 3.6 is passed the first time at the beginning, and the second time while passing through the "bay" between nodes 2 and 3. Note that the order of the leaves in the three orderings is

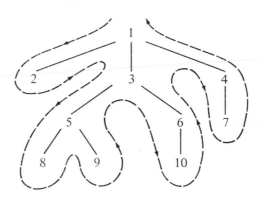

Fig. 3.6. Traversal of a tree.

always the same left-to-right ordering of the leaves. It is only the ordering of the interior nodes and their relationship to the leaves that vary among the three. □

Labeled Trees and Expression Trees

Often it is useful to associate a *label*, or value, with each node of a tree, in the same spirit with which we associated a value with a list element in the previous chapter. That is, the label of a node is not the name of the node, but a value that is "stored" at the node. In some applications we shall even change the label of a node, while the label of a node remains the same. A useful analogy is tree:list = label:element = node:position.

Example 3.4. Figure 3.7 shows a labeled tree representing the arithmetic expression $(a+b) * (a+c)$, where $n_1, \ldots, n_7$ are the names of the nodes, and the labels, by convention, are shown next to the nodes. The rules whereby a labeled tree represents an expression are as follows:

1. Every leaf is labeled by an operand and consists of that operand alone. For example, node n_4 represents the expression a.
2. Every interior node n is labeled by an operator. Suppose n is labeled by a binary operator θ, such as + or *, and that the left child represents expression E_1 and the right child E_2. Then n represents expression $(E_1) \theta (E_2)$. We may remove the parentheses if they are not necessary.

For example, node n_2 has operator +, and its left and right children represent the expressions a and b, respectively. Therefore, n_2 represents $(a)+(b)$, or just $a+b$. Node n_1 represents $(a+b)*(a+c)$, since * is the label at n_1, and $a+b$ and $a+c$ are the expressions represented by n_2 and n_3, respectively. □

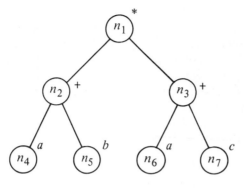

Fig. 3.7. Expression tree with labels.

Often, when we produce the preorder, inorder, or postorder listing of a tree, we prefer to list not the node names, but rather the labels. In the case of an expression tree, the preorder listing of the labels gives us what is known as the *prefix* form of an expression, where the operator precedes its left operand and its right operand. To be precise, the prefix expression for a single operand a is a itself. The prefix expression for $(E_1) \theta (E_2)$, with θ a binary operator, is $\theta P_1 P_2$, where P_1 and P_2 are the prefix expressions for E_1 and E_2. Note that no parentheses are necessary in the prefix expression, since we can scan the prefix expression $\theta P_1 P_2$ and uniquely identify P_1 as the shortest (and only) prefix of $P_1 P_2$ that is a legal prefix expression.

For example, the preorder listing of the labels of Fig. 3.7 is $*+ab+ac$. The prefix expression for n_2, which is $+ab$, is the shortest legal prefix of $+ab+ac$.

Similarly, a postorder listing of the labels of an expression tree gives us what is known as the *postfix* (or *Polish*) representation of an expression. The expression $(E_1) \theta (E_2)$ is represented by the postfix expression $P_1 P_2 \theta$, where P_1 and P_2 are the postfix representations of E_1 and E_2, respectively. Again, no parentheses are necessary in the postfix representation, as we can deduce what P_2 is by looking for the shortest suffix of $P_1 P_2$ that is a legal postfix expression. For example, the postfix expression for Fig. 3.7 is $ab+ac+*$. If we write this expression as $P_1 P_2 *$, then P_2 is $ac+$, the shortest suffix of $ab+ac+$ that is a legal postfix expression.

The inorder traversal of an expression tree gives the infix expression itself, but with no parentheses. For example, the inorder listing of the labels of Fig. 3.7 is $a+b*a+c$. The reader is invited to provide an algorithm for traversing an expression tree and producing an infix expression with all needed pairs of parentheses.

Computing Ancestral Information

The preorder and postorder traversals of a tree are useful in obtaining ancestral information. Suppose $postorder(n)$ is the position of node n in a postorder listing of the nodes of a tree. Suppose $desc(n)$ is the number of proper descendants of node n. For example, in the tree of Fig. 3.7 the postorder numbers of nodes n_2, n_4, and n_5 are 3, 1, and 2, respectively.

The postorder numbers assigned to the nodes have the useful property that the nodes in the subtree with root n are numbered consecutively from $postorder(n) - desc(n)$ to $postorder(n)$. To test if a vertex x is a descendant of vertex y, all we need do is determine whether

$$postorder(y) - desc(y) \leq postorder(x) \leq postorder(y).$$

A similar property holds for preorder.

3.2 The ADT TREE

In Chapter 2, lists, stacks, queues, and mappings were treated as abstract data types (ADT's). In this chapter trees will be treated both as ADT's and as data structures. One of our most important uses of trees occurs in the design of implementations for the various ADT's we study. For example, in Section 5.1, we shall see how a "binary search tree" can be used to implement abstract data types based on the mathematical model of a set, together with operations such as INSERT, DELETE, and MEMBER (to test whether an element is in a set). The next two chapters present a number of other tree implementations of various ADT's.

In this section, we shall present several useful operations on trees and show how tree algorithms can be designed in terms of these operations. As with lists, there are a great variety of operations that can be performed on trees. Here, we shall consider the following operations:

1. PARENT(n, T). This function returns the parent of node n in tree T. If n is the root, which has no parent, Λ is returned. In this context, Λ is a "null node," which is used as a signal that we have navigated off the tree.

2. LEFTMOST_CHILD(n, T) returns the leftmost child of node n in tree T, and it returns Λ if n is a leaf, which therefore has no children.

3. RIGHT_SIBLING(n, T) returns the right sibling of node n in tree T, defined to be that node m with the same parent p as n such that m lies immediately to the right of n in the ordering of the children of p. For example, for the tree in Fig. 3.7, LEFTMOST_CHILD(n_2) = n_4; RIGHT_SIBLING(n_4) = n_5, and RIGHT_SIBLING (n_5) = Λ.

4. LABEL(n, T) returns the label of node n in tree T. We do not, however, require labels to be defined for every tree.

5. CREATEi(v, T_1, T_2, . . . , T_i) is one of an infinite family of functions, one for each value of $i = 0, 1, 2, ...$ CREATEi makes a new node r with label v and gives it i children, which are the roots of trees T_1, T_2, . . . , T_i, in order from the left. The tree with root r is returned. Note that if $i = 0$, then r is both a leaf and the root.

6. ROOT(T) returns the node that is the root of tree T, or Λ if T is the null tree.

7. MAKENULL(T) makes T be the null tree.

Example 3.5. Let us write both recursive and nonrecursive procedures to take a tree and list the labels of its nodes in preorder. We assume that there are data types node and TREE already defined for us, and that the data type TREE is for trees with labels of the type labeltype. Figure 3.8 shows a recursive procedure that, given node n, lists the labels of the subtree rooted at n in preorder. We call PREORDER(ROOT(T)) to get a preorder listing of tree T.

```
procedure PREORDER ( n: node );
    { list the labels of the descendants of n in preorder }
    var
        c: node;
    begin
        print(LABEL(n, T));
        c := LEFTMOST_CHILD(n, T);
        while c <> Λ do begin
            PREORDER(c);
            c := RIGHT_SIBLING(c, T)
        end
    end; { PREORDER }
```

Fig. 3.8. A recursive preorder listing procedure.

We shall also develop a nonrecursive procedure to print a tree in preorder. To find our way around the tree, we shall use a stack S, whose type STACK is really "stack of nodes." The basic idea underlying our algorithm is that when we are at a node n, the stack will hold the path from the root to n, with the root at the bottom of the stack and node n at the top.†

---

† Recall our discussion of recursion in Section 2.6 in which we illustrated how the implementation of a recursive procedure involves a stack of activation records. If we examine Fig. 3.8, we can observe that when PREORDER(n) is called, the active procedure calls, and therefore the stack of activation records, correspond to the calls of PREORDER for all the ancestors of n. Thus our nonrecursive preorder procedure, like the example in Section 2.6, models closely the way the recursive procedure is implemented.

One way to perform a nonrecursive preorder traversal of a tree is given by the program NPREORDER shown in Fig. 3.9. This program has two modes of operation. In the first mode it descends down the leftmost unexplored path in the tree, printing and stacking the nodes along the path, until it reaches a leaf.

The program then enters the second mode of operation in which it retreats back up the stacked path, popping the nodes of the path off the stack, until it encounters a node on the path with a right sibling. The program then reverts back to the first mode of operation, starting the descent from that unexplored right sibling.

The program begins in mode one at the root and terminates when the stack becomes empty. The complete program is shown in Fig. 3.9.

3.3 Implementations of Trees

In this section we shall present several basic implementations for trees and discuss their capabilities for supporting the various tree operations introduced in Section 3.2.

An Array Representation of Trees

Let T be a tree in which the nodes are named 1, 2, . . . , n. Perhaps the simplest representation of T that supports the PARENT operation is a linear array A in which entry $A[i]$ is a pointer or a cursor to the parent of node i. The root of T can be distinguished by giving it a null pointer or a pointer to itself as parent. In Pascal, pointers to array elements are not feasible, so we shall have to use a cursor scheme where $A[i] = j$ if node j is the parent of node i, and $A[i] = 0$ if node i is the root.

This representation uses the property of trees that each node has a unique parent. With this representation the parent of a node can be found in constant time. A path going up the tree, that is, from node to parent to parent, and so on, can be traversed in time proportional to the number of nodes on the path. We can also support the LABEL operator by adding another array L, such that $L[i]$ is the label of node i, or by making the elements of array A be records consisting of an integer (cursor) and a label.

Example 3.6. The tree of Fig. 3.10(a) has the parent representation given by the array A shown in Fig. 3.10(b). □

The parent pointer representation does not facilitate operations that require child-of information. Given a node n, it is expensive to determine the children of n, or the height of n. In addition, the parent pointer representation does not specify the order of the children of a node. Thus, operations like LEFTMOST_CHILD and RIGHT_SIBLING are not well defined. We could impose an artificial order, for example, by numbering the children of each node after numbering the parent, and numbering the children in

```
procedure NPREORDER ( T: TREE );
    { nonrecursive preorder traversal of tree T }

  var
        m: node;  { a temporary }
        S: STACK;  { stack of nodes holding path from the root
              to the parent TOP(S) of the "current" node m }

  begin
        { initialize }
        MAKENULL(S);
        m := ROOT(T);

        while true do
            if m <> Λ then begin
                print(LABEL(m, T));
                PUSH(m, S);
                { explore leftmost child of m }
                m := LEFTMOST_CHILD(m, T)
            end
            else begin
                { exploration of path on stack
                    is now complete }
                if EMPTY(S) then
                    return;
                { explore right sibling of node
                    on top of stack }
                m := RIGHT_SIBLING(TOP(S), T);
                POP(S)
            end
    end;  { NPREORDER }
```

Fig. 3.9. A nonrecursive preorder procedure.

increasing order from left to right. On that assumption, we have written the function RIGHT_SIBLING in Fig. 3.11, for types node and TREE that are defined as follows:

```
type
      node = integer;
      TREE = array [1..maxnodes] of node;
```

For this implementation we assume the null node Λ is represented by 0.

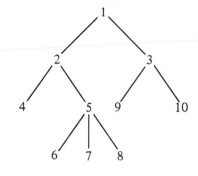

(a) a tree

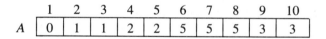

	1	2	3	4	5	6	7	8	9	10
A	0	1	1	2	2	5	5	5	3	3

(b) parent representation.

Fig. 3.10. A tree and its parent pointer representation.

```
function RIGHT_SIBLING ( n: node; T: TREE ) : node;
    { return the right sibling of node n in tree T }
    var
        i, parent: node;
    begin
        parent := T[n];
        for i := n + 1 to maxnodes do
            { search for node after n with same parent }
            if T[i] = parent then
                return (i);
            return (0)  { null node will be returned
                if no right sibling is ever found }
    end;  { RIGHT_SIBLING }
```

Fig. 3.11. Right sibling operation using array representation.

Representation of Trees by Lists of Children

An important and useful way of representing trees is to form for each node a list of its children. The lists can be represented by any of the methods suggested in Chapter 2, but because the number of children each node may have can be variable, the linked-list representations are often more appropriate.

Figure 3.12 suggests how the tree of Fig. 3.10(a) might be represented. There is an array of header cells, indexed by nodes, which we assume to be numbered 1, 2, . . . , 10. Each header points to a linked list of "elements," which are nodes. The elements on the list headed by *header*[i] are the children of node *i*; for example, 9 and 10 are the children of 3.

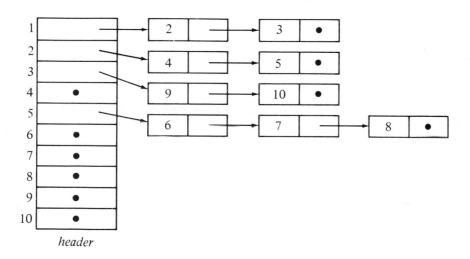

header

Fig. 3.12. A linked-list representation of a tree.

Let us first develop the data structures we need in terms of an abstract data type LIST (of nodes), and then give a particular implementation of lists and see how the abstractions fit together. Later, we shall see some of the simplifications we can make. We begin with the following type declarations:

```
type
    node = integer;
    LIST = { appropriate definition for list of nodes };
    position = { appropriate definition for positions in lists };
    TREE = record
        header: array [1..maxnodes] of LIST;
        labels: array [1..maxnodes] of labeltype;
        root: node
    end;
```

We assume that the root of each tree is stored explicitly in the *root* field. Also, 0 is used to represent the null node.

Figure 3.13 shows the code for the LEFTMOST_CHILD operation. The reader should write the code for the other operations as exercises.

```
function LEFTMOST_CHILD ( n: node; T: TREE ) : node;
    { returns the leftmost child of node n of tree T }
    var
        L: LIST; { shorthand for the list of n's children }
    begin
        L := T.header[n];
        if EMPTY(L) then   { n is a leaf }
            return (0)
        else
            return (RETRIEVE(FIRST(L), L))
    end; { LEFTMOST_CHILD }
```

Fig. 3.13. Function to find leftmost child.

Now let us choose a particular implementation of lists, in which both LIST and position are integers, used as cursors into an array *cellspace* of records:

```
    var
        cellspace[1..maxnodes] of record
            node: integer;
            next: integer
    end;
```

To simplify, we shall not insist that lists of children have header cells. Rather, we shall let *T.header*[n] point directly to the first cell of the list, as is suggested by Fig. 3.12. Figure 3.14(a) shows the function LEFTMOST_CHILD of Fig. 3.13 rewritten for this specific implementation. Figure 3.14(b) shows the operator PARENT, which is more difficult to write using this representation of lists, since a search of all lists is required to determine on which list a given node appears.

The Leftmost-Child, Right-Sibling Representation

The data structure described above has, among other shortcomings, the inability to create large trees from smaller ones, using the CREATE*i* operators. The reason is that, while all trees share *cellspace* for linked lists of children, each has its own array of headers for its nodes. For example, to implement CREATE2(v, T_1, T_2) we would have to copy T_1 and T_2 into a third tree and add a new node with label v and two children — the roots of T_1 and T_2.

If we wish to build trees from smaller ones, it is best that the representation of nodes from all trees share one area. The logical extension of Fig. 3.12 is to replace the header array by an array *nodespace* consisting of records with

```
function LEFTMOST_CHILD ( n: node; T: TREE ) : node;
    { returns the leftmost child of node n on tree T }
    var
        L: integer;  { a cursor to the beginning of the list of n's children }
    begin
        L := T.header[n];
        if L = 0 then  { n is a leaf }
            return (0)
        else
            return (cellspace[L].node)
    end;  { LEFTMOST_CHILD }
```

(a) The function LEFTMOST_CHILD.

```
function PARENT ( n: node; T: TREE ) : node;
    { returns the parent of node n in tree T }
    var
        p: node;  { runs through possible parents of n }
        i: position;  { runs down list of p's children }
    begin
        for p := 1 to maxnodes do begin
            i := T.header[p];
            while i <> 0 do  { see if n is among children of p }
                if cellspace[i].node = n then
                    return (p)
                else
                    i := cellspace[i].next
        end;
        return (0)  { return null node if parent not found }
    end;  { PARENT }
```

(b) The function PARENT.

Fig. 3.14. Two functions using linked-list representation of trees.

two fields *label* and *header*. This array will hold headers for all nodes of all
trees. Thus, we declare

```
var
    nodespace: array [1..maxnodes] of record
        label: labeltype;
        header: integer  { cursor to cellspace }
    end;
```

Then, since nodes are no longer named 1, 2, . . . , n, but are represented by arbitrary indices in *nodespace*, it is no longer feasible for the field *node* of *cellspace* to represent the "number" of a node; rather, *node* is now a cursor into *nodespace*, indicating the position of that node. The type TREE is simply a cursor into *nodespace*, indicating the position of the root.

Example 3.7. Figure 3.15(a) shows a tree, and Fig. 3.15(b) shows the data structure where we have placed the nodes labeled A, B, C, and D arbitrarily in positions 10, 5, 11, and 2 of *nodespace*. We have also made arbitrary choices for the cells of *cellspace* used for lists of children. □

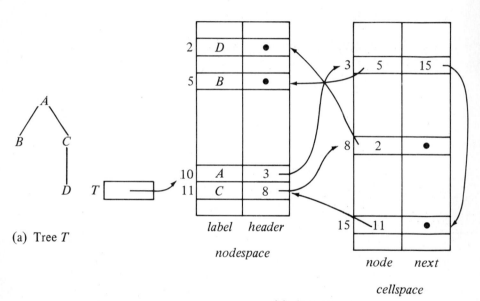

(a) Tree T

(b) Data structure

Fig. 3.15. Another linked-list structure for trees.

The structure of Fig. 3.15(b) is adequate to merge trees by the CREATEi operations. This data structure can be significantly simplified, however. First, observe that the chains of *next* pointers in *cellspace* are really right-sibling pointers.

Using these pointers, we can obtain leftmost children as follows. Suppose $cellspace[i].node = n$. (Recall that the "name" of a node, as opposed to its label, is in effect its index in *nodespace*, which is what $cellspace[i].node$ gives us.) Then $nodespace[n].header$ indicates the cell for the leftmost child of n in *cellspace*, in the sense that the *node* field of that cell is the name of that node in *nodespace*.

We can simplify matters if we identify a node not with its index in *nodespace*, but with the index of the cell in *cellspace* that represents it as a child. Then, the *next* pointers of *cellspace* truly point to right siblings, and the information contained in the *nodespace* array can be held by introducing a field *leftmost_child* in *cellspace*. The datatype TREE becomes an integer used as a cursor to *cellspace* indicating the root of the tree. We declare *cellspace* to have the following structure.

> **var**
>   *cellspace*: **array** [1..*maxnodes*] **of record**
>   *label*: labeltype;
>   *leftmost_child*: integer;
>   *right_sibling*: integer
>   **end**;

Example 3.8. The tree of Fig. 3.15(a) is represented in our new data structure in Fig. 3.16. The same arbitrary indices as in Fig. 3.15(b) have been used for the nodes. □

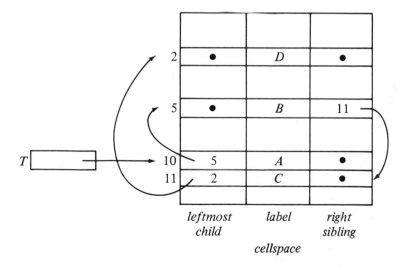

Fig. 3.16. Leftmost-child, right-sibling representation of a tree.

All operations but PARENT are straightforward to implement in the leftmost-child, right-sibling representation. PARENT requires searching the entire *cellspace*. If we need to perform the PARENT operation efficiently, we can add a fourth field to *cellspace* to indicate the parent of a node directly.

As an example of a tree operation written to use the leftmost-child, right-sibling structure as in Fig. 3.16, we give the function CREATE2 in Fig. 3.17. We assume that unused cells are linked in an available space list, headed by *avail*, and that available cells are linked by their right-sibling fields. Figure 3.18 shows the old (solid) and the new (dashed) pointers.

```
function CREATE2 ( v: labeltype; T1, T2: integer ) : integer;
    { returns new tree with root v, having T1 and T2 as subtrees }
    var
        temp: integer;   { holds index of first available cell
                for root of new tree }
    begin
        temp := avail;
        avail := cellspace[avail].right_sibling;
        cellspace[temp].leftmost_child := T1;
        cellspace[temp].label := v;
        cellspace[temp].right_sibling := 0;
        cellspace[T1].right_sibling := T2;
        cellspace[T2].right_sibling := 0;  { not necessary;
                that field should be 0 as the cell was formerly a root }
        return (temp)
    end;  { CREATE2 }
```

Fig. 3.17. The function CREATE2.

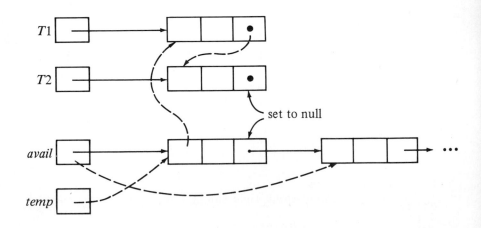

Fig. 3.18. Pointer changes produced by CREATE2.

Alternatively, we can use less space but more time if we put in the right-sibling field of the rightmost child a pointer to the parent, in place of the null pointer that would otherwise be there. To avoid confusion, we need a bit in every cell indicating whether the right-sibling field holds a pointer to the right sibling or to the parent.

Given a node, we find its parent by following right-sibling pointers until we find one that is a parent pointer. Since all siblings have the same parent, we thereby find our way to the parent of the node we started from. The time required to find a node's parent in this representation depends on the number of siblings a node has.

3.4 Binary Trees

The tree we defined in Section 3.1 is sometimes called an *ordered, oriented* tree because the children of each node are ordered from left-to-right, and because there is an oriented path (path in a particular direction) from every node to its descendants. Another useful, and quite different, notion of "tree" is the *binary tree*, which is either an empty tree, or a tree in which every node has either no children, a *left child*, a *right child*, or both a left and a right child. The fact that each child in a binary tree is designated as a left child or as a right child makes a binary tree different from the ordered, oriented tree of Section 3.1.

Example 3.9. If we adopt the convention that left children are drawn extending to the left, and right children to the right, then Fig. 3.19 (a) and (b) represent two different binary trees, even though both "look like" the ordinary (ordered, oriented) tree of Fig. 3.20. However, let us emphasize that Fig. 3.19(a) and (b) are not the same binary tree, nor are either in any sense equal to Fig. 3.20, for the simple reason that binary trees are not directly comparable with ordinary trees. For example, in Fig. 3.19(a), 2 is the left child of 1, and 1 has no right child, while in Fig. 3.19(b), 1 has no left child but has 2 as a right child. In either binary tree, 3 is the left child of 2, and 4 is 2's right child. □

The preorder and postorder listings of a binary tree are similar to those of an ordinary tree given on p. 78. The inorder listing of the nodes of a binary tree with root n, left subtree T_1 and right subtree T_2 is the inorder listing of T_1 followed by n followed by the inorder listing of T_2. For example, 35241 is the inorder listing of the nodes of Fig. 3.19(a).

Representing Binary Trees

A convenient data structure for representing a binary tree is to name the nodes 1, 2, . . . , n, and to use an array of records declared

```
var
    cellspace: array [1..maxnodes] of record
        leftchild: integer;
        rightchild: integer
    end;
```

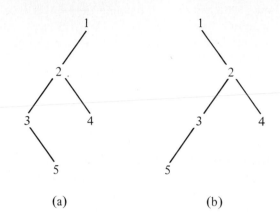

Fig. 3.19. Two binary trees.

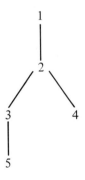

Fig. 3.20. An "ordinary" tree.

The intention is that *cellspace*[*i*].*leftchild* is the left child of node *i*, and *rightchild* is analogous. A value of 0 in either field indicates the absence of a child.

Example 3.10. The binary tree of Fig. 3.19(a) can be represented as shown in Fig. 3.21. □

An Example: Huffman Codes

Let us give an example of how binary trees can be used as a data structure. The particular problem we shall consider is the construction of "Huffman codes." Suppose we have messages consisting of sequences of characters. In each message, the characters are independent and appear with a known

	leftchild	rightchild
1	2	0
2	3	4
3	0	5
4	0	0
5	0	0

Fig. 3.21. Representation of a binary tree.

probability in any given position; the probabilities are the same for all positions. As an example, suppose we have a message made from the five characters a, b, c, d, e, which appear with probabilities .12, .4, .15, .08, .25, respectively.

We wish to encode each character into a sequence of 0's and 1's so that no code for a character is the prefix of the code for any other character. This *prefix property* allows us to decode a string of 0's and 1's by repeatedly deleting prefixes of the string that are codes for characters.

Example 3.11. Figure 3.22 shows two possible codes for our five symbol alphabet. Clearly Code 1 has the prefix property, since no sequence of three bits can be the prefix of another sequence of three bits. The decoding algorithm for Code 1 is simple. Just "grab" three bits at a time and translate each group of three into a character. Of course, sequences 101, 110, and 111 are impossible, if the string of bits really codes characters according to Code 1. For example, if we receive 001010011 we know the original message was *bcd*.

Symbol	Probability	Code 1	Code 2
a	.12	000	000
b	.40	001	11
c	.15	010	01
d	.08	011	001
e	.25	100	10

Fig. 3.22. Two binary codes.

It is easy to check that Code 2 also has the prefix property. We can decode a string of bits by repeatedly "grabbing" prefixes that are codes for characters and removing them, just as we did for Code 1. The only difference is that here, we cannot slice up the entire sequence of bits at once, because whether we take two or three bits for a character depends on the bits. For example, if a string begins 1101001, we can again be sure that the characters coded were *bcd*. The first two bits, 11, must have come from *b*, so we can remove them and worry about 01001. We then deduce that the bits 01 came from c, and so on. □

The problem we face is: given a set of characters and their probabilities, find a code with the prefix property such that the average length of a code for a character is a minimum. The reason we want to minimize the average code length is to compress the length of an average message. The shorter the average code for a character is, the shorter the length of the encoded message. For example, Code 1 has an average code length of 3. This is obtained by multiplying the length of the code for each symbol by the probability of occurrence of that symbol. Code 2 has an average length of 2.2, since symbols a and d, which together appear 20% of the time, have codes of length three, and the other symbols have codes of length two.

Can we do better than Code 2? A complete answer to this question is to exhibit a code with the prefix property having an average length of 2.15. This is the best possible code for these probabilities of symbol occurrences. One technique for finding optimal prefix codes is called *Huffman's algorithm*. It works by selecting two characters a and b having the lowest probabilities and replacing them with a single (imaginary) character, say x, whose probability of occurrence is the sum of the probabilities for a and b. We then find an optimal prefix code for this smaller set of characters, using this procedure recursively. The code for the original character set is obtained by using the code for x with a 0 appended as the code for a and with a 1 appended as a code for b.

We can think of prefix codes as paths in binary trees. Think of following a path from a node to its left child as appending a 0 to a code, and proceeding from a node to its right child as appending a 1. If we label the leaves of a binary tree by the characters represented, we can represent any prefix code as a binary tree. The prefix property guarantees no character can have a code that is an interior node, and conversely, labeling the leaves of any binary tree with characters gives us a code with the prefix property for these characters.

Example 3.12. The binary trees for Code 1 and Code 2 of Fig. 3.22 are shown in Fig. 3.23(a) and (b), respectively. □

We shall implement Huffman's algorithm using a *forest* (collection of trees), each of which has its leaves labeled by characters whose codes we desire to select and whose roots are labeled by the sum of the probabilities of all the leaf labels. We call this sum the *weight* of the tree. Initially, each character is in a one-node tree by itself, and when the algorithm ends, there will be only one tree, with all the characters at its leaves. In this tree, the path from the root to any leaf represents the code for the label of that leaf, according to the left = 0, right = 1 scheme of Fig. 3.23.

The essential step of the algorithm is to select the two trees in the forest that have the smallest weights (break ties arbitrarily). Combine these two trees into one, whose weight is the sum of the weights of the two trees. To combine the trees we create a new node, which becomes the root and has the

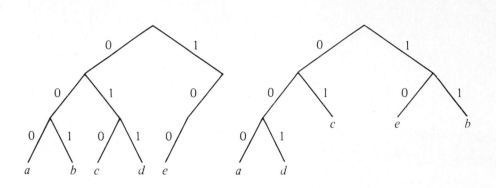

Fig. 3.23. Binary trees representing codes with the prefix property.

roots of the two given trees as left and right children (which is which doesn't matter). This process continues until only one tree remains. That tree represents a code that, for the probabilities given, has the minimum possible average code length.

Example 3.13. The sequence of steps taken for the characters and probabilities in our running example is shown in Fig. 3.24. From Fig. 3.24(e) we see the code words for a, b, c, d, and e are 1111, 0, 110, 1110, and 10. In this example, there is only one nontrivial tree, but in general, there can be many. For example, if the probabilities of b and e were .33 and .32, then after Fig. 3.24(c) we would combine b and e, rather than attaching e to the large tree as we did in Fig. 3.24(d). □

Let us now describe the needed data structures. First, we shall use an array *TREE* of records of the type

```
record
    leftchild: integer;
    rightchild: integer;
    parent: integer
end
```

to represent binary trees. Parent pointers facilitate finding paths from leaves to roots, so we can discover the code for a character. Second, we use an array *ALPHABET* of records of type

```
record
    symbol: char;
    probability: real;
    leaf: integer { cursor into tree }
end
```

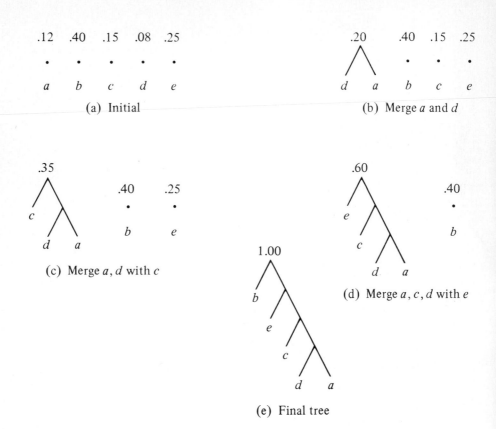

Fig. 3.24. Steps in the construction of a Huffman tree.

to associate, with each symbol of the alphabet being encoded, its corresponding leaf. This array also records the probability of each character. Third, we
need an array *FOREST* of records that represent the trees themselves. The
type of these records is

> **record**
>   *weight*: real;
>   *root*: integer { cursor into tree }
> **end**

The initial values of all these arrays, assuming the data of Fig. 3.24(a), are
shown in Fig. 3.25. A sketch of the program to build the Huffman tree is
shown in Fig. 3.26.

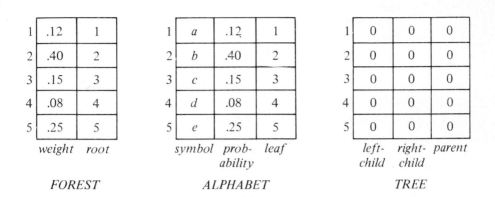

	weight	root
1	.12	1
2	.40	2
3	.15	3
4	.08	4
5	.25	5

FOREST

	symbol	prob-ability	leaf
1	a	.12	1
2	b	.40	2
3	c	.15	3
4	d	.08	4
5	e	.25	5

ALPHABET

	left-child	right-child	parent
1	0	0	0
2	0	0	0
3	0	0	0
4	0	0	0
5	0	0	0

TREE

Fig. 3.25. Initial contents of arrays.

(1)   **while** there is more then one tree in the forest **do**
              begin
(2) i := index of the tree in *FOREST* with smallest weight;
(3) j := index of the tree in *FOREST* with second smallest weight;
(4)   create a new node with left child *FOREST*[i].*root* and
                      right child *FOREST*[j].*root*;
(5)   replace tree i in *FOREST* by a tree whose root
                      is the new node and whose weight is
                      FOREST[i].*weight* + *FOREST*[j].*weight*;
(6)   delete tree j from *FOREST*
          end;

Fig. 3.26. Sketch of Huffman tree construction.

To implement line (4) of Fig. 3.26, which increases the number of cells of the *TREE* array used, and lines (5) and (6), which decrease the number of utilized cells of *FOREST*, we shall introduce cursors *lasttree* and *lastnode*, pointing to *FOREST* and *TREE*, respectively. We assume that cells 1 to *lasttree* of *FOREST* and 1 to *lastnode* of *TREE* are occupied.† We assume that arrays of Fig. 3.25 have some declared lengths, but in what follows we omit comparisons between these limits and cursor values.

---

† For the data reading phase, which we omit, we also need a cursor for the array *ALPHABET* as it fills with symbols and their probabilities.

```
procedure lightones ( var least, second: integer );
    { sets least and second to the indices in FOREST of
        the trees of smallest weight.  We assume lasttree ≥2. }
    var
        i: integer;
    begin    { initialize least and second, considering first two trees }
        if FOREST[1].weight <= FOREST[2].weight then
            begin least := 1; second := 2 end
        else
            begin least := 2; second := 1 end;
        { Now let i run from 3 to lasttree.  At each iteration
            least is the tree of smallest weight among the first i trees
            in FOREST, and second is the next smallest of these }
        for i := 3 to lasttree do
            if FOREST[i].weight < FOREST[least].weight then
                begin second := least; least := i end
            else if FOREST[i].weight < FOREST[second].weight then
                second := i
    end;  { lightones }


function create ( lefttree, righttree: integer ) : integer;
    { returns new node whose left and right children are
        FOREST[lefttree].root and FOREST[righttree].root }
    begin
        lastnode := lastnode + 1;
            { cell for new node is TREE[lastnode] }
        TREE[lastnode].leftchild := FOREST[lefttree].root;
        TREE[lastnode].rightchild := FOREST[righttree].root;
        { now enter parent pointers for new node and its children }
        TREE[lastnode].parent := 0;
        TREE[FOREST[lefttree].root].parent := lastnode;
        TREE[FOREST[righttree].root].parent := lastnode;
        return (lastnode)
    end;  { create }
```

Fig. 3.27. Two procedures.

Figure 3.27 shows two useful procedures. The first implements lines (2) and (3) of Fig. 3.26 to select indices of the two trees of smallest weight. The second is the command $create(n_1, n_2)$ that creates a new node and makes n_1 and n_2 its left and right children.

Now the steps of Fig. 3.26 can be described in greater detail. A procedure *Huffman*, which has no input or output, but works on the global structures of Fig. 3.25, is shown in Fig. 3.28.

```
procedure Huffman;
    var
        i, j: integer; { the two trees of least weight in FOREST }
        newroot: integer;
    begin
        while lasttree > 1 do begin
            lightones(i, j);
            newroot := create(i, j);
            { Now replace tree i by the tree whose root is newroot }
            FOREST[i].weight := FOREST[i].weight + FOREST[j].weight;
            FOREST[i].root := newroot;
            { next, replace tree j, which is no longer needed, by lasttree,
                and shrink FOREST by one }
            FOREST[j] := FOREST[lasttree];
            lasttree := lasttree − 1
        end
    end; { Huffman }
```

Fig. 3.28. Huffman's algorithm.

Figure 3.29 shows the data structure of Fig. 3.25 after *lasttree* has been reduced to 3, that is, when the forest looks like Fig. 3.24(c).

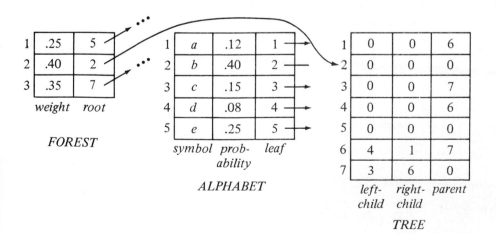

Fig. 3.29. Tree data structure after two iterations.

After completing execution of the algorithm, the code for each symbol can be determined as follows. Find the symbol in the *symbol* field of the *ALPHA-BET* array. Follow the *leaf* field of the same record to get to a record of the *TREE* array; this record corresponds to the leaf for that symbol. Repeatedly follow the *parent* pointer from the "current" record, say for node *n*, to the record of the *TREE* array for its parent *p*. Remember node *n*, so it is possible to examine the *leftchild* and *rightchild* pointers for node *p* and see which is *n*. In the former case, print 0, in the latter print 1. The sequence of bits printed is the code for the symbol, in reverse. If we wish the bits printed in the correct order, we could push each onto a stack as we go up the tree, and then repeatedly pop the stack, printing symbols as we pop them.

Pointer-Based Implementations of Binary Trees

Instead of using cursors to point to left and right children (and parents if we wish), we can use true Pascal pointers. For example, we might declare

```
type
    node = record
        leftchild: ↑ node;
        rightchild: ↑ node;
        parent: ↑ node
    end
```

For example, if we used this type for nodes of a binary tree, the function *create* of Fig. 3.27 could be written as in Fig. 3.30.

```
function create ( lefttree, righttree: ↑ node ) : ↑ node;
    var
        root: ↑ node;
    begin
        new(root);
        root↑.leftchild := lefttree;
        root↑.rightchild := righttree;
        root↑.parent := 0;
        lefttree↑.parent := root;
        righttree↑.parent := root;
        return (root)
    end; { create }
```

Fig. 3.30. Pointer-based implementation of binary trees.

Exercises

3.1 Answer the following questions about the tree of Fig. 3.31.

    a) Which nodes are leaves?

    b) Which node is the root?

    c) What is the parent of node C?

    d) Which nodes are children of C?

    e) Which nodes are ancestors of E?

    f) Which nodes are descendants of E?

    g) What are the right siblings of D and E?

    h) Which nodes are to the left and to the right of G?

    i) What is the depth of node C?

    j) What is the height of node C?

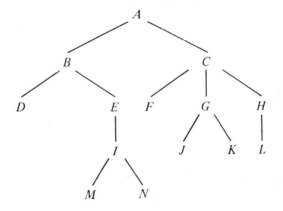

Fig. 3.31. A tree.

3.2 In the tree of Fig. 3.31 how many different paths of length three are there?

3.3 Write programs to compute the height of a tree using each of the three tree representations of Section 3.3.

3.4 List the nodes of Fig. 3.31 in

    a) preorder,

b) postorder, and

c) inorder.

3.5 If m and n are two different nodes in the same tree, show that exactly one of the following statements is true:

a) m is to the left of n

b) m is to the right of n

c) m is a proper ancestor of n

d) m is a proper descendant of n.

3.6 Place a check in row i and column j if the two conditions represented by row i and column j can occur simultaneously.

	preorder(n) < preorder(m)	inorder(n) < inorder(m)	postorder(n) < postorder(m)
n is to the left of m			
n is to the right of m			
n is a proper ancestor of m			
n is a proper descendant of m			

For example, put a check in row 3 and column 2 if you believe that n can be a proper ancestor of m and at the same time n can precede m in inorder.

3.7 Suppose we have arrays PREORDER[n], INORDER[n], and POSTORDER[n] that give the preorder, inorder, and postorder positions, respectively, of each node n of a tree. Describe an algorithm that tells whether node i is an ancestor of node j, for any pair of nodes i and j. Explain why your algorithm works.

*3.8 We can test whether a node m is a proper ancestor of a node n by testing whether m precedes n in X-order but follows n in Y-order, where X and Y are chosen from {pre, post, in}. Determine all those pairs X and Y for which this statement holds.

3.9 Write programs to traverse a binary tree in

a) preorder,

b) postorder,

c) inorder.

3.10 The *level-order* listing of the nodes of a tree first lists the root, then all nodes of depth 1, then all nodes of depth 2, and so on. Nodes at the same depth are listed in left-to-right order. Write a program to list the nodes of a tree in level-order.

3.11 Convert the expression $((a+b) + c * (d+e) + f) * (g+h)$ to a

   a) prefix expression

   b) postfix expression.

3.12 Draw tree representations for the expressions

   a) $*a+b*c+de$

   b) $*a+*b+cde$

3.13 Let T be a tree in which every nonleaf node has two children. Write a program to convert

   a) a preorder listing of T into a postorder listing,

   b) a postorder listing of T into a preorder listing,

   c) a preorder listing of T into an inorder listing.

3.14 Write a program to evaluate

   a) preorder

   b) postorder

arithmetic expressions.

3.15 We can define a binary tree as an ADT with the binary tree structure as a mathematical model and with operations such as LEFTCHILD(n), RIGHTCHILD(n), PARENT(n), and NULL(n). The first three operations return the left child, the right child, and the parent of node n (Λ if there is none) and the last returns true if and only if n is Λ. Implement these procedures using the binary tree representation of Fig. 3.21.

3.16 Implement the seven tree operations of Section 3.2 using the following tree implementations:

   a) parent pointers

   b) lists of children

   c) leftmost-child, right-sibling pointers.

3.17 The *degree* of a node is the number of children it has. Show that in any binary tree the number of leaves is one more than the number of nodes of degree two.

3.18 Show that the maximum number of nodes in a binary tree of height h is $2^{h+1}-1$. A binary tree of height h with the maximum number of nodes is called a *full* binary tree.

*3.19 Suppose trees are implemented by leftmost-child, right-sibling and parent pointers. Give nonrecursive preorder, postorder, and inorder traversal algorithms that do not use "states" or a stack, as Fig. 3.9 does.

3.20 Suppose characters a, b, c, d, e, f have probabilities .07, .09, .12, .22, .23, .27, respectively. Find an optimal Huffman code and draw the Huffman tree. What is the average code length?

*3.21 Suppose T is a Huffman tree, and that the leaf for symbol a has greater depth than the leaf for symbol b. Prove that the probability of symbol b is no less than that of a.

*3.22 Prove that Huffman's algorithm works, i.e., it produces an optimal code for the given probabilities. *Hint*: Use Exercise 3.21.

Bibliographic Notes

Berge [1958] and Harary [1969] discuss the mathematical properties of trees. Knuth [1973] and Nievergelt [1974] contain additional information on binary search trees. Many of the works on graphs and applications referenced in Chapter 6 also cover material on trees.

The algorithm given in Section 3.4 for finding a tree with a minimal weighted path length is from Huffman [1952]. Parker [1980] gives some more recent explorations into that algorithm.

CHAPTER 4

Basic
Operations
on
Sets

The set is the basic structure underlying all of mathematics. In algorithm design, sets are used as the basis of many important abstract data types, and many techniques have been developed for implementing set-based abstract data types. In this chapter we review the basic operations on sets and introduce some simple implementations for sets. We present the "dictionary" and "priority queue," two abstract data types based on the set model. Implementations for these abstract data types are covered in this and the next chapter.

4.1 Introduction to Sets

A set is a collection of *members* (or *elements*); each member of a set either is itself a set or is a primitive element called an *atom*. All members of a set are different, which means no set can contain two copies of the same element.

When used as tools in algorithm and data structure design, atoms usually are integers, characters, or strings, and all elements in any one set are usually of the same type. We shall often assume that atoms are linearly ordered by a relation, usually denoted "<" and read "less than" or "precedes." A *linear order* < on a set S satisfies two properties:

1. For any a and b in S, exactly one of $a < b$, $a = b$, or $b < a$ is true.
2. For all a, b, and c in S, if $a < b$ and $b < c$, then $a < c$ (transitivity).

Integers, reals, characters, and character strings have a natural linear ordering for which < is used in Pascal. A linear ordering can be defined on objects that consist of sets of ordered objects. We leave as an exercise how one develops such an ordering. For example, one question to be answered in constructing a linear order for a set of integers is whether the set consisting of integers 1 and 4 should be regarded as being less than or greater than the set consisting of 2 and 3.

Set Notation

A set of atoms is generally exhibited by putting curly brackets around its members, so $\{1, 4\}$ denotes the set whose only members are 1 and 4. We should bear in mind that a set is not a list, even though we represent sets in this manner as if they were lists. The order in which the elements of a set are listed is irrelevant, and we could just as well have written $\{4, 1\}$ in place of $\{1, 4\}$. Note also that in a set each element appears exactly once, so $\{1, 4, 1\}$ is not a set.†

Sometimes we represent sets by *set formers*, which are expressions of the form

$$\{\, x \mid \text{statement about } x \,\}$$

where the statement about x is a predicate that tells us exactly what is needed for an arbitrary object x to be in the set. For example, $\{x \mid x$ is a positive integer and $x \le 1000\}$ is another way of representing $\{1, 2, \ldots , 1000\}$, and $\{x \mid$ for some integer y, $x = y^2\}$ denotes the set of perfect squares. Note that the set of perfect squares is infinite and cannot be represented by listing its members.

The fundamental relationship of set theory is membership, which is denoted by the symbol $\in$. That is, $x \in A$ means that element x is a member of set A; the element x could be an atom or another set, but A cannot be an atom. We use $x \notin A$ for "x is not a member of A." There is a special set, denoted $\varnothing$ and called the *null set* or *empty set*, that has no members. Note that $\varnothing$ is a set, not an atom, even though the set $\varnothing$ does not have any members. The distinction is that $x \in \varnothing$ is false for every x, whereas if y is an atom, then $x \in y$ doesn't even make sense; it is syntactically meaningless rather than false.

We say set A is *included* (or *contained*) in set B, written $A \subseteq B$, or $B \supseteq A$, if every member of A is also a member of B. We also say A is a *subset of B* and B is a *superset* of A, if $A \subseteq B$. For example, $\{1, 2\} \subseteq \{1, 2, 3\}$, but $\{1, 2, 3\}$ is not a subset of $\{1, 2\}$ since 3 is a member of the former but not the latter. Every set is included in itself, and the empty set is included in every set. Two sets are equal if each is included in the other, that is, if their members are the same. Set A is a *proper subset* or *proper superset* of set B if $A \ne B$, and $A \subseteq B$ or $A \supseteq B$, respectively.

The most basic operations on sets are union, intersection, and difference. If A and B are sets, then $A \cup B$, the *union* of A and B, is the set of elements that are members of A or B or both. The *intersection* of A and B, written $A \cap B$, is the set of elements in both A and B, and the difference, $A - B$, is the set of elements in A that are not in B. For example, if $A = \{a, b, c\}$ and $B = \{b, d\}$, then $A \cup B = \{a, b, c, d\}$, $A \cap B = \{b\}$, and $A - B = \{a, c\}$.

---

† Sometimes the term *multiset* is used for a "set with repetitions." That is, the multiset $\{1, 4, 1\}$ has 1 (twice) and 4 (once) as members. A multiset is not a list any more than a set is, and thus this multiset could also have been written $\{4, 1, 1\}$ or $\{1, 1, 4\}$.

Abstract Data Types Based on Sets

We shall consider ADT's that incorporate a variety of set operations. Some collections of these operations have been given special names and have special implementations of high efficiency. Some of the more common set operations are the following.

1.–3. The three procedures UNION(A, B, C), INTERSECTION(A, B, C), and DIFFERENCE(A, B, C) take set-valued arguments A and B, and assign the result, $A \cup B$, $A \cap B$, or $A - B$, respectively, to the set variable C.

4. We shall sometimes use an operation called *merge*, or *disjoint set union*, that is no different from union, but that assumes its operands are *disjoint* (have no members in common). The procedure MERGE(A, B, C) assigns to the set variable C the value $A \cup B$, but is not defined if $A \cap B \neq \emptyset$, i.e., if A and B are not disjoint.

5. The function MEMBER(x, A) takes set A and object x, whose type is the type of elements of A, and returns a boolean value — true if $x \in A$ and false if $x \notin A$.

6. The procedure MAKENULL(A) makes the null set be the value for set variable A.

7. The procedure INSERT(x, A), where A is a set-valued variable, and x is an element of the type of A's members, makes x a member of A. That is, the new value of A is $A \cup \{x\}$. Note that if x is already a member of A, then INSERT(x, A) does not change A.

8. DELETE(x, A) removes x from A, i.e., A is replaced by $A - \{x\}$. If x is not in A originally, DELETE(x, A) does not change A.

9. ASSIGN(A, B) sets the value of set variable A to be equal to the value of set variable B.

10. The function MIN(A) returns the least element in set A. This operation may be applied only when the members of the parameter set are linearly ordered. For example, MIN($\{2, 3, 1\}$) $= 1$ and MIN($\{'a', 'b', 'c'\}$) $= 'a'$. We also use a function MAX with the obvious meaning.

11. EQUAL(A, B) is a function whose value is true if and only if sets A and B consist of the same elements.

12. The function FIND(x) operates in an environment where there is a collection of disjoint sets. FIND(x) returns the name of the (unique) set of which x is a member.

4.2 An ADT with Union, Intersection, and Difference

We begin by defining an ADT for the mathematical model "set" with the three basic set-theoretic operations, union, intersection, and difference. First we give an example where such an ADT is useful and then we discuss several simple implementations of this ADT.

Example 4.1. Let us write a program to do a simple form of "data-flow analysis" on flowcharts that represent procedures. The program will use variables of an abstract data type SET, whose operations are UNION, INTERSECTION, DIFFERENCE, EQUAL, ASSIGN, and MAKENULL, as defined in the previous section.

In Fig. 4.1 we see a flowchart whose boxes have been named $B_1, \ldots, B_8$, and for which the *data definitions* (read and assignment statements) have been numbered $1, 2, \ldots, 9$. This flowchart happens to implement the Euclidean algorithm, to compute the greatest common divisor of inputs p and q, but the details of the algorithm are not relevant to the example.

In general, *data-flow analysis* refers to that part of a compiler that examines a flowchart-like representation of a source program, such as Fig. 4.1, and collects information about what can be true as control reaches each box of the flowchart. The boxes are often called *blocks* or *basic blocks*, and they represent collections of statements through which the flow-of-control proceeds sequentially. The information collected during data-flow analysis is used to help improve the code generated by the compiler. For example, if data-flow analysis told us that each time control reached block B, variable x had the value 27, then we could substitute 27 for all uses of x in block B, unless x were assigned a new value within block B. If constants can be accessed more quickly than variables, this change could speed up the code produced by the compiler.

In our example, we want to determine where a variable could last have been given a new value. That is, we want to compute for each block B_i the set $DEFIN[i]$ of data definitions d such that there is a path from B_1 to B_i in which d appears, but is not followed by any other definition of the same variable as d defines. $DEFIN[i]$ is called the set of *reaching definitions* for B_i.

To see how such information could be useful, consider Fig. 4.1. The first block B_1 is a "dummy" block of three data definitions, making the three variables t, p, and q have "undefined" values. If we discover, for example, that $DEFIN[7]$ includes definition 3, which gives q an undefined value, then the program might contain a bug, as apparently it could print q without first assigning a valid value to q. Fortunately, we shall discover that it is impossible to reach block B_7 without assigning to q; that is, 3 is not in $DEFIN[7]$.

The computation of the $DEFIN[i]$'s is aided by several rules. First, we precompute for each block i two sets $GEN[i]$ and $KILL[i]$. $GEN[i]$ is the set of data definitions in block i, with the exception that if B_i contains two or more definitions of variable x, then only the last is in $GEN[i]$. Thus, $GEN[i]$ is the set of definitions in B_i that are "generated" by B_i; they reach the end of B_i without having their variables redefined.

The set $KILL[i]$ is the set of definitions d not in B_i such that B_i has a definition of the same variable as d. For example, in Fig. 4.1, $GEN[4] = \{6\}$, since definition 6 (of variable t) is in B_4 and there are no subsequent definitions of t in B_4. $KILL[4] = \{1, 9\}$, since these are the definitions of variable t that are not in B_4.

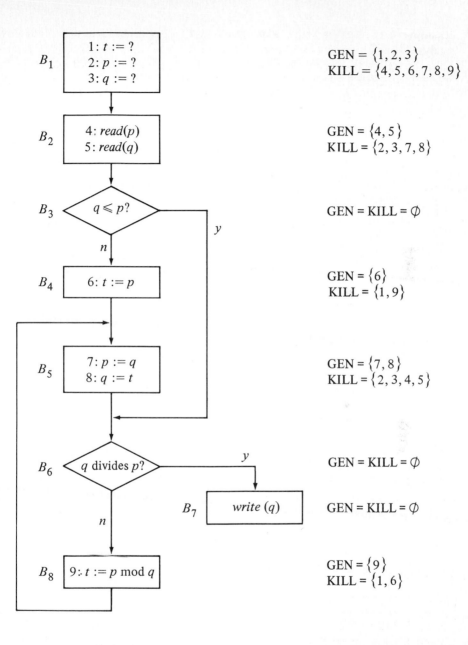

Fig. 4.1. A flowchart of the Euclidean algorithm.

In addition to the *DEFIN*[*i*]'s, we compute the set *DEFOUT*[*i*] for each block B_i. Just as *DEFIN*[*i*] is the set of definitions that reach the beginning of B_i, *DEFOUT*[*i*] is the set of definitions reaching the end of B_i. There is a simple formula relating *DEFIN* and *DEFOUT*, namely

$$DEFOUT[i] = (DEFIN[i] - KILL[i]) \bigcup GEN[i] \qquad (4.1)$$

That is, definition d reaches the end of B_i if and only if it either reaches the beginning of B_i and is not killed by B_i, or it is generated in B_i. The second rule relating $DEFIN$ and $DEFOUT$ is that $DEFIN[i]$ is the union, over all predecessors p of B_i, of $DEFOUT[p]$, that is:

$$DEFIN[i] = \bigcup_{B_p \text{ a predecessor of } B_i} DEFOUT[p] \qquad (4.2)$$

Rule (4.2) says that a data definition enters B_i if and only if it reaches the end of one of B_i's predecessors. As a special case, if B_i has no predecessors, as B_1 in Fig. 4.1, then $DEFIN[i] = \varnothing$.

Because we have introduced a variety of new concepts in this example, we shall not try to complicate matters by writing a general algorithm for computing the reaching definitions of an arbitrary flowgraph. Rather, we shall write a part of a program that assumes $GEN[i]$ and $KILL[i]$ are available for $i = 1, \ldots, 8$ and computes $DEFIN[i]$ and $DEFOUT[i]$ for $1, \ldots, 8$, assuming the particular flowgraph of Fig. 4.1. This program fragment assumes the existence of an ADT SET with operations UNION, INTERSECTION, DIFFERENCE, EQUAL, ASSIGN, and MAKENULL; we shall give alternative implementations of this ADT later.

The procedure *propagate*($GEN, KILL, DEFIN, DEFOUT$) applies rule (4.1) to compute $DEFOUT$ for a block, given $DEFIN$. If a program were loop-free, then the calculation of $DEFOUT$ would be straightforward. The presence of a loop in the program fragment of Fig. 4.2 necessitates an iterative procedure. We approximate $DEFIN[i]$ by starting with $DEFIN[i] = \varnothing$ and $DEFOUT[i] = GEN[i]$ for all i, and then repeatedly apply (4.1) and (4.2) until no more changes to $DEFIN$'s and $DEFOUT$'s occur. Since each new value assigned to $DEFIN[i]$ or $DEFOUT[i]$ can be shown to be a superset (not necessarily proper) of its former value, and there are only a finite number of data definitions in any program, the process must converge eventually to a solution to (4.1) and (4.2).

The successive values of $DEFIN[i]$ after each iteration of the repeat-loop are shown in Fig. 4.3. Note that none of the dummy assignments 1, 2, and 3 reaches a block where their variable is used, so there are no undefined variable uses in the program of Fig. 4.1. Also note that by deferring the application of (4.2) for B_i until just before we apply (4.1) for B_i would make the process of Fig. 4.2 converge in fewer iterations in general. □

4.3 A Bit-Vector Implementation of Sets

The best implementation of a SET ADT depends on the operations to be performed and on the size of the set. When all sets in our domain of discourse are subsets of a small "universal set" whose elements are the integers $1, \ldots, N$ for some fixed N, then we can use a *bit-vector* (boolean array) implementation. A set is represented by a bit vector in which the ith bit is true if i is an element of the set. The major advantage of this representation

```
var
     GEN, KILL, DEFIN, DEFOUT: array[1..8] of SET;
          { we assume GEN and KILL are computed externally }
     i: integer;
     changed: boolean;

procedure propagate ( G, K, I: SET; var O: SET );
     { apply (4.1) and set changed to true if
          a change in DEFOUT is detected }
     var
          TEMP: SET;
     begin
          DIFFERENCE(I, K, TEMP);
          UNION(TEMP, G, TEMP);
          if not EQUAL(TEMP, O) do begin
               ASSIGN(O, TEMP);
               changed := true
          end
     end;  { propagate }

begin
     for i := 1 to 8 do
          ASSIGN(DEFOUT[i], GEN[i]);
     repeat
          changed := false;
          {the next 8 statements apply (4.2) for the
               graph of Fig. 4.1 only}
          MAKENULL(DEFIN[1]);
          ASSIGN(DEFIN[2], DEFOUT[1]);
          ASSIGN(DEFIN[3], DEFOUT[2]);
          ASSIGN(DEFIN[4], DEFOUT[3]);
          UNION(DEFOUT[4], DEFOUT[8], DEFIN[5]);
          UNION(DEFOUT[3], DEFOUT[5], DEFIN[6]);
          ASSIGN(DEFIN[7], DEFOUT[6]);
          ASSIGN(DEFIN[8], DEFOUT[6]);
          for i := 1 to 8 do
               propagate(GEN[i], KILL[i], DEFIN[i], DEFOUT[i]);
     until
          not changed
end.
```

Fig. 4.2. Program to compute reaching definitions.

i	pass 1	pass 2	pass 3	pass 4
1	∅	∅	∅	∅
2	{1,2,3}	{1,2,3}	{1,2,3}	{1,2,3}
3	{4,5}	{1,4,5}	{1,4,5}	{1,4,5}
4	∅	{4,5}	{1,4,5}	{1,4,5}
5	{6,9}	{6,9}	{4,5,6,7,8,9}	{4,5,6,7,8,9}
6	{7,8}	{4,5,6,7,8,9}	{1,4,5,6,7,8,9}	{1,4,5,6,7,8,9}
7	∅	{7,8}	{4,5,6,7,8,9}	{1,4,5,6,7,8,9}
8	∅	{7,8}	{4,5,6,7,8,9}	{1,4,5,6,7,8,9}

Fig. 4.3. *DEFIN*[i] after each iteration.

is that MEMBER, INSERT, and DELETE operations can be performed in constant time by directly addressing the appropriate bit. UNION, INTERSECTION, and DIFFERENCE can be performed in time proportional to the size of the universal set.

If the universal set is sufficiently small so that a bit vector fits in one computer word, then UNION, INTERSECTION, and DIFFERENCE can be performed by single logical operations in the language of the underlying machine. Certain small sets can be represented directly in Pascal using the **set** construct. The maximum size of such sets depends on the particular compiler used and, unfortunately, it is often too small to be used in typical set problems. However, in writing our own programs, we need not be constrained by any limit on our set size, as long as we can treat our sets as subsets of some universal set {1, . . . ,N}. We intend that if A is a set represented as a boolean array, then $A[i]$ is true if and only if element i is a member of A. Thus, we can define an ADT SET by the PASCAL declaration

const
    N = { whatever value is appropriate };
type
    SET = **packed array**[1..N] **of** boolean;

We can then implement the procedure UNION as shown in Fig. 4.4. To implement INTERSECTION and DIFFERENCE, we replace "or" in Fig. 4.4 by "and" and "and not," respectively. The reader can implement the other operations mentioned in Section 4.1 (except MERGE and FIND, which make little sense in this context) as easy exercises.

It is possible to use the bit-vector implementation of sets when the universal set is a finite set other than a set of consecutive integers. Normally, we would then need a way to translate between members of the universal set and the integers 1, . . . ,N. Thus, in Example 4.1 we assumed that the data definitions were assigned numbers from 1 to 9. In general, the translations in

```
procedure UNION ( A, B: SET; var C: SET );
    var
        i: integer;
    begin
        for i := 1 to N do
            C[i] := A[i] or B[i]
    end
```

Fig. 4.4. Implementation of UNION.

both directions could be performed by the MAPPING ADT described in Chapter 2. However, the inverse translation from integers to elements of the universal set can be accomplished better using an array A, where $A[i]$ is the element corresponding to integer i.

4.4 A Linked-List Implementation of Sets

It should also be evident that sets can be represented by linked lists, where the items in the list are the elements of the set. Unlike the bit-vector representation, the list representation uses space proportional to the size of the set represented, not the size of the universal set. Moreover, the list representation is somewhat more general since it can handle sets that need not be subsets of some finite universal set.

When we have operations like INTERSECTION on sets represented by linked lists, we have several options. If the universal set is linearly ordered, then we can represent a set by a sorted list. That is, we assume all set members are comparable by a relation "$<$" and the members of a set appear on a list in the order $e_1, e_2, \ldots, e_n$, where $e_1 < e_2 < e_3 < \cdots < e_n$. The advantage of a sorted list is that we do not need to search the entire list to determine whether an element is on the list.

An element is in the intersection of lists L_1 and L_2 if and only if it is on both lists. With unsorted lists we must match each element on L_1 with each element on L_2, a process that takes $O(n^2)$ steps on lists of length n. The reason that sorting the lists makes intersection and some other operations easy is that if we wish to match an element e on one list L_1 with the elements of another list L_2, we have only to look down L_2 until we either find e, or find an element greater than e; in the first case we have found the match, while in the second case we know none exists. More importantly, if d is the element on L_1 that immediately precedes e, and we have found on L_2 the first element, say f, such that $d \leq f$, then to search L_2 for an occurrence of e we can begin with f. The conclusion from this reasoning is that we can find matches for all the elements of L_1, if they exist, by scanning L_1 and L_2 once, provided we advance the position markers for the two lists in the proper order, always advancing the one with the smaller element. The routine to implement INTERSECTION is shown in Fig. 4.5. There, sets are represented by linked lists of "cells" whose type is defined

```
type
    celltype = record
        element: elementtype;
        next: ↑ celltype
    end
```

Figure 4.5 assumes elementtype is a type, such as integer, that can be compared by $<$. If not, we have to write a function that determines which of two elements precedes the other.

The linked lists in Fig. 4.5 are headed by empty cells that serve as entry points to the lists. The reader may, as an exercise, write this program in a more general abstract form using list primitives. The program in Fig. 4.5, however, may be more efficient than the more abstract program. For example, Fig. 4.5 uses pointers to particular cells rather than "position" variables that point to previous cells. We can do so because we only append to list C, and A and B are only scanned, with no insertions or deletions done on those lists.

The operations of UNION and DIFFERENCE can be written to look surprisingly like the INTERSECTION procedure of Fig. 4.5. For UNION, we must attach all elements from either the A or B list to the C list, in their proper, sorted order, so when the elements are unequal (lines 12–14), we add the smaller to the C list just as we do when the elements are equal. We also append to list C all elements on the list not exhausted when the test of line (5) fails. For DIFFERENCE we do not add an element to the C list when equal elements are found. We only add the current A list element to the C list when it is smaller than the current B list element; for then we know the former cannot be found on the B list. Also, we add to C those elements on A when and if the test of line (5) fails because B is exhausted.

The operator ASSIGN(A, B) copies list A into list B. Note that, this operator cannot be implemented simply by making the header cell of A point to the same place as the header cell of B, because in that case, subsequent changes to B would cause unexpected changes to A. The MIN operator is especially easy; just return the first element on the list. DELETE and FIND can be implemented by finding the target item as discussed for general lists and in the case of a DELETE, disposing of its cell.

Lastly, insertion is not difficult to implement, but we must arrange to insert the new element into the proper position. Figure 4.6 shows a procedure INSERT that takes as parameters an element and a pointer to the header cell of a list, and inserts the element into the list. Figure 4.7 shows the crucial cells and pointers just before (solid) and after (dashed) insertion.

```
        procedure INTERSECTION ( ahead, bhead: ↑ celltype;
                    var pc: ↑ celltype );
            { computes the intersection of sorted lists A and B with
                header cells ahead and bhead, leaving the result
                as a sorted list whose header is pointed to by pc }
        var
            acurrent, bcurrent, ccurrent: ↑ celltype;
            { the current cells of lists A and B, and the last cell added list C }
        begin
(1)         new(pc);  { create header for list C }
(2)         acurrent := ahead↑.next;
(3)         bcurrent := bhead↑.next;
(4)         ccurrent := pc;
(5)         while (acurrent <> nil) and (bcurrent <> nil) do begin
                { compare current elements on lists A and B }
(6)             if acurrent↑.element = bcurrent↑.element then begin
                    { add to intersection }
(7)                 new(ccurrent↑.next);
(8)                 ccurrent := ccurrent↑.next;
(9)                 ccurrent↑.element := acurrent↑.element;
(10)                acurrent := acurrent↑.next;
(11)                bcurrent := bcurrent↑.next
                end
                else { elements unequal }
(12)                if acurrent↑.element < bcurrent↑.element then
(13)                    acurrent := acurrent↑.next
                    else
(14)                    bcurrent := bcurrent↑.next
            end;
(15)        ccurrent↑.next := nil
        end;  { INTERSECTION }
```

Fig. 4.5. Intersection procedure using sorted lists.

4.5 The Dictionary

When we use a set in the design of an algorithm, we may not need powerful operations like union and intersection. Often, we only need to keep a set of "current" objects, with periodic insertions and deletions from the set. Also, from time to time we may need to know whether a particular element is in the set. A set ADT with the operations INSERT, DELETE, and MEMBER has been given the name *dictionary*. We shall also include MAKENULL as a dictionary operation to initialize whatever data structure is used in the implementation. Let us consider an example application of the dictionary, and then discuss implementations that are well suited for representing dictionaries.

```
procedure INSERT ( x: elementtype; p: ↑ celltype );
    { inserts x onto list whose header is pointed to by p }
    var
        current, newcell: ↑ celltype;
    begin
        current := p;
        while current↑.next <> nil do begin
            if current↑.next↑.element = x then
                return;  { if x is already on the list, return }
            if current↑.next↑.element > x then
                goto add;  { break }
            current := current↑.next
        end;
    add: { current is now the cell after which x is to be inserted }
        new(newcell);
        newcell↑.element := x;
        newcell↑.next := current↑.next;
        current↑.next := newcell
    end;  { INSERT }
```

Fig. 4.6. Insertion procedure.

Example 4.2. The Society for the Prevention of Injustice to Tuna (SPIT) keeps a database recording the most recent votes of legislators on issues of importance to tuna lovers. This database is, conceptually, two sets of legislators' names, called *goodguys* and *badguys*. The society is very forgiving of past mistakes, but also tends to forget former friends easily. For example, if a vote to declare Lake Erie off limits to tuna fishermen is taken, all legislators voting in favor will be inserted into *goodguys* and deleted from *badguys*, while the opposite will happen to those voting against. Legislators not voting remain in whatever set they were in, if any.

In operation, the database system accepts three commands, each represented by a single character, followed by a ten character string denoting the name of a legislator. Each command is on a separate line. The commands are:

1. F (a legislator voting favorably follows)
2. U (a legislator voting unfavorably follows)
3. ? (determine the status of the legislator that follows).

We also allow the character 'E' on the input line to signal the end of processing. Figure 4.8 shows the sketch of the program, written in terms of the as-yet-undefined ADT DICTIONARY, which in this case is intended to be a set of strings of length 10. □

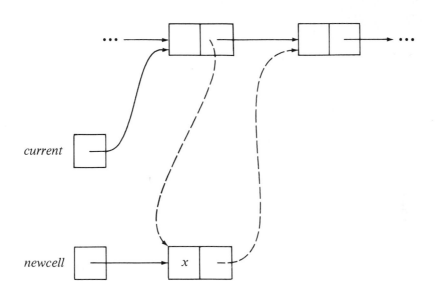

Fig. 4.7. The insertion picture.

4.6 Simple Dictionary Implementations

A dictionary can be implemented by a sorted or unsorted linked list. Another possible implementation of a dictionary is by a bit vector, provided the elements of the underlying set are restricted to the integers $1, \ldots, N$ for some N, or are restricted to a set that can be put in correspondence with such a set of integers.

A third possible implementation of a dictionary is to use a fixed-length array with a pointer to the last entry of the array in use. This implementation is only feasible if we can assume our sets never get larger than the length of the array. It has the advantage of simplicity over the linked-list representation, while its disadvantages are that (1) sets cannot grow arbitrarily, (2) deletion is slower, and (3) space is not utilized efficiently if sets are of varying sizes.

It is for the last of these reasons that we did not discuss the array implementation in connection with sets whose unions and intersections were taken frequently. Since arrays as well as lists can be sorted, however, the reader could consider the array implementation we now describe for dictionaries as a possible implementation for sets in general. Figure 4.9 shows the declarations and procedures necessary to supplement Fig. 4.8 to make it a working program.

```
program tuna ( input, output );
    { legislative database }
    type
        nametype = array[1..10] of char;
    var
        command: char;
        legislator: nametype;
        goodguys, badguys: DICTIONARY;

    procedure favor ( friend: nametype );
        begin
            INSERT(friend, goodguys);
            DELETE(friend, badguys)
        end; { favor }

    procedure unfavor ( foe: nametype );
        begin
            INSERT(foe, badguys);
            DELETE(foe, goodguys)
        end; { unfavor }

    procedure report ( subject: nametype );
        begin
            if MEMBER(subject, goodguys) then
                writeln(subject, ' is a friend')
            else if MEMBER(subject, badguys) then
                writeln(subject, ' is a foe')
            else
                writeln('we have no information about ', subject)
        end; { report }

begin  { main program }
    MAKENULL(goodguys);
    MAKENULL(badguys);
    read(command);
    while command <> 'E' do begin
        readln(legislator);
        if command = 'F' then
            favor(legislator)
        else if command = 'U' then
            unfavor(legislator)
        else if command = '?' then
            report(legislator)
        else
            error('unknown command');
        read(command)
    end
end. { tuna }
```

Fig. 4.8. Outline of the SPIT database program.

```
const
    maxsize = { some suitable number };
type
    DICTIONARY = record
        last: integer;
        data: array[1..maxsize] of nametype
    end;
procedure MAKENULL ( var A: DICTIONARY );
    begin
        A.last := 0
    end;  { MAKENULL }
function MEMBER ( x: nametype; var A: DICTIONARY ) : boolean;
    var
        i: integer;
    begin
        for i := 1 to A.last do
            if A.data[i] = x then return (true);
        return (false)  { if x is not found }
    end;  { MEMBER }
procedure INSERT ( x: nametype; var A: DICTIONARY );
    begin
        if not MEMBER(x, A) then
            if A.last < maxsize then begin
                A.last := A.last + 1;
                A.data[A.last] := x
            end
            else error('database is full')
    end;  { INSERT }
procedure DELETE ( x: nametype; var A: DICTIONARY );
    var
        i: integer;
    begin
        if A.last > 0 then begin
            i := 1;
            while (A.data[i] <> x) and (i < A.last) do
                i := i + 1;
            { when we reach here, either x has been found, or
                we are at the last element in set A, or both }
            if A.data[i] = x then begin
                A.data[i] := A.data[A.last];
                    { move the last element into the place of x;
                        Note that if i = A.last, this step does
                        nothing, but the next step will delete x }
                A.last := A.last - 1
            end
        end
    end;  { DELETE }
```

Fig. 4.9. Type and procedure declarations for array dictionary.

4.7 The Hash Table Data Structure

The array implementation of dictionaries requires, on the average, $O(N)$ steps to execute a single INSERT, DELETE, or MEMBER instruction on a dictionary of N elements; we get a similar speed if a list implementation is used. The bit-vector implementation takes constant time to do any of these three operations, but we are limited to sets of integers in some small range for that implementation to be feasible.

There is another important and widely useful technique for implementing dictionaries called "hashing." Hashing requires constant time per operation, on the average, and there is no requirement that sets be subsets of any particular finite universal set. In the worst case, this method requires time proportional to the size of the set for each operation, just as the array and list implementations do. By careful design, however, we can make the probability of hashing requiring more than a constant time per operation be arbitrarily small.

We shall consider two somewhat different forms of hashing. One, called *open* or *external* hashing, allows the set to be stored in a potentially unlimited space, and therefore places no limit on the size of the set. The second, called *closed* or *internal* hashing, uses a fixed space for storage and thus limits the size of sets.†

Open Hashing

In Fig. 4.10 we see the basic data structure for open hashing. The essential idea is that the (possibly infinite) set of potential set members is partitioned into a finite number of classes. If we wish to have B classes, numbered $0, 1, \ldots, B-1$, then we use a *hash function* h such that for each object x of the data type for members of the set being represented, $h(x)$ is one of the integers 0 through $B-1$. The value of $h(x)$, naturally, is the class to which x belongs. We often call x the *key* and $h(x)$ the *hash value* of x. The "classes" we shall refer to as *buckets*, and we say that x belongs to bucket $h(x)$.

In an array called the *bucket table*, indexed by the *bucket numbers* $0, 1, \ldots, B-1$, we have headers for B lists. The elements on the ith list are the members of the set being represented that belong to class i, that is, the elements x in the set such that $h(x) = i$.

It is our hope that the buckets will be roughly equal in size, so the list for each bucket will be short. If there are N elements in the set, then the average bucket will have N/B members. If we can estimate N and choose B to be roughly as large, then the average bucket will have only one or two members, and the dictionary operations take, on the average, some small constant number of steps, independent of what N (or equivalently B) is.

---

† Changes to the data structure can be made that will speed up open hashing and allow closed hashing to handle larger sets. We shall describe these techniques after giving the basic methods.

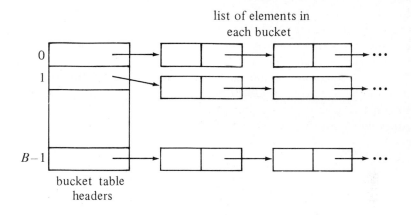

list of elements in
each bucket

Fig. 4.10. The open hashing data organization.

It is not always clear that we can select h so that a typical set will have its members distributed fairly evenly among the buckets. We shall later have more to say about selecting h so that it truly "hashes" its argument, that is, so that $h(x)$ is a "random" value that is not dependent on x in any trivial way.

```
function h ( x: nametype ) : 0..B−1;
    var
        i, sum: integer;
    begin
        sum := 0;
        for i := 1 to 10 do
            sum := sum + ord(x[i]);
        h := sum mod B
    end;  { h }
```

Fig. 4.11. A simple hash function h.

Let us here, for specificity, introduce one hash function on character strings that is fairly good, although not perfect. The idea is to treat the characters as integers, using the character code of the machine to define the correspondence. Pascal provides the *ord* built-in function, where $ord(c)$ is the integer code for character c. Thus, if x is a key and the type of keys is **array**[1..10] **of** char (what we called nametype in Example 4.2), we might declare the hash function h as in Fig. 4.11. In that function we sum the integers for each character and divide the result by B, taking the remainder, which is an integer from 0 to $B-1$.

In Fig. 4.12 we see the declarations of the data structures for an open hash table and the procedures implementing the operations for a dictionary. The member type for the dictionary is assumed to be nametype (a character array), so these declarations can be used in Example 4.2 directly. We should note that in Fig. 4.12 we have made the headers of the bucket lists be pointers to cells, rather than complete cells. We did so because the bucket table, where the headers are, could take as much space as the lists themselves if we made the bucket be an array of cells rather than pointers. Notice, however, that we pay a price for saving this space. The code for the DELETE procedure now must distinguish the first cell from the remaining ones.

```
const
    B = { suitable constant }
type
    celltype = record
        element: nametype;
        next: ↑ celltype
    end;
    DICTIONARY = array[0..B-1] of ↑ celltype;

procedure MAKENULL ( var A: DICTIONARY );
    var
        i: integer;
    begin
        for i := 0 to B-1 do
            A[i] := nil;
    end;  { MAKENULL }

function MEMBER ( x: nametype; var A: DICTIONARY ) : boolean;
    var
        current: ↑ celltype;
    begin
        current := A[h(x)];
        { initially current is the header of x's bucket }
        while current <> nil do
            if current↑.element = x then
                return (true)
            else
                current := current↑.next;
        return (false)  { if x is not found }
    end;  { MEMBER }

procedure INSERT ( x: nametype; var A: DICTIONARY );
    var
        bucket: integer;
        oldheader: ↑ celltype;
    begin
```

```
              if not MEMBER(x, A) then begin
                  bucket := h(x);
                  oldheader := A[bucket];
                  new(A[bucket]);
                  A[bucket]↑.element := x;
                  A[bucket]↑.next := oldheader
              end
          end; { INSERT }

    procedure DELETE ( x: nametype; var A: DICTIONARY );
    var
          current: ↑ celltype;
          bucket: integer;
    begin
          bucket := h(x);
          if A[bucket] <> nil then begin
              if A[bucket]↑.element = x then   { x is in first cell }
                  A[bucket] := A[bucket]↑.next  { remove x from list }
              else begin  { x, if present at all, is not in first cell of bucket }
                  current := A[bucket];  { current points to the previous cell }
                  while current↑.next <> nil do
                      if current↑.next↑.element = x then begin
                          current↑.next := current↑.next↑.next;
                              { remove x from list }
                          return { break }
                      end
                      else  { x not yet found }
                          current := current↑.next
              end
          end
    end; { DELETE }
```

Fig. 4.12. Dictionary implementation by an open hash table.

Closed Hashing

A closed hash table keeps the members of the dictionary in the bucket table itself, rather than using that table to store list headers. As a consequence, it appears that we can put only one element in any bucket. However, associated with closed hashing is a *rehash strategy*. If we try to place x in bucket $h(x)$ and find it already holds an element, a situation called a *collision*, the rehash strategy chooses a sequence of alternative locations, $h_1(x), h_2(x), \ldots$ within the bucket table, in which we could place x. We try each of these locations, in order, until we find an empty one. If none is empty, then the table is full, and we cannot insert x.

Example 4.3. Suppose $B = 8$ and keys a, b, c, and d have hash values

$h(a) = 3$, $h(b) = 0$, $h(c) = 4$, $h(d) = 3$. We shall use the simplest rehash strategy, called *linear hashing*, where $h_i(x) = (h(x) + i)$ **mod** B. Thus, for example, if we wished to insert a and found bucket 3 already filled, we would try buckets 4, 5, 6, 7, 0, 1, and 2, in that order.

Initially, we assume the table is empty, that is, each bucket holds a special value *empty*, which is not equal to any value we might try to insert.† If we insert a, b, c, and d, in that order, into an initially empty table, we find a goes into bucket 3, b into 0 and c into 4. When we try to insert d, we first try $h(d) = 3$ and find it filled. Then we try $h_1(d) = 4$ and find it filled as well. Finally, when we try $h_2(d) = 5$, we find an empty space and put d there. The resulting positions are shown in Fig. 4.13. □

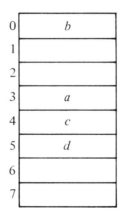

Fig. 4.13. Partially filled hash table.

The membership test for an element x requires that we examine $h(x)$, $h_1(x)$, $h_2(x)$, . . . , until we either find x or find an empty bucket. To see why we can stop looking if we reach an empty bucket, suppose first that there are no deletions permitted. If, say, $h_3(x)$ is the first empty bucket found in the series, it is not possible that x is in bucket $h_4(x)$, $h_5(x)$, or further along the sequence, because x could not have been placed there unless $h_3(x)$ were filled at the time we inserted x.

Note, however, that if we allow deletions, then we can never be sure, if we reach an empty bucket without finding x, that x is not somewhere else, and the now empty bucket was occupied when x was inserted. When we must do deletions, the most effective approach is to place a constant called *deleted* into a bucket that holds an element that we wish to delete. It is important

---

† If the type of members of the dictionary does not suggest an appropriate value for *empty*, let each bucket have an extra one-bit field to tell whether or not it is empty.

that we can distinguish between *deleted* and *empty*, the constant found in all buckets that have never been filled. In this way it is possible to permit deletions without having to search the entire table during a MEMBER test. When inserting we can treat *deleted* as an available space, so with luck the space of a deleted element will eventually be reused. However, as the space of a deleted element is not immediately reclaimable as it is with open hashing, we may prefer the open to the closed scheme.

Example 4.4. Suppose we wish to test if e is in the set represented by Fig. 4.13. If $h(e) = 4$, we try buckets 4, 5, and then 6. Bucket 6 is empty and since we have not found e, we conclude e is not in the set.

If we delete c, we must put the constant *deleted* in bucket 4. In that way, when we look for d, and we begin at $h(d) = 3$, we shall scan 4 and 5 to find d, and not stop at 4 as we would if we put *empty* in that bucket. □

In Fig 4.14 we see the type declarations and operations for the DICTIONARY ADT with set members of type nametype and the closed hash table as the underlying implementation. We use an arbitrary hash function h, of which Fig. 4.11 is one possibility, and we use the linear hashing strategy to rehash in case of collisions. For convenience, we have identified *empty* with a string of ten blanks and *deleted* with a string of 10 *'s, assuming that neither of these strings could be real keys. (In the SPIT database these strings are unlikely to be names of legislators.) The procedure INSERT(x, A) first uses *locate* to determine if x is already in A, and, if not, it uses a special routine *locate*1 to find a location in which to insert x. Note that *locate*1 searches for *deleted* as well as *empty* locations.

```
const
    empty = '          ';  { 10 blanks }
    deleted = '**********';  { 10 *'s }

type
    DICTIONARY = array[0..B−1] of nametype;

procedure MAKENULL ( var A: DICTIONARY );
    var
        i: integer;
    begin
        for i := 0 to B−1 do
            A[i] := empty
    end;  { MAKENULL }

function locate ( x: nametype ) : integer;
    { locate scans DICTIONARY from the bucket for h(x) until
        either x is found, or an empty bucket is found, or it has
        scanned completely around the table, thereby determining
        that the table does not contain x. locate returns the index
        of the bucket at which it stops for any of these reasons }
```

```
    var
        initial, i: integer;
        { initial holds h(x); i counts the number of
            buckets thus far scanned when looking for x }
    begin
        initial := h(x);
        i := 0;
        while (i < B) and (A[(initial + i) mod B] <> x) and
            (A[(initial + i) mod B] <> empty) do
                i := i + 1;
        return ((initial + i) mod B)
    end; { locate }

function locate1 ( x: nametype ) : integer;
    { like locate, but it will also stop at and return a deleted entry }

function MEMBER ( x: nametype; var A: DICTIONARY ) : boolean;
    begin
        if A[locate(x)] = x then
            return (true)
        else
            return (false)
    end; { MEMBER }

procedure INSERT ( x: nametype; var A: DICTIONARY );
    var
        bucket: integer;
    begin
        if A[locate(x)] = x then
            return; { x is already in A }
        bucket := locate1(x);
        if (A[bucket] = empty) or (A[bucket] = deleted) then
            A[bucket] := x;
        error('INSERT failed: table is full')
    end; { INSERT }

procedure DELETE ( x: nametype; var A: DICTIONARY );
    var
        bucket: integer;
    begin
        bucket := locate(x);
        if A[bucket] = x then
            A[bucket] := deleted
    end; { DELETE }
```

Fig. 4.14. Dictionary implementation by a closed hash table.

4.8 Estimating the Efficiency of Hash Functions

As we have mentioned, hashing is an efficient way to represent dictionaries and some other abstract data types based on sets. In this section we shall examine the average time per dictionary operation in an open hash table. If there are B buckets and N elements stored in the hash table, then the average bucket has N/B members, and we expect that an average INSERT, DELETE, or MEMBER operation will take $O(1 + N/B)$ time. The constant 1 represents the time to find the bucket, and N/B the time to search the bucket. If we can choose B to be about N, this time becomes a constant per operation. Thus the average time to insert, delete, or test an element for membership, assuming the element is equally likely to be hashed to any bucket, is a constant independent of N.

Suppose we are given a program written in some language like Pascal, and we wish to insert all identifiers appearing in the program into a hash table. Whenever a declaration of a new identifier is encountered, the identifier is inserted into the hash table after checking that it is not already there. During this phase it is reasonable to assume that an identifier is equally likely to be hashed into any given bucket. Thus we can construct a hash table with N elements in time $O(N(1 + N/B))$. By choosing B equal to N this becomes $O(N)$.

In the next phase identifiers are encountered in the body of the program. We must locate the identifiers in the hash table to retrieve information associated with them. But what is the expected time to locate an identifier? If the element searched for is equally likely to be any element in the table, then the expected time to search for it is just the average time spent inserting an element. To see this, observe that in searching once for each element in the table, the time spent is exactly the same as the time spent inserting each element, assuming that elements are always appended to the end of the list for the appropriate bucket. Thus the expected time for a search is also $O(1 + N/B)$.

The above analysis assumes that a hash function distributes elements uniformly over the buckets. Do such functions exist? We can regard a function such as that of Fig. 4.11 (convert characters to integers, sum and take the remainder modulo B) as a typical hash function. The following example examines the performance of this hash function.

Example 4.5. Suppose we use the hash function of Fig. 4.11 to hash the 100 keys consisting of the character strings A0, A1, . . . , A99 into a table of 100 buckets. On the assumption that $ord(0)$, $ord(1)$, . . . , $ord(9)$ form an arithmetic progression, as they do in the common character codes ASCII and EBCDIC, it is easy to check that the keys hash into at most 29 of the 100 buckets,[†] and the largest bucket contains A18, A27, A36, . . . , A90, or 9 of the 100 elements. If we compute the average number of steps for insertion,

---

† Note that A2 and A20 do not necessarily hash to the same bucket, but A23 and A41, for example, must do so.

using the fact that the insertion of the ith element into a bucket takes $i+1$ steps, we get 395 steps for these 100 keys. In comparison, our estimate $N(1 + N/B)$ would suggest 200 steps. □

The simple hashing function of Fig. 4.11 may treat certain sets of inputs, such as the consecutive strings in Example 4.5, in a nonrandom manner. "More random" hash functions are available. As an example, we can use the idea of squaring and taking middle digits. Thus, if we have a 5-digit number n as a key and square it, we get a 9 or 10 digit number. The "middle digits," such as the 4th through 7th places from the right, have values that depend on almost all the digits of n. For example, digit 4 depends on all but the leftmost digit of n, and digit 5 depends on all digits of n. Thus, if $B = 100$, we might choose to take digits 6 and 5 to form the bucket number.

This idea can be generalized to situations where B is not a power of 10. Suppose keys are integers in the range 0, 1, . . . , K. If we pick an integer C such that BC^2 is about equal to K^2, then the function

$$h(n) = \lfloor n^2/C \rfloor \bmod B$$

effectively extracts a base-B digit from the middle of n^2.

Example 4.6. If $K = 1000$ and $B = 8$, we might choose $C = 354$. Then

$$h(456) = \lfloor \frac{207936}{354} \rfloor \bmod 8 = 587 \bmod 8 = 3.$$

□

To use the "square and take the middle" strategy when keys are character strings, first group the characters in the string from the right into blocks of a fixed number of characters, say 4, padding the left with blanks, if necessary. Treat each block as a single integer, formed by concatenating the binary codes for the characters. For example, ASCII uses a 7-bit character code, so characters may be viewed as "digits" in base 2^7, or 128. Thus we can regard the character string $abcd$ as the integer $(128)^3a + (128)^2b + (128)c + d$. After converting all blocks to integers, add them† and proceed as we have suggested previously for integers.

Analysis of Closed Hashing

In a closed hashing scheme the speed of insertion and other operations depends not only on how randomly the hash function distributes the elements into buckets, but also on how well the rehashing strategy avoids additional collisions when a bucket is already filled. For example, the linear strategy for

---

† If strings can be extremely long, this addition will have to be performed modulo some constant c. For example, c could be one more than the largest integer obtainable from a single block.

resolving collisions is not as good as possible. While the analysis is beyond the scope of this book, we can observe the following. As soon as a few consecutive buckets are filled, any key that hashes to one of them will be sent, by the rehash strategy, to the end of the group, thereby increasing the length of that consecutive group. We are thus likely to find more long runs of consecutive buckets that are filled than if elements filled buckets at random. Moreover, runs of filled blocks cause some long sequences of tries before a newly inserted element finds an empty bucket, so having unusually large runs of filled buckets slows down insertion and other operations.

We might wish to know how many tries (or *probes*) are necessary on the average to insert an element when N out of B buckets are filled, assuming all combinations of N out of B buckets are equally likely to be filled. It is generally assumed, although not proved, that no closed hashing strategy can give a better average time performance than this for the dictionary operations. We shall then derive a formula for the cost of insertion if the alternative locations used by the rehashing strategy are chosen at random. Finally, we shall consider some strategies for rehashing that approximate random behavior.

The probability of a collision on our initial probe is N/B. Assuming a collision, our first rehash will try one of $B - 1$ buckets, of which $N - 1$ are filled, so the probability of at least two collisions is $\dfrac{N(N-1)}{B(B-1)}$. Similarly, the probability of at least i collisions is

$$\frac{N(N-1) \ \cdots \ (N-i+1)}{B(B-1) \ \cdots \ (B-i+1)}. \tag{4.3}$$

If B and N are large, this probability approximates $(N/B)^i$. The average number of probes is one (for the successful insertion) plus the sum over all $i \geq 1$ of the probability of at least i collisions, that is, approximately $1 + \sum_{i=1}^{\infty}(N/B)^i$, or $\dfrac{B}{B-N}$. It can be shown that the exact value of the summation when formula (4.3) is substituted for $(N/B)^i$ is $\dfrac{B+1}{B+1-N}$, so our approximation is a good one except when N is very close to B.

Observe that $\dfrac{B+1}{B+1-N}$ grows very slowly as N begins to grow from 0 to $B - 1$, the largest value of N for which another insertion is possible. For example, if N is half B, then about two probes are needed for the next insertion, on the average. The average insertion cost per bucket to fill M of the B buckets is $\dfrac{1}{M}\sum_{N=0}^{M-1}\dfrac{B+1}{B+1-N}$, which is approximately $\dfrac{1}{M}\int_{0}^{M-1}\dfrac{B}{B-x}dx$, or $\dfrac{B}{M}\log_e(\dfrac{B}{B-M+1})$. Thus, to fill the table completely ($M = B$) requires an average of $\log_e B$ per bucket, or $B\log_e B$ probes in total. However, to fill the table to 90% of capacity ($M = .9B$) only requires $B((10/9)\log_e 10)$ or approximately $2.56B$ probes.

The average cost of the membership test for a nonexistent element is the same as the cost of inserting the next element, but the cost of the membership test for an element in the set is the average cost of all insertions made so far, which is substantially less if the table is fairly full. Deletions have the same average cost as membership testing. But unlike open hashing, deletions from a closed hash table do not help speed up subsequent insertions or membership tests. It should be emphasized that if we never fill closed hash tables to more than any fixed fraction less than one, the average cost of operations is a constant; the constant grows as the permitted fraction of capacity grows. Figure 4.15 graphs the cost of insertions, deletions and membership tests, as a function of the percentage of the table that is full at the time the operation is performed.

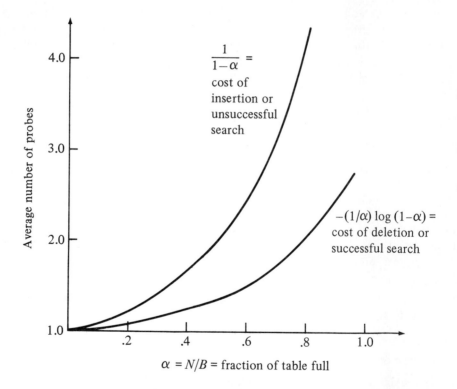

Fig. 4.15. Average operation cost.

"Random" Strategies for Collision Resolution

We have observed that the linear rehashing strategy tends to group full buckets into large consecutive blocks. Perhaps we could get more "random" behavior if we probed at a constant interval greater than one. That is, let $h_i(x) = (h(x) + ci) \bmod B$ for some $c > 1$. For example, if $B = 8$, $c = 3$, and $h(x) = 4$, we would probe buckets 4, 7, 2, 5, 0, 3, 6, and 1, in that order. Of course, if c and B have a common factor greater than one, this strategy doesn't even allow us to search all buckets; try $B = 8$ and $c = 2$, for example. But more significantly, even if c and B are *relatively prime* (have no common factors), we have the same "bunching up" problem as with linear hashing, although here it is sequences of full buckets separated by difference c that tend to occur. This phenomenon slows down operations as for linear hashing, since an attempted insertion into a full bucket will tend to travel down a chain of full buckets separated by distance c, and the length of this chain will increase by one.

In fact, any rehashing strategy where the target of a probe depends only on the target of the previous probe (as opposed to depending on the number of unsuccessful probes so far, the original bucket $h(n)$, or the value of the key x itself) will exhibit the bunching property of linear hashing. Perhaps the simplest strategy in which the problem does not occur is to let $h_i(x) = (h(x)+d_i) \bmod B$ where $d_1, d_2, \ldots, d_{B-1}$ is a "random" permutation of the integers $1, 2, \ldots, B-1$. Of course, the same sequence $d_1, \ldots, d_{B-1}$ is used for all insertions, deletions and membership tests; the "random" shuffle of the integers is decided upon once, when we design the rehash algorithm.

The generation of "random" numbers is a complicated subject, but fortunately, many common methods do produce a sequence of "random" numbers that is actually a permutation of integers from 1 up to some limit. These random number generators, if reset to their initial values for each operation on the hash table, serve to generate the desired sequence $d_1, \ldots, d_{B-1}$.

One effective approach uses "shift register sequences." Let B be a power of 2 and k a constant between 1 and $B-1$. Start with some number d_1 in the range 1 to $B - 1$, and generate successive numbers in the sequence by taking the previous value, doubling it, and if the result exceeds B, subtracting B and taking the *bitwise modulo 2 sum* of the result and the selected constant k. The bitwise modulo 2 sum of x and y, written $x \oplus y$, is computed by writing x and y in binary, with leading 0's if necessary so both are of the same length, and forming the numbers whose binary representation has a 1 in those positions where one, but not both, of x and y have a 1.

Example 4.7. $25 \oplus 13$ is computed by taking

$$
\begin{array}{rcl}
25 & = & 11001 \\
13 & = & \underline{01101} \\
25 \oplus 13 & = & 10100
\end{array}
$$

Note that this "addition" can be thought of as ordinary binary addition with carries from place to place ignored. □

Not every value of k will produce a permutation of $1, 2, \ldots, B-1$; sometimes a number repeats before all are generated. However, for given B, there is a small but finite chance that any particular k will work, and we need only find one k for each B.

Example 4.8. Let $B = 8$. If we pick $k = 3$, we succeed in generating all of 1, 2, . . . , 7. For example, if we start with $d_1 = 5$, then we compute d_2 by first doubling d_1 to get 10. Since $10 > 8$, we subtract 8 to get 2, and then compute $d_2 = 2 \oplus 3 = 1$. Note that $x \oplus 3$ can be computed by complementing the last two bits of x.

It is instructive to see the 3-bit binary representations of $d_1, d_2, \ldots, d_7$. These are shown in Fig. 4.16, along with the method of their calculation. Note that multiplication by 2 corresponds to a shift left in binary. Thus we have a hint of the origin of the term "shift register sequence."

$$
\begin{array}{ll}
                        & d_1 = 101 = 5 \\
\text{shift}   & \quad\;\; 1010 \\
\text{delete leading 1} & \quad\;\;\; 010 \\
\oplus\ 3   & d_2 = 001 = 1 \\
\text{shift}   & d_3 = 010 = 2 \\
\text{shift}   & d_4 = 100 = 4 \\
\text{shift}   & \quad\; 1000 \\
\text{delete leading 1} & \quad\;\;\; 000 \\
\oplus\ 3   & d_5 = 011 = 3 \\
\text{shift}   & d_6 = 110 = 6 \\
\text{shift}   & \quad\; 1100 \\
\text{delete leading 1} & \quad\;\; 100 \\
\oplus\ 3   & d_7 = 111 = 7 \\
\end{array}
$$

Fig. 4.16. Calculating a shift register sequence.

The reader should check that we also generate a permutation of 1, 2, . . . , 7 if we choose $k = 5$, but we fail for other values of k. $\square$

Restructuring Hash Tables

If we use an open hash table, the average time for operations increases as N/B, a quantity that grows rapidly as the number of elements exceeds the number of buckets. Similarly, for a closed hash table, we saw from Fig. 4.15 that efficiency goes down as N approaches B, and it is not possible that N exceeds B.

To retain the constant time per operation that is theoretically possible with hash tables, we suggest that if N gets too large, for example $N \geq .9B$ for a closed table or $N \geq 2B$ for an open one, we simply create a new hash table with twice as many buckets. The insertion of the current members of the set into the new table will, on the average, take less time than it took to insert them into the smaller table, and we more than make up this cost when doing subsequent dictionary operations.

4.9 Implementation of the Mapping ADT

Recall our discussion of the MAPPING ADT from Chapter 2 in which we defined a mapping as a function from domain elements to range elements. The operations for this ADT are:

1. MAKENULL(A) initializes the mapping A by making each domain element have no assigned range value.
2. ASSIGN(A, d, r) defines $A(d)$ to be r.
3. COMPUTE(A, d, r) returns true and sets r to $A(d)$ if $A(d)$ is defined; false is returned otherwise.

The hash table is an effective way to implement a mapping. The operations ASSIGN and COMPUTE are implemented much like INSERT and MEMBER operations for a dictionary. Let us consider an open hash table first. We assume the hash function $h(d)$ hashes domain elements to bucket numbers. While for the dictionary, buckets consisted of a linked list of elements, for the mapping we need a list of domain elements paired with their range values. That is, we replace the cell definition in Fig. 4.12 by

```
type
    celltype = record
        domainelement: domaintype;
        range: rangetype;
        next: ↑ celltype
    end
```

where domaintype and rangetype are whatever types domain and range elements have in this mapping. The declaration of a MAPPING is

```
type
    MAPPING = array[0..B − 1] of ↑ celltype
```

This array is the bucket array for a hash table. The procedure ASSIGN is written in Fig. 4.17. The code for MAKENULL and COMPUTE are left as an exercise.

Similarly, we can use a closed hash table as a mapping. Define cells to consist of domain element and range fields and declare a MAPPING to be an array of cells. As for open hash tables, let the hash function apply to domain elements, not range elements. We leave the implementation of the mapping operations for a closed hash table as an exercise.

4.10 Priority Queues

The priority queue is an ADT based on the set model with the operations INSERT and DELETEMIN, as well as the usual MAKENULL for initialization of the data structure. To define the new operation, DELETEMIN, we first assume that elements of the set have a "priority" function defined on

```
procedure ASSIGN ( var A: MAPPING; d: domaintype; r: rangetype );
    var
        bucket: integer;
        current: ↑ celltype;
    begin
        bucket := h(d);
        current := A[bucket];
        while current <> nil do
            if current↑.domainelement = d then begin
                current↑.range := r;  { replace old value for d }
                return
            end
            else
                current := current↑.next;
        { at this point, d was not found on the list }
        current := A[bucket];  { use current to remember first cell }
        new(A[bucket]);
        A[bucket]↑.domainelement := d;
        A[bucket]↑.range := r;
        A[bucket]↑.next := current
    end;  { ASSIGN }
```

Fig. 4.17. The procedure ASSIGN for an open hash table.

them; for each element a, $p(a)$, the *priority* of a, is a real number or, more generally, a member of some linearly ordered set. The operation INSERT has the usual meaning, while DELETEMIN is a function that returns some element of smallest priority and, as a side effect, deletes it from the set. Thus, as its name implies, DELETEMIN is a combination of the operations DELETE and MIN discussed earlier in the chapter.

Example 4.9. The term "priority queue" comes from the following sort of use for this ADT. The word "queue" suggests that people or entities are waiting for some service, and the word "priority" suggests that the service is given not on a "first-come-first-served" basis as for the QUEUE ADT, but rather that each person has a priority based on the urgency of need. An example is a hospital waiting room, where patients having potentially fatal problems will be taken before any others, no matter how long the respective waits are.

As a more mundane example of the use of priority queues, a time-shared computing system needs to maintain a set of processes waiting for service. Usually, the system designers want to make short processes appear to be instantaneous (in practice, response within a second or two appears instantaneous), so these are given priority over processes that have already consumed substantial time. A process that requires several seconds of computing time cannot be made to appear instantaneous, so it is sensible strategy to defer these until all processes that have a chance to appear instantaneous have been done. However, if we are not careful, processes that have taken substantially more time than average may never get another time slice and will wait forever.

One possible way to favor short processes, yet not lock out long ones is to give process P a priority $100t_{used}(P) - t_{init}(P)$. The parameter t_{used} gives the amount of time consumed by the process so far, and t_{init} gives the time at which the process initiated, measured from some "time zero." Note that priorities will generally be large negative integers, unless we choose to measure t_{init} from a time in the future. Also note that 100 in the above formula is a "magic number"; it is selected to be somewhat larger than the largest number of processes we expect to be active at once. The reader may observe that if we always pick the process with the smallest priority number, and there are not too many short processes in the mix, then in the long run, a process that does not finish quickly will receive 1% of the processor's time. If that is too much or too little, another constant can replace 100 in the priority formula.

We shall represent processes by records consisting of a process identifier and a priority number. That is, we define

```
type
    processtype = record
        id: integer;
        priority: integer
    end;
```

The priority of a process is the value of the priority field, which here we have defined to be an integer. We can define the priority function as follows.

```
function p ( a: processtype ) : integer;
    begin
        return (a.priority)
    end;
```

In selecting processes to receive a time slice, the system maintains a priority queue WAITING of processtype elements and uses two procedures, *initial* and *select*, to manipulate the priority queue by the operations INSERT and DELETEMIN. Whenever a process is initiated, procedure *initial* is called to place a record for that process in WAITING. Procedure *select* is called when the system has a time slice to award to some process. The record for the selected process is deleted from WAITING, but retained by *select* for reentry into the queue with a new priority; the priority is increased by 100 times the amount of time used.

We make use of function *currenttime*, which returns the current time, in whatever time units are used by the system, say microseconds, and we use procedure *execute(P)* to cause the process with identifier P to execute for one time slice. Figure 4.18 shows the procedures *initial* and *select*. □

```
        procedure initial ( P: integer );
            { initial places process with id P on the queue }
            var
                process: processtype;
            begin
                process.id := P;
                process.priority := −currenttime;
                INSERT (process, WAITING)
            end; { initial }

        procedure select;
            { select allocates a time slice to process with highest priority }
            var
                begintime, endtime: integer;
                process: processtype;
            begin
                process := ↑ DELETEMIN(WAITING);
                    { DELETEMIN returns a pointer to the deleted element }
                begintime := currenttime;
                execute(process.id);
                endtime := currenttime;
                process.priority := process.priority + 100*(endtime−begintime);
                    { adjust priority to incorporate amount of time used }
                INSERT (process, WAITING)
                    { put selected process back on queue with new priority }
            end; { select }
```

Fig. 4.18. Allocating time to processes.

4.11 Implementations of Priority Queues

With the exception of the hash table, the set implementations we have studied so far are also appropriate for priority queues. The reason the hash table is inappropriate is that there is no convenient way to find the minimum element, so hashing merely adds complications, and does not improve performance over, say, a linked list.

If we use a linked list, we have a choice of sorting it or leaving it unsorted. If we sort the list, finding a minimum is easy — just take the first element on the list. However, insertion requires scanning half the list on the average to maintain the sorted list. On the other hand, we could leave the list unsorted, which makes insertion easy and selection of a minimum more difficult.

Example 4.10. We shall implement DELETEMIN for an unsorted list of elements of type processtype, as defined in Example 4.9. The list is headed by an empty cell. The implementations of INSERT and MAKENULL are straightforward, and we leave the implementation using sorted lists as an exercise. Figure 4.19 gives the declaration for cells, for the type PRIORITYQUEUE, and for the procedure DELETEMIN.

Partially Ordered Tree Implementation of Priority Queues

Whether we choose sorted or unsorted lists to represent priority queues, we must spend time proportional to n to implement one or the other of INSERT or DELETEMIN on sets of size n. There is another implementation in which DELETEMIN and INSERT each require $O(\log n)$ steps, a substantial improvement for large n (say $n \geq 100$). The basic idea is to organize the elements of the priority queue in a binary tree that is as balanced as possible; an example is in Fig. 4.20. At the lowest level, where some leaves may be missing, we require that all missing leaves are to the right of all leaves that are present on the lowest level.

Most importantly, the tree is *partially ordered*, meaning that the priority of node v is no greater than the priority of the children of v, where the priority of a node is the priority number of the element stored at the node. Note from Fig. 4.20 that small priority numbers need not appear at higher levels than larger priority numbers. For example, level three has 6 and 8, which are less than the priority number 9 appearing on level 2. However, the parent of 6 and 8 has priority 5, which is, and must be, at least as small as the priorities of its children.

To execute DELETEMIN, we return the minimum-priority element, which, it is easily seen, must be at the root. However, if we simply remove the root, we no longer have a tree. To maintain the property of being a partially ordered tree, as balanced as possible, with leaves at the lowest level as far left as can be, we take the rightmost leaf at the lowest level and temporarily put it at the root. Figure 4.21(a) shows this change from Fig. 4.20. Then we push this element as far down the tree as it will go, by exchanging it with the one of its children having smaller priority, until the element is either at a leaf or at a position where it has priority no larger than either of its children.

In Fig. 4.21(a) we must exchange the root with its smaller-priority child, which has priority 5. The result is shown in Fig. 4.21(b). The element being pushed down is still larger than its children, with priorities 6 and 8. We exchange it with the smaller of these, 6, to reach the tree of Fig. 4.21(c). This tree has the partially ordered property.

In this percolation, if a node v has an element with priority a, if its children have elements with priorities b and c, and if at least one of b and c is smaller than a, then exchanging a with the smaller of b and c results in v having an element whose priority is smaller than either of its children. To prove this, suppose $b \leq c$. After the exchange, node v acquires b, and its

```
type
    celltype = record
        element: processtype;
        next: ↑ celltype
    end;
    PRIORITYQUEUE = ↑ celltype;
        { cell pointed to is a list header }


function DELETEMIN ( var A: PRIORITYQUEUE ) : ↑ celltype;
    var
        current: ↑ celltype;  { cell before one being "scanned" }
        lowpriority: integer;  { smallest priority found so far }
        prewinner: ↑ celltype;  { cell before one with element
                of smallest priority }
    begin
        if A↑.next = nil then
            error('cannot find minimum of empty list')
        else begin
            lowpriority := p(A↑.next↑.element);
                { p returns priority of the first element. Note A points
                    to a header cell that does not contain an element }
            prewinner := A;
            current := A↑.next;
            while current↑.next <> nil do begin
                { compare priorities of current winner
                    and next element }
                if p(current↑.next↑.element) < lowpriority then begin
                    prewinner := current;
                    lowpriority := p(current↑.next↑.element)
                end;
                current := current↑.next
            end;
            DELETEMIN := prewinner↑.next;
                { return pointer to the winner }
            prewinner↑.next := prewinner↑.next↑.next
                { remove winner from list }
        end
    end;  { DELETEMIN }
```

Fig. 4.19. Linked list priority queue implementation.

children hold a and c. We assumed $b \leq c$, and we were told that a is larger than at least one of b and c. Therefore $b \leq a$ surely holds. Thus the insertion procedure outlined above percolates an element down the tree until there

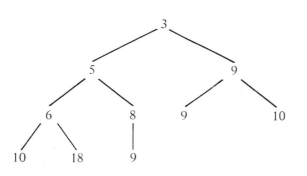

Fig. 4.20. A partially ordered tree.

are no more violations of the partially ordered property.

Let us also observe that DELETEMIN applied to a set of n elements takes $O(\log n)$ time. This follows since no path in the tree has more than $1 + \log n$ nodes, and the process of forcing an element down the tree takes a constant time per node. Observe that for any constant c, the quantity $c(1+\log n)$ is at most $2c\log n$ for $n \geq 2$. Thus $c(1+\log n)$ is $O(\log n)$.

Now let us consider how INSERT should work. First, we place the new element as far left as possible on the lowest level, starting a new level if the current lowest level is all filled. Fig. 4.22(a) shows the result of placing an element with priority 4 into Fig. 4.20. If the new element has priority lower than its parent, exchange it with its parent. The new element is now at a position where it is of lower priority than either of its children, but it may also be of lower priority than its parent, in which case we must exchange it with its parent, and repeat the process, until the new element is either at the root or has larger priority than its parent. Figures 4.22(b) and (c) show how 4 moves up by this process.

We could prove that the above steps result in a partially ordered tree. While we shall not attempt a rigorous proof, let us observe that an clement with priority a can become the parent of an element with priority b in three ways. (In what follows, we identify an element with its priority.)

1. a is the new element and moves up the tree replacing the old parent of b. Let the old parent of b have priority c. Then $a < c$, else the exchange would not takė place. But $c \leq b$, since the original tree was partially ordered. Thus $a < b$. For example in Fig. 4.22(c) 4 became the parent of 6, replacing a parent of larger priority, 5.

2. a was pushed down the tree due to an exchange with the new element. In this case a must have been an ancestor of b in the original partially ordered tree. Thus $a \leq b$. For example, in Fig. 4.22(c), 5 becomes the

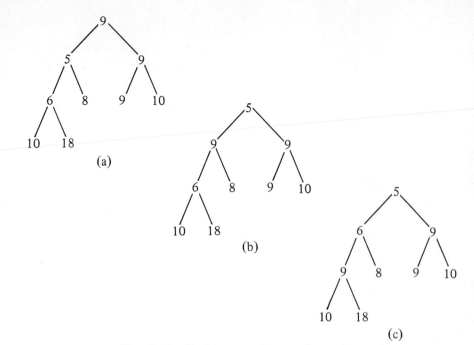

(a)

(b)

(c)

Fig. 4.21. Pushing an element down the tree.

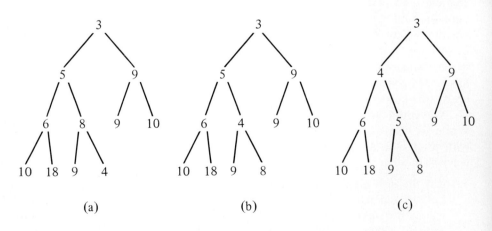

(a)   (b)   (c)

Fig. 4.22. Inserting an element.

parent of elements with priority 8 and 9. Originally, 5 was the parent of the former and the "grandparent" of the latter.

3. b might be the new element, and it moves up to become a child of a. If $a > b$, then a and b will be exchanged at the next step, so this violation of the partially ordered property will be removed.

The time to perform an insertion is proportional to the distance up the tree that the new element travels. As for DELETEMIN, we observe that this distance can be no greater than $1 + \log n$, so both INSERT and DELETEMIN take $O(\log n)$ steps.

An Array Implementation of Partially Ordered Trees

The fact that the trees we have been considering are binary, balanced as much as possible, and have leaves at the lowest level pushed to the left means that we can use a rather unusual representation for these trees, called a *heap*. If there are n nodes, we use the first n positions of an array A. $A[1]$ holds the root. The left child of the node in $A[i]$, if it exists, is at $A[2i]$, and the right child, if it exists, is at $A[2i+1]$. An equivalent viewpoint is that the parent of $A[i]$ is $A[i \textbf{ div } 2]$, for $i > 1$. Still another equivalent observation is that the nodes of a tree fill up $A[1], A[2], \ldots, A[n]$ level-by-level, from the top, and within a level, from the left. For example, Fig. 4.20 corresponds to an array containing the elements 3, 5, 9, 6, 8, 9, 10, 10, 18, 9.

We can declare a priority queue of elements of some type, say processtype as in Example 4.9, to consist of an array of processtype and an integer *last*, indicating the currently last element of the array that is in use. If we assume *maxsize* is the desired size of arrays for a priority queue, we can declare:

```
type
    PRIORITYQUEUE = record
        contents := array[1..maxsize] of processtype;
        last: integer
    end;
```

The priority queue operations are implemented in Fig. 4.23.

```
procedure MAKENULL ( var A: PRIORITYQUEUE );
    begin
        A.last := 0
    end;  { MAKENULL }

procedure INSERT ( x: processtype; var A: PRIORITYQUEUE );
    var
        i: integer;
        temp: processtype;
    begin
        if A.last >= maxsize then
```

```
        error('priority queue is full')
    else begin
        A.last := A.last + 1;
        A.contents[A.last] := x;
        i := A.last; { i is index of current position of x }
        while (i > 1) and (p(A.contents[i]) < p(A.contents[i div 2])) do
            begin
                { push x up the tree by exchanging it with its parent of
                    larger priority.  Recall p computes the priority of a
                    processtype element }
                temp := A.contents[i];
                A.contents[i] := A.contents[i div 2];
                A.contents[i div 2] := temp;
                i := i div 2
            end
    end
end; { INSERT }


function DELETEMIN ( var A: PRIORITYQUEUE ) : ↑ processtype;
    var
        i, j: integer;
        temp: processtype;
        minimum: ↑ processtype;
    begin
        if A.last = 0 then
            error('priority queue is empty')
        else begin
            new(minimum);
            minimum↑ := A.contents[1];
                { we shall return a pointer to a copy of the root of A }
            A.contents[1] := A.contents[A.last];
            A.last := A.last - 1;
                { move the last element to the beginning }
            i := 1; { i is the current position of the old last element }
            while i <= A.last div 2 do begin
                { push old last element down tree }
                if (p(A.contents[2*i]) < p(A.contents[2*i+1]))
                        or (2*i = A.last) then
                            j := 2 * i
                else
                        j := 2 * i + 1;
                { j will be the child of i having the smaller priority
                    or if 2*i = A.last, then j is the only child of i }
                if p(A.contents[i]) > p(A.contents[j]) then begin
                    { exchange old last element with smaller priority child }
```

```
                temp := A.contents[i];
                A.contents[i] := A.contents[j];
                A.contents[j] := temp;
                i := j
        end
        else
                return (minimum)  { cannot push further }
    end;
        return (minimum)  { pushed all the way to a leaf }
    end
end;  { DELETEMIN }
```

Fig. 4.23. Array implementation of priority queue.

4.12 Some Complex Set Structures

In this section we consider two more complex uses of sets to represent data. The first problem is that of representing many-many relationships, as might occur in a database system. A second case study exhibits how a pair of data structures representing the same object (a mapping in our example) can be a more efficient representation than either acting alone.

Many-Many Relationships and the Multilist Structure

An example of a many-many relationship between students and courses is suggested in Fig. 4.24. It is called a "many-many" relationship because there can be many students taking each course and many courses taken by each student.

	CS101	CS202	CS303
Alan			X
Alex	X	X	
Alice			X
Amy	X		
Andy		X	X
Ann	X		X

Enrollment

Fig. 4.24. An example relationship between students and courses.

From time to time the registrar may wish to insert or delete students from courses, to determine which students are taking a given course, or to know which courses a given student is taking. The simplest data structure with which these questions can be answered is obvious; just use the 2-dimensional array suggested by Fig. 4.24, where the value 1 (or true) replaces the X's and

0 (or false) replaces the blanks.

For example, to insert a student into a course, we need a mapping, MS, perhaps implemented as a hash table, that translates student names into array indices, and another, MC, that translates course names into array indices. Then, to insert student s into course c, we simply set

$$Enrollment[MS(s), MC(c)] := 1.$$

Deletions are performed by setting this element to 0. To find the courses taken by the student with name s, run across the row $MS(s)$, and similarly run down the column $MC(c)$ find the students in course c.

Why might we wish to look further for a more appropriate data structure? Consider a large university with perhaps 1000 courses and 20,000 students, taking an average of three courses each. The array suggested by Fig. 4.24 would have 20,000,000 elements, of which 60,000, or 0.3%, would be 1.[†] Such an array, which is called *sparse* to indicate that almost all its elements are zero, can be represented in far less space if we simply list the nonzero entries. Moreover, we can spend a great deal of time scanning a column of 20,000 entries searching for, on the average, 60 that are nonzero; row scans take considerable time as well.

One way to do better is to begin by phrasing the problem as one of maintaining a collection of sets. Two of these sets are S and C, the sets of all students and all courses. Each element of S is actually a record of a type such as

```
type
    studenttype = record
        id: integer;
        name: array[1..30] of char;
    end
```

and we would invent a similar record type for courses. To implement the structure we have in mind, we need a third set of elements, E, standing for enrollments. The elements of E each represent one of the boxes in the array of Fig. 4.24 in which there is an X. The E elements are records of a fixed type. As of now, we don't know any fields that belong in these records[‡], although we shall soon learn of them. For the moment, let us simply postulate that there is an enrollment record for each X-entry in the array and that enrollment records are distinguishable from one another somehow.

We also need sets that represent the answers to the crucial questions: given a student or course, what are the related courses or students, respectively. It would be nice if for each student s there were a set C_s of all the courses s was taking and conversely, a set S_c of the students taking course c.

[†] If this were a real database system, the array would be kept in secondary storage. However, this data structure would waste a great deal of space.

[‡] It would in practice be useful to place fields like grade or status (credit or audit) in enrollment records, but our original problem statement does not require this.

Such sets would be hard to implement because there would be no limit on the number of sets any element could be in, forcing us to complicate student and course records. We could instead let S_c and C_s be sets of pointers to course and student records, rather than the records themselves, but there is a method that saves a significant fraction of the space and allows equally fast answers to questions about students and courses.

Let us make each set C_s be the set of enrollment records corresponding to student s and some course c. That is, if we think of an enrollment as a pair (s, c), then

$$C_s = \{(s, c) \mid s \text{ is taking course } c\}$$

We can similarly define

$$S_c = \{(s, c) \mid s \text{ is taking course } c\}$$

Note the only difference in the meaning of the two set formers above is that in the first case s is constant and in the second c is. For example, based on Fig. 4.24, $C_{\text{Alex}} = \{(\text{Alex}, \text{CS101}), (\text{Alex}, \text{CS202})\}$ and $S_{\text{CS101}} = \{(\text{Alex}, \text{CS101}), (\text{Amy}, \text{CS101}), (\text{Ann}, \text{CS101})\}$.

Multilist Structures

In general, a multilist structure is any collection of cells some of which have more than one pointer and can therefore be on more than one list simultaneously. For each type of cell in a multilist structure, it is important to distinguish among pointer fields so we can follow one particular list and not get confused about which of several pointers in a particular cell we are following.

As a case in point, we can put one pointer field in each student and course record, pointing to the first enrollment record in the set C_s or S_c, respectively. Each enrollment record needs two pointer fields, one, which we shall call *cnext*, for the next enrollment on the list for the C_s set to which it belongs and the other, *snext*, for the S_c set to which it belongs.

It turns out that an enrollment record indicates explicitly neither the student nor the course that it represents. The information is implicit in the lists on which the enrollment record appears. Call the student and course records heading these lists the *owners* of the enrollment record. Thus, to tell what courses student s is taking, we must scan the enrollment records in C_s and find for each one the owner course record. We could do that by placing a pointer in each enrollment record to the owning course record, and we would also need a pointer to the owning student record.

While we might wish to use these pointers and thereby answer questions in the minimum possible time, we can save substantial space† at the cost of slowing down some computations, if we eliminate these pointers and instead place at the end of each S_c list a pointer to the owning course and at the end of each C_s list a pointer to the owning student. Thus, each student and course

---

† Note that there are probably many more enrollment records than course or student

record becomes part of a ring that includes all the enrollment records it owns. These rings are depicted in Fig. 4.25 for the data of Fig. 4.24. Note that each enrollment record has its *cnext* pointer first and its *snext* pointer second.

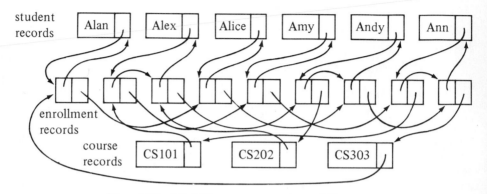

Fig. 4.25. Multilist representation of Fig. 4.24.

Example 4.11. To answer a question like "which students are taking CS101," we find the course record for CS101. How we find such a record depends on how the set of courses is maintained. For example, there might be a hash table containing all such records, and we obtain the desired record by applying some hash function to "CS101".

We follow the pointer in the CS101 record to the first enrollment record in the ring for CS101. This is the second enrollment record from the left. We must then find the student owner of this enrollment record, which we do by following *cnext* pointers (the first pointer in enrollment records) until we reach a student record.† In this case, we reach the third enrollment record, then the student record for Alex; we now know that Alex is taking CS101.

Now we must find the next student in CS101. We do so by following the *snext* pointer (second pointer) in the second enrollment record, which leads to the fifth enrollment record. The *cnext* pointer in that record leads us directly to the owner, Amy, so Amy is in CS101. Finally, we follow the *snext* pointer in the fifth enrollment record to the eighth enrollment record. The ring of *cnext* pointers from that record leads to the ninth enrollment record, then to the student record for Ann, so Ann is in CS101. The *snext* pointer in the eighth enrollment record leads back to CS101, so we are done. □

---

records, so shrinking enrollment records shrinks the total space requirement almost as much.

† We must have some way of identifying the type of records; we shall discuss a way to do this momentarily.

Abstractly, we can express the operation in Example 4.11 above as

for each enrollment record in the set for CS101 **do begin**
    $s :=$ the student owner of the enrollment record;
    print(s)
end

The above assignment to s can be elaborated as

$f := e$;
repeat
    $f := f{\uparrow}.cnext$
until
    f is a pointer to a student record;
$s := studentname$ field of record pointed to by f;

where e is a pointer to the first enrollment record in the set for CS101.

In order to implement a structure like Fig. 4.25 in Pascal, we need to have only one record type, with variants to cover the cases of student, course, and enrollment records. Pascal forces this arrangement on us since the fields *cnext* and *snext* each have the ability to point to different record types. This arrangement, however, does solve one of our other problems. It is now easy to tell what type of record we have reached as we travel around a ring. Figure 4.26 shows a possible declaration of the records and a procedure that prints the students taking a particular class.

Dual Data Structures for Efficiency

Often, a seemingly simple representation problem for a set or mapping presents a difficult problem of data structure choice. Picking one data structure for the set makes certain operations easy, but others take too much time, and it seems that there is no one data structure that makes all the operations easy. In that case, the solution often turns out to be the use of two or more different structures for the same set or mapping.

Suppose we wish to maintain a "tennis ladder," in which each player is on a unique "rung." New players are added to the bottom, that is, the highest-numbered rung. A player can challenge the player on the rung above, and if the player below wins the match, they trade rungs. We can represent this situation as an abstract data type, where the underlying model is a mapping from names (character strings) to rungs (integers 1, 2, . . .). The three operations we perform are

1. ADD(*name*) adds the named person to the ladder at the highest-numbered rung.
2. CHALLENGE(*name*) is a function that returns the name of the person on rung $i-1$ if the named person is on rung i, $i > 1$.
3. EXCHANGE(i) swaps the names of the players on rungs i and $i-1$, $i > 1$.

```
type
    stype = array[1..20] of char;
    ctype = array[1..5] of char;
    recordkinds = (student, course, enrollment);
    recordtype = record
        case kind : recordkinds of
            student : (studentname: stype;
                        firstcourse: ↑ recordtype);
            course: (coursename: ctype;
                        firststudent: ↑ recordtype);
            enrollment: (cnext, snext: ↑ recordtype)
        end;

procedure printstudents ( cname: ctype );
    var
        c, e, f: ↑ recordtype;
    begin
        c := pointer to course record with c↑.coursename = cname;
        { above depends on how course set is implemented }
        e := c↑.firststudent;
        { e runs around the ring of enrollments pointed to by c }
        while e↑.kind = enrollment do begin
            f := e;
            repeat
                f := f↑.cnext
            until
                f↑.kind = student;
            { f now points to student owner of enrollment e↑ }
            writeln(f↑.studentname);
            e := e↑.snext
        end
    end
```

Fig. 4.26. Implementation of search through a multilist.

Notice that we have chosen to pass only the higher rung number to EXCHANGE, while the other two operations take a name as argument.

We might, for example, choose an array LADDER, where LADDER[i] is the name of the person on rung i. If we also keep a count of the number of players, adding a player to the first unoccupied rung can be done in some small constant number of steps.

EXCHANGE is also easy, as we simply swap two elements of the array. However, CHALLENGE(name) requires that we examine the entire array for the name, which takes $O(n)$ time, if n is the number of players on the ladder.

On the other hand, we might consider a hash table to represent the mapping from names to rungs. On the assumption that we can keep the number of buckets roughly proportional to the number of players, ADD takes $O(1)$ time, on the average. Challenging takes $O(1)$ time on the average to look up the given name, but $O(n)$ time to find the name on the next lower-numbered rung, since the entire hash table may have to be searched. Exchanging requires $O(n)$ time to find the players on rungs i and $i-1$.

Suppose, however, that we combine the two structures. The cells of the hash table will contain pairs consisting of a name and a rung, while the array will have in $LADDER[i]$ a pointer to the cell for the player on rung i as suggested in Fig. 4.27.

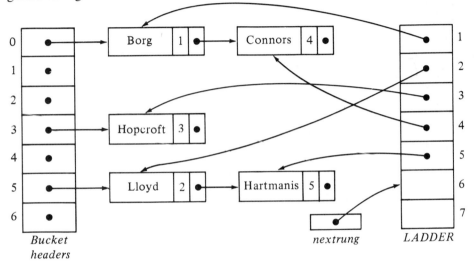

Fig. 4.27. Combined structure for high performance.

We can add a name by inserting into the hash table in $O(1)$ time on the average, and also placing a pointer to the newly created cell into the array *LADDER* at a position marked by the cursor *nextrung* in Fig. 4.27. To challenge, we look up the name in the hash table, taking $O(1)$ time on the average, get the rung i for the given player, and follow the pointer in $LADDER[i-1]$ to the cell of the hash table for the player to be challenged. Consulting $LADDER[i-1]$ takes constant time in the worst case, and the lookup in the hash table takes $O(1)$ time on the average, so CHALLENGE is $O(1)$ in the average case.

EXCHANGE(i) takes $O(1)$ time to find the cells for the players on rungs i and $i-1$, swap the rung numbers in those cells, and swap the pointers to the two cells in *LADDER*. Thus EXCHANGE requires constant time in even the worst case.

Exercises

4.1 If $A = \{1, 2, 3\}$ and $B = \{3, 4, 5\}$, what are the results of

a) UNION(A, B, C)

b) INTERSECTION(A, B, C)

c) DIFFERENCE(A, B, C)

d) MEMBER($1, A$)

e) INSERT($1, A$)

f) DELETE($1, A$)

g) MIN(A)?

*__4.2__ Write a procedure in terms of the basic set operations to print all the elements of a (finite) set. You may assume that a procedure to print an object of the element type is available. You must not destroy the set you are printing. What data structures would be most suitable for implementing sets in this case?

4.3 The bit-vector implementation of sets can be used whenever the "universal set" can be translated into the integers 1 through N. Describe how this translation would be done if the universal set were

a) the integers $0, 1, \ldots, 99$

b) the integers n through m for any $n \le m$

c) the integers $n, n+2, n+4, \ldots, n+2k$ for any n and k

d) the characters $'a', 'b', \ldots, 'z'$

e) arrays of two characters, each chosen from $'a'$ through $'z'$.

4.4 Write MAKENULL, UNION, INTERSECTION, MEMBER, MIN, INSERT, and DELETE procedures for sets represented by linked lists using the abstract operations for the sorted list ADT. Note that Fig. 4.5 is a procedure for INTERSECTION using a specific implementation of the list ADT.

4.5 Repeat Exercise 4.4 for the following set implementations:

a) open hash table (use abstract list operations within buckets).

b) closed hash table with linear resolution of collisions.

c) unsorted list (use abstract list operations).

d) fixed length array and pointer to the last position used.

4.6 For each of the operations and each of the implementations in Exercises 4.4 and 4.5, give the order of magnitude for the running time of the operations on sets of size n.

4.7 Suppose we are hashing integers with a 7-bucket hash table using the hash function $h(i) = i \bmod 7$.

a) Show the resulting open hash table if the perfect cubes 1, 8, 27, 64, 125, 216, 343 are inserted.

b) Repeat part (a) using a closed hash table with linear resolution of collisions.

4.8 Suppose we are using a closed hash table with 5 buckets and the hashing function $h(i) = i \bmod 5$. Show the hash table that results with linear resolution of collisions if the sequence 23, 48, 35, 4, 10 is inserted into an initially empty table.

4.9 Implement the mapping ADT operations using open and closed hash tables.

4.10 To improve the speed of operations, we may wish to replace an open hash table with B_1 buckets holding many more than B_1 elements by another hash table with B_2 buckets. Write a procedure to construct the new hash table from the old, using the list ADT operations to process each bucket.

4.11 In Section 4.8 we discussed "random" hash functions, where $h_i(X)$, the bucket to be tried after i collisions, is $(h(x) + d_i) \bmod B$ for some sequence $d_1, d_2, \ldots, d_{B\text{-}1}$. We also suggested that one way to compute a suitable sequence $d_1, d_2, \ldots, d_{B-1}$ was to pick a constant k, an arbitrary $d_1 > 0$, and let

$$d_i = \begin{cases} 2d_{i-1} & \text{if } 2d_{i-1} < B \\ (2d_{i-1}-B) \oplus k & \text{if } 2d_{i-1} \geq B \end{cases}$$

where $i > 1$, B is a power of 2, and $\oplus$ stands for the bitwise modulo 2 sum. If $B = 16$, find those values of k for which the sequence $d_1, d_2, \ldots, d_{15}$ includes all the integers 1, 2, $\ldots$, 15.

4.12 a) Show the partially ordered tree that results if the integers 5, 6, 4, 9, 3, 1, 7 are inserted into an empty tree.

b) What is the result of three successive DELETEMIN operations on the tree from (a)?

4.13 Suppose we represent the set of courses by

a) a linked list

b) a hash table

c) a binary search tree

Modify the declarations in Fig. 4.26 for each of these structures.

4.14 Modify the data structure of Fig. 4.26 to give each enrollment record a direct pointer to its student and course owners. Rewrite the procedure *printstudents* of Fig. 4.26 to take advantage of this structure.

4.15 Assuming 20,000 students, 1,000 courses, and each student in an average of three courses, compare the data structure of Fig. 4.26 with its modification suggested in Exercise 4.14 as to

    a) the amount of space required

    b) the average time to execute *printstudents*

    c) the average time to execute the analogous procedure that prints the courses taken by a given student.

4.16 Assuming the data structure of Fig. 4.26, given course record c and student record s, write procedures to insert and to delete the fact that s is taking c.

4.17 What, if anything, is the difference between the data structure of Exercise 4.14 and the structure in which sets C_s and S_c are represented by lists of pointers to course and student records, respectively?

4.18 Employees of a certain company are represented in the company database by their name (assumed unique), employee number, and social security number. Suggest a data structure that lets us, given one representation of an employee, find the other two representations of the same individual. How fast, on the average, can you make each such operation?

Bibliographic Notes

Knuth [1973] is a good source for additional information on hashing. Hashing was developed in the mid-to-late 1950's, and Peterson [1957] is a fundamental early paper on the subject. Morris [1968] and Maurer and Lewis [1975] are good surveys of hashing.

    The multilist is a central data structure of the network-based database systems proposed in DBTG [1971]. Ullman [1982] provides additional information on database applications of structures such as these.

    The heap implementation of partially ordered trees is based on an idea in Williams [1964]. Priority queues are discussed further in Knuth [1973].

    Reingold [1972] discusses the computational complexity of basic set operations. Set-based data flow analysis techniques are treated in detail in Cocke and Allen [1976] and Aho and Ullman [1977].

CHAPTER 5

Advanced
Set
Representation
Methods

This chapter introduces data structures for sets that permit more efficient implementation of common collections of set operations than those of the previous chapter. These structures, however, are more complex and are often only appropriate for large sets. All are based on various kinds of trees, such as binary search trees, tries, and balanced trees.

5.1 Binary Search Trees

We shall begin with binary search trees, a basic data structure for representing sets whose elements are ordered by some linear order. We shall, as usual, denote that order by $<$. This structure is useful when we have a set of elements from a universe so large that it is impractical to use the elements of the set themselves as indices into arrays. An example of such a universe would be the set of possible identifiers in a Pascal program. A binary search tree can support the set operations INSERT, DELETE, MEMBER, and MIN, taking $O(\log n)$ steps per operation on the average for a set of n elements.

A *binary search tree* is a binary tree in which the nodes are labeled with elements of a set. The important property of a binary search tree is that all elements stored in the left subtree of any node x are all less than the element stored at x, and all elements stored in the right subtree of x are greater than the element stored at x. This condition, called the *binary search tree property*, holds for every node of a binary search tree, including the root.

Figure 5.1 shows two binary search trees representing the same set of integers. Note the interesting property that if we list the nodes of a binary search tree in inorder, then the elements stored at those nodes are listed in sorted order.

Suppose a binary search tree is used to represent a set. The binary search tree property makes testing for membership in the set simple. To determine whether x is a member of the set, first compare x with the element r at the root of the tree. If $x = r$ we are done and the answer to the membership query is "true." If $x < r$, then by the binary search tree property, x can only

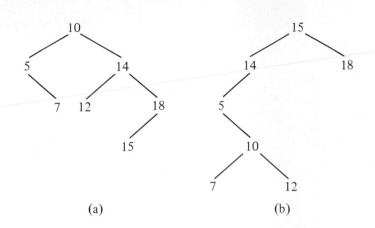

Fig. 5.1. Two binary search trees.

be a descendant of the left child of the root, if x is present at all.† Similarly, if $x > r$, then x could only be at a descendant of the right child of the root.

We shall write a simple recursive function MEMBER(x, A) to implement this membership test. We assume the elements of the set are of an unspecified type that will be called elementtype. For convenience, we assume elementtype is a type for which $<$ and $=$ are defined. If not, we must define functions LT(a, b) and EQ(a, b), where a and b are of type elementtype, such that LT(a, b) is true if and only if a is "less than" b, and EQ(a, b) is true if and only if a and b are the same.

The type for nodes consists of an element and two pointers to other nodes:

```
type
    nodetype = record
        element: elementtype;
        leftchild, rightchild: ↑ nodetype
    end;
```

Then we can define the type SET as a pointer to a node, which we take to be the root of the binary search tree representing the set. That is:

---

† Recall the left child of the root is a descendant of itself, so we have not ruled out the possibility that x is at the left child of the root.

type
    SET = ↑ nodetype;

Now we can specify fully the function MEMBER, in Fig. 5.2. Notice that since SET and "pointer to nodetype" are synonymous, MEMBER can call itself on subtrees as if those subtrees represented sets. In effect, the set can be partitioned into the subset of members less than x and the subset of members greater than x.

```
function MEMBER ( x: elementtype; A: SET ) : boolean;
    { returns true if x is in A, false otherwise }
    begin
        if A = nil then
            return (false) { x is never in ∅ }
        else if x = A↑.element then
            return (true)
        else if x < A↑.element then
            return (MEMBER(x, A↑.leftchild))
        else { x > A↑.element }
            return (MEMBER(x, A↑.rightchild))
    end; { MEMBER }
```

Fig. 5.2. Testing membership in a binary search tree.

The procedure INSERT(x, A), which adds element x to set A, is also easy to write. The first action INSERT must do is test whether A = nil, that is, whether the set is empty. If so, we create a new node to hold x and make A point to it. If the set is not empty, we search for x more or less as MEMBER does, but when we find a **nil** pointer during our search, we replace it by a pointer to a new node holding x. Then x will be in the right place, namely, the place where the function MEMBER will find it. The code for INSERT is shown in Fig. 5.3.

Deletion presents some problems. First, we must locate the element x to be deleted in the tree. If x is at a leaf, we can delete that leaf and be done. However, x may be at an interior node *inode*, and if we simply deleted *inode*, we would disconnect the tree.

If *inode* has only one child, as the node numbered 14 in Fig. 5.1(b), we can replace *inode* by that child, and we shall be left with the appropriate binary search tree. If *inode* has two children, as the node numbered 10 in Fig. 5.1(a), then we must find the lowest-valued element among the descendants of the right child.† For example, in the case the element 10 is deleted from Fig. 5.1(a), we must replace it by 12, the minimum-valued descendant of

_____
† The highest-valued node among the descendants of the left child would do as well.

```
procedure INSERT ( x: elementtype; var A: SET );
    { add x to set A }
    begin
        if A = nil then begin
            new(A);
            A↑.element := x
            A↑.leftchild := nil;
            A↑.rightchild := nil
        end
        else if x < A↑.element then
            INSERT(x, A↑.leftchild)
        else if x > A↑.element then
            INSERT (x, A↑.rightchild)
            { if x = A↑.element, we do nothing; x is already in the set }
    end; { INSERT }
```

Fig. 5.3. Inserting an element into a binary search tree.

the right child of 10.

To write DELETE, it is useful to have a function DELETEMIN(A) that removes the smallest element from a nonempty tree and returns the value of the element removed. The code for DELETEMIN is shown in Fig. 5.4. The code for DELETE uses DELETEMIN and is shown in Fig. 5.5.

```
function DELETEMIN ( var A: SET ) : elementtype;
    { returns and removes the smallest element from set A }
    begin
        if A↑.leftchild = nil then begin
            { A points to the smallest element }
            DELETEMIN := A↑.element;
            A := A↑.rightchild;
                { replace the node pointed to by A by its right child }
        end
        else { the node pointed to by A has a left child }
            DELETEMIN := DELETEMIN(A↑.leftchild)
    end; { DELETEMIN }
```

Fig. 5.4. Deleting the smallest element.

Example 5.1. Suppose we try to delete 10 from Fig. 5.1(a). Then, in the last statement of DELETE we call DELETEMIN with argument a pointer to node 14. That pointer is the *rightchild* field in the root. That call results in another call to DELETEMIN. The argument is then a pointer to node 12; this pointer is found in the *leftchild* field of node 14. We find that 12 has no

```
procedure DELETE ( x: elementtype; var A: SET );
    { remove x from set A }
begin
    if A <> nil then
        if x < A↑.element then
            DELETE(x, A↑.leftchild)
        else if x > A↑.element then
            DELETE(x, A↑.rightchild)
        { if we reach here, x is at the node pointed to by A }
        else if (A↑.leftchild = nil) and (A↑.rightchild = nil) then
            A := nil  { delete the leaf holding x }
        else if A↑.leftchild = nil then
            A := A↑.rightchild
        else if A↑.rightchild = nil then
            A := A↑.leftchild
        else { both children are present }
            A↑.element := DELETEMIN(A↑.rightchild)
end; { DELETE }
```

Fig. 5.5. Deletion from a binary search tree.

left child, so we return element 12 and make the left child for 14 be the right child of 12, which happens to be **nil**. Then DELETE takes the value 12 returned by DELETEMIN and replaces 10 by it. The resulting tree is shown in Fig. 5.6. □

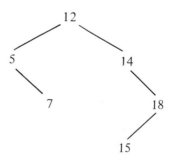

Fig. 5.6. Tree of Fig. 5.1(a) after deleting 10.

5.2 Time Analysis of Binary Search Tree Operations

In this section we analyze the average behavior of various binary search tree operations. We show that if we insert n random elements into an initially empty binary search tree, then the average path length from the root to a leaf is $O(\log n)$. Testing for membership, therefore, takes $O(\log n)$ time.

It is easy to see that if a binary tree of n nodes is complete (all nodes, except those at the lowest level have two children), then no path has more than $1 + \log n$ nodes.† Thus, each of the procedures MEMBER, INSERT, DELETE, and DELETEMIN takes $O(\log n)$ time. To see this, observe that they all take a constant amount of time at a node, then may call themselves recursively on at most one child. Therefore, the sequence of nodes at which calls are made forms a path from the root. Since that path is $O(\log n)$ in length, the total time spent following the path is $O(\log n)$.

However, when we insert n elements in a "random" order, they do not necessarily arrange themselves into a complete binary tree. For example, if we happen to insert smallest first, in sorted order, then the resulting tree is a chain of n nodes, in which each node except the lowest has a right child but no left child. In this case, it is easy to show that, as it takes $O(i)$ steps to insert the ith element, and $\sum_{i=1}^{n} i = n(n+1)/2$, the whole process of n insertions takes $O(n^2)$ steps, or $O(n)$ steps per operation.

We must determine whether the "average" binary search tree of n nodes is closer to the complete tree in structure than to the chain, that is, whether the average time per operation on a "random" tree takes $O(\log n)$ steps, $O(n)$ steps, or something in between. As we cannot know the true frequency of insertions and deletions, or whether deleted elements have some special property (e.g., do we always delete the minimum?), we can only analyze the average path length of "random" trees if we make some assumptions. The particular assumptions we make are that trees are formed by insertions only, and all orders of the n inserted elements are equally likely.

Under these fairly natural assumptions, we can calculate $P(n)$, the average number of nodes on the path from the root to some node (not necessarily a leaf). We assume the tree was formed by inserting n random elements into an initially empty tree. Clearly $P(0) = 0$ and $P(1) = 1$. Suppose we have a list of $n \geq 2$ elements to insert into an empty tree. The first element on the list, call it a, is equally likely to be first, second, or nth in the sorted order. Suppose that i elements on the list are less than a, so $n - i - 1$ are greater than a. When we build the tree, a will appear at the root, the i smaller elements will be descendants of the left child of the root, and the remaining $n - i - 1$ will be descendants of the right child. This tree is sketched in Fig. 5.7.

As all orders for the i small elements and for the $n - i - 1$ large elements are equally likely, we expect that the left and right subtrees of the root will

---

† Recall that all logarithms are to the base 2 unless otherwise noted.

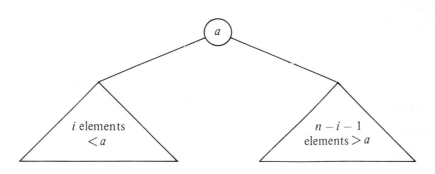

Fig. 5.7. Binary search tree.

have average path lengths $P(i)$ and $P(n-i-1)$, respectively. Since these elements are reached through the root of the complete tree, we must add 1 to the number of nodes on every path. Thus $P(n)$ can be calculated by averaging, for all i between 0 and $n-1$, the sum

$$\frac{i}{n}(P(i)+1) \; + \; \frac{(n-i-1)}{n} \, (P(n-i-1)+1) \; + \; \frac{1}{n}$$

The first term is the average path length in the left subtree, weighted by its size. The second term is the analogous quantity for the right subtree, and the $1/n$ term represents the contribution of the root. By averaging the above sum for all i between 1 and n, we obtain the recurrence

$$P(n) = 1 + \frac{1}{n^2} \sum_{i=0}^{n-1} \left(iP(i) \; + \; (n-i-1) \, P(n-i-1) \right) \qquad (5.1)$$

The first part of the summation (5.1), $\sum_{i=0}^{n-1} iP(i)$, can be made identical to the second part $\sum_{i=0}^{n-1} (n-i-1)P(n-i-1)$ if we substitute i for $n-i-1$ in the second part. Also, the term for $i=0$ in the summation $\sum_{i=0}^{n-1} iP(i)$ is zero, so we can begin the summation at 1. Thus (5.1) can be written

$$P(n) = 1 + \frac{2}{n^2} \sum_{i=1}^{n-1} iP(i) \qquad \text{for } n \geq 2 \qquad (5.2)$$

We shall show by induction on n, starting at $n=1$, that $P(n) \leq 1 + 4\log n$. Surely this statement is true for $n = 1$, since $P(1) = 1$. Suppose it is true for all $i < n$. Then by (5.2)

$$P(n) \leq 1 + \frac{2}{n^2} \sum_{i=1}^{n-1} (4i \log i + i)$$

$$\leq 1 + \frac{2}{n^2} \sum_{i=1}^{n-1} 4i \log i + \frac{2}{n^2} \sum_{i=1}^{n-1} i$$

$$\leq 2 + \frac{8}{n^2} \sum_{i=1}^{n-1} i \log i \tag{5.3}$$

The last step is justified, since $\sum_{i=1}^{n-1} i \leq n^2/2$, and therefore, the last term of the second line is at most 1. We shall divide the terms in the summation of (5.3) into two parts, those for $i \leq \lceil n/2 \rceil - 1$, which do not exceed $i \log(n/2)$, and those for $i > \lceil n/2 \rceil - 1$, which do not exceed $i \log n$. Thus (5.3) can be rewritten

$$P(n) \leq 2 + \frac{8}{n^2} \left[\sum_{i=1}^{\lceil n/2 \rceil - 1} i \log(n/2) + \sum_{i=\lceil n/2 \rceil}^{n-1} i \log n \right] \tag{5.4}$$

Whether n is even or odd, one can show that the first sum of (5.4) does not exceed $(n^2/8) \log(n/2)$, which is $(n^2/8) \log n - (n^2/8)$, and the second sum does not exceed $(3n^2/8) \log n$. Thus, we can rewrite (5.4) as

$$P(n) \leq 2 + \frac{8}{n^2} \left[\frac{n^2}{2} \log n - \frac{n^2}{8} \right]$$

$$\leq 1 + 4 \log n$$

as we wished to prove. This step completes the induction and shows that the average time to follow a path from the root to a random node of a binary search tree constructed by random insertions is $O(\log n)$, that is to within a constant factor as good as if the tree were complete. A more careful analysis shows that the constant 4 above is really about 1.4.

We can conclude from the above that the time of membership testing for a random member of the set takes $O(\log n)$ time. A similar analysis shows that if we include in our average path length only those nodes that are missing both children, or only those missing left children, then the average path length still obeys an equation similar to (5.1), and is therefore $O(\log n)$. We may then conclude that testing membership of a random element not in the set, inserting a random new element, and deleting a random element also all take $O(\log n)$ time on the average.

Evaluation of Binary Search Tree Performance

Hash table implementations of dictionaries require constant time per operation on the average. Although this performance is better than that for a binary search tree, a hash table requires $O(n)$ steps for the MIN operation, so if MIN is used frequently, the binary search tree will be the better choice; if MIN is not needed, we would probably prefer the hash table.

The binary search tree should also be compared with the partially ordered tree used for priority queues in Chapter 4. A partially ordered tree with n

elements requires only $O(\log n)$ steps for each INSERT and DELETEMIN operation not only on the average, but also in the worst case. Moreover, the actual constant of proportionality in front of the $\log n$ factor will be smaller for a partially ordered tree than for a binary search tree. However, the binary search tree permits general DELETE and MIN operations, as well as the combination DELETEMIN, while the partially ordered tree permits only the latter. Moreover, MEMBER requires $O(n)$ steps on a partially ordered tree but only $O(\log n)$ steps on a binary search tree. Thus, while the partially ordered tree is well suited to implementing priority queues, it cannot do as efficiently any of the additional operations that the binary search tree can do.

5.3 Tries

In this section we shall present a special structure for representing sets of character strings. The same method works for representing data types that are strings of objects of any type, such as strings of integers. This structure is known as the *trie*, derived from the middle letters of the word "retrieval."† By way of introduction, consider the following use of a set of character strings.

Example 5.2. As indicated in Chapter 1, one way to implement a spelling checker is to read a text file, break it into words (character strings separated by blanks and new lines), and find those words not in a standard dictionary of words in common use. Words in the text but not in the dictionary are printed out as possible misspellings. In Fig. 5.8 we see a sketch of one possible program *spell*. It makes use of a procedure $getword(x, f)$ that sets x to be the next word in text file f; variable x is of a type called wordtype, which we shall define later. The variable A is of type SET; the SET operations we shall need are INSERT, DELETE, MAKENULL, and PRINT. The PRINT operator prints the members of the set. □

The trie structure supports these set operations when the elements of the set are words, i.e., character strings. It is appropriate when many words begin with the same sequences of letters, that is, when the number of distinct prefixes among all the words in the set is much less than the total length of all the words.

In a trie, each path from the root to a leaf corresponds to one word in the represented set. This way, the nodes of the trie correspond to the prefixes of words in the set. To avoid confusion between words like THE and THEN, let us add a special *endmarker* symbol, $, to the ends of all words, so no prefix of a word can be a word itself.

Example 5.3. In Fig. 5.9 we see a trie representing the set of words {THE, THEN, THIN, THIS, TIN, SIN, SING}. That is, the root corresponds to the empty string, and its two children correspond to the prefixes T and S. The

---

† Trie was originally intended to be a homonym of "tree" but to distinguish these two terms many people prefer to pronounce trie as though it rhymes with "pie."

```
program spell ( input, output, dictionary );
    type
        wordtype = { to be defined };
        SET = { to be defined, using the trie structure };
    var
        A: SET;  { holds input words not yet found in the dictionary }
        nextword: wordtype;
        dictionary: file of char;

    procedure getword ( var x: wordtype; f: file of char );
        { a procedure to be defined that sets x
            to be the next word in file f }

    procedure INSERT ( x: wordtype; var S: SET );
        { to be defined }

    procedure DELETE (x: wordtype; var S: SET );
        { to be defined }

    procedure MAKENULL ( var S: SET );
        { to be defined }

    procedure PRINT ( var S: SET );
        { to be defined }

    begin
        MAKENULL(A);
        while not eof(input) do begin
            getword(nextword, input);
            INSERT(nextword, A)
        end
        while not eof(dictionary) do begin
            getword(nextword, dictionary);
            DELETE(nextword, A)
        end;
        PRINT(A);
    end;  { spell }
```

Fig. 5.8. Sketch of spelling checker.

leftmost leaf represents the word THE, the next leaf the word THEN, and so on. □

We can make the following observations about the trie in Fig. 5.9.

1. Each node has at most 27 children, one for each letter and $.

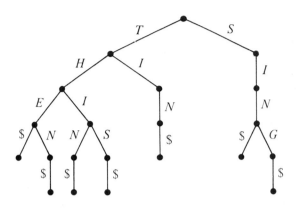

Fig. 5.9. A trie.

2. Most nodes will have many fewer than 27 children.
3. A leaf reached by an edge labeled $ cannot have any children, and may as well not be there.

Trie Nodes as ADT's

We can view a node of a trie as a mapping whose domain is {A, B, . . . , Z, $} (or whatever alphabet we choose) and whose value set is the type "pointer to trie node." Moreover, the trie itself can be identified with its root, so the ADT's TRIE and TRIENODE have the same data type, although their operations are substantially different. On TRIENODE we need the following operations:

1. procedure ASSIGN(*node*, *c*, *p*) that assigns value *p* (a pointer to a node) to character *c* in node *node*,
2. function VALUEOF(*node*, *c*) that produces the value associated with character *c* in *node*,† and
3. procedure GETNEW(*node*, *c*) to make the value of *node* for character *c* be a pointer to a new node.

Technically, we also need a procedure MAKENULL(*node*) to make *node* be the null mapping. One simple implementation of trie nodes is an array *node* of pointers to nodes, with the index set being {A, B, . . . , Z, $}. That is, we define

---

† VALUEOF is a function version of COMPUTE in Section 2.5.

```
type
    chars = ('A', 'B', . . . ,'Z', '$');
    TRIENODE = array[chars] of ↑ TRIENODE;
```

If *node* is a trie node, *node*[c] is VALUEOF(*node*, c) for any c in the set
chars. To avoid creating many leaves that are children corresponding to '$',
we shall adopt the convention that *node*['$'] is either **nil** or a pointer to the
node itself. In the former case, *node* has no child corresponding to '$', and in
the latter case it is deemed to have such a child, although we never create it.
Then we can write the procedures for trie nodes as in Fig. 5.10.

```
procedure MAKENULL ( var node: TRIENODE );
    { makes node a leaf, i.e., a null mapping }
    var
        c: char;
    begin
        for c := 'A' to '$' do
            node[c] := nil
    end; { MAKENULL }

procedure ASSIGN ( var node: TRIENODE; c: char; p: ↑ TRIENODE );
    begin
        node[c] := p
    end; { ASSIGN }

function VALUEOF ( var node: TRIENODE; c: char ) : ↑ TRIENODE;
    begin
        return (node[c])
    end; { VALUEOF }

procedure GETNEW ( var node: TRIENODE; c: char );
    begin
        new(node[c]);
        MAKENULL(node[c])
    end; { GETNEW }
```

Fig. 5.10. Operations on trie nodes.

Now let us define

```
type
    TRIE = ↑ TRIENODE;
```

We shall assume wordtype is an array of characters of some fixed length. The value of such an array will always be assumed to have at least one '$'; we take the end of the represented word to be the first '$', no matter what follows (presumably more '$''s). On this assumption, we can write the procedure INSERT(x, *words*) to insert x into set *words* represented by a trie, as shown in Fig. 5.11. We leave the writing of MAKENULL, DELETE, and PRINT for tries represented as arrays for exercises.

```
procedure INSERT ( x: wordtype; var words: TRIE );
    var
        i: integer;  { counts positions in word x }
        t: TRIE;  { used to point to trie nodes
              corresponding to prefixes of x }
    begin
        i := 1;
        t := words;
        while x[i] <> '$' do begin
            if VALUEOF(t↑, x[i]) = nil then
                { if current node has no child for character x[i],
                    create one }
                GETNEW(t↑, x[i]);
            t := VALUEOF(t↑, x[i]);
                { proceed to the child of t for character x[i],
                    whether or not that child was just created }
            i := i+1  { move along the word x }
        end;
        { now we have reached the first '$' in x }
        ASSIGN(t↑, '$', t)
            { make loop for '$' to represent a leaf }
    end; {INSERT}
```

Fig. 5.11. The procedure INSERT.

A List Representation for Trie Nodes

The array representation of trie nodes takes a collection of words, having among them p different prefixes, and represents them with $27p$ bytes of storage. That amount of space could far exceed the total length of the words in the set. However, there is another implementation of tries that may save space. Recall that each trie node is a mapping, as discussed in Section 2.6. In principle, any implementation of mappings would do. Yet in practice, we want a representation suitable for mappings with a small domain and for mappings defined for comparatively few members of that domain. The linked list representation for mappings satisfies these requirements nicely. We may represent the mapping that is a trie node by a linked list of the characters for which the associated value is not the **nil** pointer. That is, a trie node is a linked list of cells of the type

```
    type
        celltype = record
            domain: char;
            value: ↑ celltype;
                { pointer to first cell on list for the child node }
            next: ↑ celltype;
                { pointer to next cell on the list }
        end;
```

We shall leave the procedures ASSIGN, VALUEOF, MAKENULL, and GETNEW for this implementation of trie nodes as exercises. After writing these procedures, the INSERT operations on tries, in Fig. 5.11, and the other operations on tries that we left as exercises, should work correctly.

Evaluation of the Trie Data Structure

Let us compare the time and space needed to represent n words with a total of p different prefixes and a total length of l using a hash table and a trie. In what follows, we shall assume that pointers require four bytes. Perhaps the most space-efficient way to store words and still support INSERT and DELETE efficiently is in a hash table. If the words are of varying length, the cells of the buckets should not contain the words themselves; rather, the cells consist of two pointers, one to link the cells of the bucket and the other pointing to the beginning of a word belonging in the bucket.

The words themselves are stored in a large character array, and the end of each word is indicated by an endmarker character such as '$'. For example, the words THE, THEN, and THIN could be stored as

THE$THEN$THIN$ . . .

The pointers for the three words are cursors to positions 1, 5, and 10 of the array. The amount of space used in the buckets and character array is

1. $8n$ bytes for the cells of the buckets, there being one cell for each of the n words, and a cell has two pointers or 8 bytes,
2. $l + n$ bytes for the character array to store the n words of total length l and their endmarkers.

The total space is thus $9n + l$ bytes plus whatever amount is used for the bucket headers.

In comparison, a trie with nodes implemented by linked lists requires $p + n$ cells, one cell for each prefix and one cell for the end of each word. Each trie cell has a character and two pointers, and needs nine bytes, for a total space of $9n + 9p$. If l plus the space for the bucket headers exceeds $9p$, the trie uses less space. However, for applications such as storing a dictionary where l/p is typically less than 3, the hash table would use less space.

In favor of the trie, however, let us note that we can travel down a trie, and thus perform operations INSERT, DELETE, and MEMBER in time proportional to the length of the word involved. A hash function to be truly "random" should involve each character of the word being hashed. It is fair, therefore, to state that computing the hash function takes roughly as much time as performing an operation like MEMBER on the trie. Of course the time spent computing the hash function does not include the time spent resolving collisions or performing the insertion, deletion, or membership test on the hash table, so we can expect tries to be considerably faster than hash tables for dictionaries whose elements are character strings.

Another advantage to the trie is that it supports the MIN operation efficiently, while hash tables do not. Further, in the hash table organization described above, we cannot easily reuse the space in the character array when a word is deleted (but see Chapter 12 for methods of handling such a problem).

5.4 Balanced Tree Implementations of Sets

In Sections 5.1 and 5.2 we saw how sets could be implemented by binary search trees, and we saw that operations like INSERT could be performed in time proportional to the average depth of nodes in that tree. Further, we discovered that this average depth is $O(\log n)$ for a "random" tree of n nodes. However, some sequences of insertions and deletions can produce binary search trees whose average depth is proportional to n. This suggests that we might try to rearrange the tree after each insertion and deletion so that it is always complete; then the time for INSERT and similar operations would always be $O(\log n)$.

In Fig. 5.12(a) we see a tree of six nodes that becomes the complete tree of seven nodes in Fig. 5.12(b), when element 1 is inserted. Every element in Fig. 5.12(a), however, has a different parent in Fig. 5.12(b), so we must take n steps to insert 1 into a tree like Fig. 5.12(a), if we wish to keep the tree as balanced as possible. It is thus unlikely that simply insisting that the binary search tree be complete will lead to an implementation of a dictionary, priority queue, or other ADT that includes INSERT among its operations, in $O(\log n)$ time.

There are several other approaches that do yield worst case $O(\log n)$ time per operation for dictionaries and priority queues, and we shall consider one of them, called a "2-3 tree," in detail. A *2-3 tree* is a tree with the following two properties.

1. Each interior node has two or three children.
2. Each path from the root to a leaf has the same length.

We shall also consider a tree with zero nodes or one node as special cases of a 2-3 tree.

We represent sets of elements that are ordered by some linear order $<$, as follows. Elements are placed at the leaves; if element a is to the left of

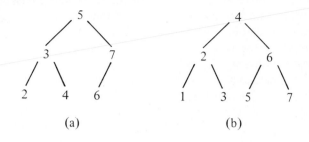

(a)   (b)

Fig. 5.12. Complete trees.

element b, then $a < b$ must hold. We shall assume that the "$<$" ordering of elements is based on one field of a record that forms the element type; this field is called the *key*. For example, elements might represent people and certain information about people, and in that case, the key field might be "social security number."

At each interior node we record the key of the smallest element that is a descendant of the second child and, if there is a third child, we record the key of the smallest element descending from that child as well.† Figure 5.13 is an example of a 2-3 tree. In that and subsequent examples, we shall identify an element with its key field, so the order of elements becomes obvious.

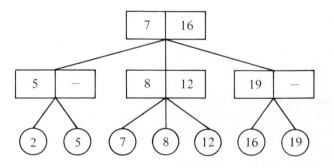

Fig. 5.13. A 2-3 tree.

---

† There is another version of 2-3 trees that places whole records at interior nodes, as a binary search tree does.

Observe that a 2-3 tree of k levels has between 2^{k-1} and 3^{k-1} leaves. Put another way, a 2-3 tree representing a set of n elements requires at least $1 + \log_3 n$ levels and no more than $1 + \log_2 n$ levels. Thus, path lengths in the tree are $O(\log n)$.

We can test membership of a record with key x in a set represented by a 2-3 tree in $O(\log n)$ time by simply moving down the tree, using the values of the elements recorded at the interior nodes to guide our path. At a node *node*, compare x with the value y that represents the smallest element descending from the second child of *node*. (Recall we are treating elements as if they consisted solely of a key field.) If $x < y$, move to the first child of *node*. If $x \geq y$, and *node* has only two children, move to the second child of *node*. If *node* has three children and $x \geq y$, compare x with z, the second value recorded at *node*, the value that indicates the smallest descendant of the third child of *node*. If $x < z$, go to the second child, and if $x \geq z$, go to the third child. In this manner, we find ourselves at a leaf eventually, and x is in the represented set if and only if x is at the leaf. Evidently, if during this process we find $x = y$ or $x = z$, we can stop immediately. However, we stated the algorithm as we did because in some cases we shall wish to find the leaf with x as well as to verify its existence.

Insertion into a 2-3 Tree

To insert a new element x into a 2-3 tree, we proceed at first as if we were testing membership of x in the set. However, at the level just above the leaves, we shall be at a node *node* whose children, we discover, do not include x. If *node* has only two children, we simply make x the third child of *node*, placing the children in the proper order. We then adjust the two numbers at *node* to reflect the new situation.

For example, if we insert 18 into Fig. 5.13, we wind up with *node* equal to the rightmost node at the middle level. We place 18 among the children of *node*, whose proper order is 16, 18, 19. The two values recorded at *node* become 18 and 19, the elements at the second and third children. The result is shown in Fig. 5.14.

Suppose, however, that x is the fourth, rather than the third child of *node*. We cannot have a node with four children in a 2-3 tree, so we split *node* into two nodes, which we call *node* and *node'*. The two smallest elements among the four children of *node* stay with *node*, while the two larger elements become children of *node'*. Now, we must insert *node'* among the children of p, the parent of *node*. This part of the insertion is analogous to the insertion of a leaf as a child of *node*. That is, if p had two children, we make *node'* the third and place it immediately to the right of *node*. If p had three children before *node'* was created, we split p into p and p', giving p the two leftmost children and p' the remaining two, and then we insert p' among the children of p's parent, recursively.

One special case occurs when we wind up splitting the root. In that case we create a new root, whose two children are the two nodes into which the

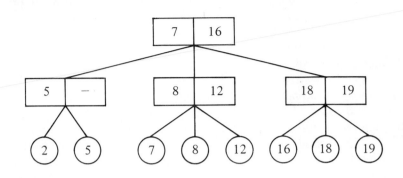

Fig. 5.14. 2-3 tree with 18 inserted.

old root was split. This is how the number of levels in a 2-3 tree increases.

Example 5.4. Suppose we insert 10 into the tree of Fig. 5.14. The intended parent of 10 already has children 7, 8, and 12, so we split it into two nodes. The first of these has children 7 and 8; the second has 10 and 12. The result is shown in Fig. 5.15(a). We must now insert the new node with children 10 and 12 in its proper place as a child of the root of Fig. 5.15(a). Doing so gives the root four children, so we split it, and create a new root, as shown in Fig. 5.15(b). The details of how information regarding smallest elements of subtrees is carried up the tree will be given when we develop the program for the command INSERT. □

Deletion in a 2-3 tree

When we delete a leaf, we may leave its parent *node* with only one child. If *node* is the root, delete *node* and let its lone child be the new root. Otherwise, let *p* be the parent of *node*. If *p* has another child, adjacent to *node* on either the right or the left, and that child of *p* has three children, we can transfer the proper one of those three to *node*. Then *node* has two children, and we are done.

If the children of *p* adjacent to *node* have only two children, transfer the lone child of *node* to an adjacent sibling of *node*, and delete *node*. Should *p* now have only one child, repeat all the above, recursively, with *p* in place of *node*.

Example 5.5. Let us begin with the tree of Fig. 5.15(b). If 10 is deleted, its parent has only one child. But the grandparent has another child that has three children, 16, 18, and 19. This node is to the right of the deficient node, so we pass the deficient node the smallest element, 16, leaving the 2-3 tree in Fig. 5.16(a).

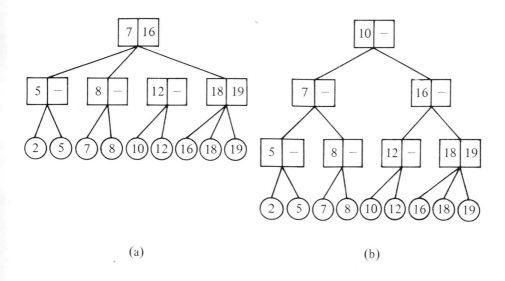

(a)   (b)

Fig. 5.15. Insertion of 10 into the tree of Fig. 5.14.

Next suppose we delete 7 from the tree of Fig. 5.16(a). Its parent now has only one child, 8, and the grandparent has no child with three children. We therefore make 8 be a sibling of 2 and 5, leaving the tree of Fig. 5.16(b). Now the node starred in Fig. 5.16(b) has only one child, and its parent has no other child with three children. Thus we delete the starred node, making its child be a child of the sibling of the starred node. Now the root has only one child, and we delete it, leaving the tree of Fig. 5.16(c).

Observe in the above examples, the frequent manipulation of the values at interior nodes. While we can always calculate these values by walking the tree, it can be done as we manipulate the tree itself, provided we remember the smallest value among the descendants of each node along the path from the root to the deleted leaf. This information can be computed by a recursive deletion algorithm, with the call at each node being passed, from above, the correct quantity (or the value "minus infinity" if we are along the leftmost path). The details require careful case analysis and will be sketched later when we consider the program for the DELETE operation. □

Data Types for 2-3 Trees

Let us restrict ourselves to representing by 2-3 trees sets of elements whose keys are real numbers. The nature of other fields that go with the field key, to make up a record of type elementtype, we shall leave unspecified, as it has no bearing on what follows.

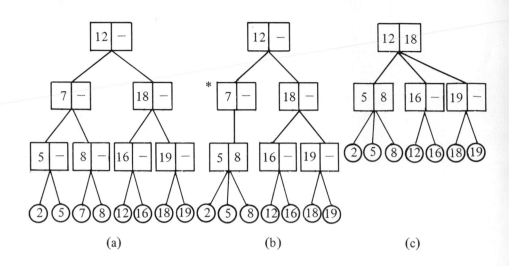

Fig. 5.16. Deletion in a 2-3 tree.

In Pascal, the parents of leaves must be records consisting of two reals (the keys of the smallest elements in the second and third subtrees) and of three pointers to elements. The parents of these nodes are records consisting of two reals and of three pointers to parents of leaves. This progression continues indefinitely; each level in a 2-3 tree is of a different type from all other levels. Such a situation would make programming 2-3 tree operations in Pascal impossible, but fortunately, Pascal provides a mechanism, the variant record structure, that enables us to regard all 2-3 tree nodes as having the same type, even though some are elements, and some are records with pointers and reals.† We can define nodes as in Fig. 5.17. Then we declare a set, represented by a 2-3 tree, to be a pointer to the root as in Fig. 5.17.

Implementation of INSERT

The details of operations on 2-3 trees are quite involved, although the principles are simple. We shall therefore describe only one operation, insertion, in detail; the others, deletion and membership testing, are similar in spirit, and finding the minimum is a trivial search down the leftmost path. We shall write the insertion routine as main procedure, INSERT, which we call at the root, and a procedure insert1, which gets called recursively down the tree.

---

† All nodes, however, take the largest amount of space needed for any of the variant types, so Pascal is not really the best language for implementing 2-3 trees in practice.

```
type
    elementtype = record
        key: real;
        {other fields as warranted}
    end;
    nodetypes = (leaf, interior);
    twothreenode = record
        case kind: nodetypes of
            leaf: (element: elementtype);
            interior: (firstchild, secondchild, thirdchild: ↑ twothreenode;
                lowofsecond, lowofthird: real)
    end;
    SET = ↑ twothreenode;
```

Fig. 5.17. Definition of a node in a 2-3 tree.

For convenience, we assume that a 2-3 tree is not a single node or empty. These two cases require a straightforward sequence of steps which the reader is invited to provide as an exercise.

```
procedure insert1 ( node: ↑ twothreenode;
    x: elementtype; { x is to be inserted into the subtree of node }
    var pnew: ↑ twothreenode; { pointer to new node created to right of node }
    var low: real ); { smallest element in the subtree pointed to by pnew }

begin
    pnew := nil;
    if node is a leaf then begin
        if x is not the element at node then begin
            create new node pointed to by pnew;
            put x at the new node;
            low := x.key
        end
    end
    else begin { node is an interior node }
        let w be the child of node to whose subtree x belongs;
        insert1(w, x, pback, lowback);
        if pback <> nil then begin
            insert pointer pback among the children of node just
                to the right of w;
            if node has four children then begin
                create new node pointed to by pnew;
                give the new node the third and fourth children of node;
                adjust lowofsecond and lowofthird in node and the new node,
```

```
                    set low to be the lowest key among the
                         children of the new node
              end
          end
     end
end;  { insert1 }
```

Fig. 5.18. Sketch of 2-3 tree insertion program.

We would like *insert*1 to return both a pointer to a new node, if it must create one, and the key of the smallest element descended from that new node. As the mechanism in Pascal for creating such a function is awkward, we shall instead declare *insert*1 to be a procedure that assigns values to parameters *pnew* and *low* in the case that it must "return" a new node. We sketch *insert*1 in Fig. 5.18. The complete procedure is shown in Fig. 5.19; some comments in Fig. 5.18 have been omitted from Fig. 5.19 to save space.

```
procedure insert1 ( node: ↑ twothreenode; x: elementtype;
     var pnew: ↑ twothreenode; var low: real );

     var
          pback: ↑ twothreenode;
          lowback: real;
          child: 1..3;  { indicates which child of node is followed
               in recursive call (cf. w in Fig. 5.18) }
          w: ↑ twothreenode;  { pointer to the child }

     begin
          pnew := nil;
          if node↑.kind = leaf then begin
               if node↑.element.key <> x.key then begin
                    { create new leaf holding x.key and "return" this node }
                    new(pnew, leaf);
                    if (node↑.element.key < x.key) then
                         { place x in new node to right of current node }
                         begin pnew↑.element := x; low := x.key end
                    else begin  { x belongs to left of element at current node }
                         pnew↑.element := node↑.element;
                         node↑.element := x;
                         low := pnew↑.element.key
                    end
               end
          end
          else begin  { node is an interior node }
               { select the child of node that we must follow }
               if x.key < node↑.lowofsecond then
                    begin child := 1; w := node↑.firstchild end
```

```
    else if (node↑.thirdchild = nil) or (x.key < node↑.lowofthird) then begin
        { x is in second subtree }
        child := 2;
        w := node↑.secondchild
end
else begin  { x is in third subtree }
    child := 3;
    w := node↑.thirdchild
end;
insert1(w, x, pback, lowback);
if pback <> nil then
    { a new child of node must be inserted }
    if node↑.thirdchild = nil then
        { node had only two children, so insert new node in proper place }
        if child = 2 then begin
            node↑.thirdchild := pback;
            node↑.lowofthird := lowback
        end
        else begin  { child = 1 }
            node↑.thirdchild := node↑.secondchild;
            node↑.lowofthird := node↑.lowofsecond;
            node↑.secondchild := pback;
            node↑.lowofsecond := lowback
        end
    else begin  { node already had three children }
        new(pnew, interior);
        if child = 3 then begin
            { pback and third child become children of new node }
            pnew↑.firstchild := node↑.thirdchild;
            pnew↑.secondchild := pback;
            pnew↑.thirdchild := nil;
            pnew↑.lowofsecond := lowback;
                { lowofthird is undefined for pnew }
            low := node↑.lowofthird;
            node↑.thirdchild := nil
        end
        else begin  { child ≤ 2; move third child of node to pnew}
            pnew↑.secondchild := node↑.thirdchild;
            pnew↑.lowofsecond := node↑.lowofthird;
            pnew↑.thirdchild := nil;
            node↑.thirdchild := nil
        end
        if child = 2 then begin
            { pback becomes first child of pnew }
            pnew↑.firstchild := pback;
```

```
                    low := lowback
                end
                if child = 1 then begin
                    { second child of node is moved to pnew;
                        pback becomes second child of node }
                    pnew↑.firstchild := node↑.secondchild;
                    low := node↑.lowofsecond;
                    node↑.secondchild := pback;
                    node↑.lowofsecond := lowback
                end
            end
        end
end; { insert1 }
```

Fig. 5.19. The procedure *insert1*.

Now we can write the procedure INSERT, which calls *insert1*. If *insert1* "returns" a new node, then INSERT must create a new root. The code is shown in Fig. 5.20 on the assumption that the type SET is ↑ twothreenode, i.e., a pointer to the root of a 2-3 tree whose leaves contain the members of the set.

```
procedure INSERT ( x: elementtype; var S: SET );
    var
        pback: ↑ twothreenode; { pointer to new node returned by insert1 }
        lowback: real; { low value in subtree of pback }
        saveS: SET; { place to store a temporary copy of the pointer S }
    begin
        { checks for S being empty or a single node should occur here,
            and an appropriate insertion procedure should be included }
        insert1(S, x, pback, lowback);
        if pback <> nil then begin
            { create new root; its children are now pointed to by S and pback }
            saveS := S;
            new(S);
            S↑.firstchild := saveS;
            S↑.secondchild := pback;
            S↑.lowofsecond := lowback;
            S↑.thirdchild := nil;
        end
end; { INSERT }
```

Fig. 5.20. INSERT for sets represented by 2-3 trees.

Implementation of DELETE

We shall sketch a function *delete*1 that takes a pointer to a node *node* and an element *x*, and deletes a leaf descended from *node* having value *x*, if there is one.† Function *delete*1 returns true if after deletion *node* has only one child, and it returns false if *node* still has two or three children. A sketch of the code for *delete*1 is shown in Fig. 5.21.

We leave the detailed code for function *delete*1 for the reader. Another exercise is to write a procedure DELETE(*S*, *x*) that checks for the special cases that the set *S* consists only of a single leaf or is empty, and otherwise calls *delete*1(*S*, *x*); if *delete*1 returns true, the procedure removes the root (the node pointed to by *S*) and makes *S* point to its lone child.

```
function delete1 ( node: ↑ twothreenode; x: elementtype ) : boolean;
    var
        onlyone: boolean;  { to hold the value returned by a call to delete1 }
    begin
        delete := false;
        if the children of node are leaves then begin
            if x is among those leaves then begin
                remove x;
                shift children of node to the right of x one position left;
                if node now has one child then
                    delete1 := true
            end
        end
        else begin  { node is at level two or higher }
            determine which child of node could have x as a descendant;
            onlyone := delete1(w, x);  { w stands for node↑.firstchild,
                node↑.secondchild, or node↑.thirdchild, as appropriate }
            if onlyone then begin  { fix children of node }
                if w is the first child of node then
                    if y, the second child of node, has three children then
                        make the first child of y be the second child of w
                    else begin  { y has two children }
                        make the child of w be the first child of y;
                        remove w from among the children of node;
                        if node now has one child then
                            delete1 := true
                    end
                if w is the second child of node then
                    if y, the first child of node, has three children then
                        make the third child of y be the first child of w
```

---

† A useful variant would take only a key value and delete any element with that key.

                else { y has two children }
                    if z, the third child of *node*, exists
                        and has three children **then**
                            make first child of z be the second child of w
                else begin { no other child of *node* has three children }
                    make the child of w be the third child of y;
                    remove w from among the children of *node*;
                    if *node* now has one child **then**
                        *delete*1 := true
              end;
            if w is the third child of *node* **then**
                if y, the second child of *node*, has three children **then**
                    make the third child of y be the second child of w
                else begin { y has two children }
                  make the child of w be the third child of y;
                  remove w from among the children of *node*
              end { note *node* surely has two children left in this case }
        end
      end
    end; { *delete*1 }

Fig. 5.21. Recursive deletion procedure.

5.5 Sets with the MERGE and FIND Operations

In certain problems we start with a collection of objects, each in a set by itself; we then combine sets in some order, and from time to time ask which set a particular object is in. These problems can be solved using the operations MERGE and FIND. The operation MERGE(A, B, C) makes C equal to the union of sets A and B, provided A and B are disjoint (have no member in common); MERGE is undefined if A and B are not disjoint. FIND(x) is a function that returns the set of which x is a member; in case x is in two or more sets, or in no set, FIND is not defined.

Example 5.6. An *equivalence relation* is a reflexive, symmetric, and transitive relation. That is, if $\equiv$ is an equivalence relation on set S, then for any (not necessarily distinct) members a, b, and c in S, the following properties hold:

1. $a \equiv a$ (*reflexivity*)
2. If $a \equiv b$, then $b \equiv a$ (*symmetry*).
3. If $a \equiv b$ and $b \equiv c$, then $a \equiv c$ (*transitivity*).

The relation "is equal to" ($=$) is the paradigm equivalence relation on any set S. For a, b, and c in S, we have (1) $a = a$, (2) if $a = b$, then $b = a$, and (3) if $a = b$ and $b = c$, then $a = c$. There are many other equivalence relations, however, and we shall shortly see several additional examples.

In general, whenever we partition a collection of objects into disjoint groups, the relation $a \equiv b$ if and only if a and b are in the same group is an equivalence relation. "Is equal to" is the special case where every element is in a group by itself.

More formally, if a set S has an equivalence relation defined on it, then the set S can be partitioned into disjoint subsets $S_1, S_2, \ldots$, called *equivalence classes,* whose union is S. Each subset S_i consists of equivalent members of S. That is, $a \equiv b$ for all a and b in S_i, and $a \not\equiv b$ if a and b are in different subsets. For example, the relation congruence modulo n† is an equivalence relation on the set of integers. To check that this is so, note that $a-a = 0$, which is a multiple of n (reflexivity); if $a-b = dn$, then $b-a = (-d)n$ (symmetry); and if $a-b = dn$ and $b-c = en$, then $a-c = (d+e)n$ (transitivity). In the case of congruence modulo n there are n equivalence classes, which are the set of integers congruent to 0, the set of integers congruent to 1, . . . , the set of integers congruent to $n-1$.

The *equivalence problem* can be formulated in the following manner. We are given a set S and a sequence of statements of the form "a is equivalent to b." We are to process the statements in order in such a way that at any time we are able to determine in which equivalence class a given element belongs. For example, suppose $S = \{1, 2, \ldots, 7\}$ and we are given the sequence of statements

$$1 \equiv 2 \quad 5 \equiv 6 \quad 3 \equiv 4 \quad 1 \equiv 4$$

to process. The following sequence of equivalence classes needs to be constructed, assuming that initially each element of S is in an equivalence class by itself.

$1 \equiv 2$	$\{1,2\}$ $\{3\}$ $\{4\}$ $\{5\}$ $\{6\}$ $\{7\}$
$5 \equiv 6$	$\{1,2\}$ $\{3\}$ $\{4\}$ $\{5,6\}$ $\{7\}$
$3 \equiv 4$	$\{1,2\}$ $\{3,4\}$ $\{5,6\}$ $\{7\}$
$1 \equiv 4$	$\{1,2,3,4\}$ $\{5,6\}$ $\{7\}$

We can "solve" the equivalence problem by starting with each element in a named set. When we process statement $a \equiv b$, we FIND the equivalence classes of a and b and then MERGE them. We can at any time use FIND to tell us the current equivalence class of any element.

The equivalence problem arises in several areas of computer science. For example, one form occurs when a Fortran compiler has to process "equivalence declarations" such as

---

† We say a is *congruent to b modulo n* if a and b have the same remainders when divided by n, or put another way, $a-b$ is a multiple of n.

EQUIVALENCE (A(1),B(1,2),C(3)), (A(2),D,E)), (F,G)

Another example, presented in Chapter 6, uses solutions to the equivalence problem to help find minimum-cost spanning trees. □

A Simple Implementation of MFSET

Let us begin with a simplified version of the MERGE-FIND ADT. We shall define an ADT, called MFSET, consisting of a set of subsets, which we shall call *components*, together with the following operations:

1. MERGE(A, B) takes the union of the components A and B and calls the result either A or B, arbitrarily.

2. FIND(x) is a function that returns the name of the component of which x is a member.

3. INITIAL(A, x) creates a component named A that contains only the element x.

To make a reasonable implementation of MFSET, we must restrict our underlying types, or alternatively, we should recognize that MFSET really has two other types as "parameters" — the type of set names and the type of members of these sets. In many applications we can use integers as set names. If we take n to be the number of elements, we may also use integers in the range $[1..n]$ for the members of components. For the implementation we have in mind, it is important that the type of set members be a subrange type, because we want to index into an array defined over that subrange. The type of set names is not important, as this type is the type of array elements, not their indices. Observe, however, that if we wanted the member type to be other than a subrange type, we could create a mapping, with a hash table, for example, that assigned these to unique integers in a subrange. We only need to know the total number of elements in advance.

The implementation we have in mind is to declare

>   **const**
> n = { number of elements };
>   **type**
>   MFSET = **array**[1..*n*] **of** integer;

as a special case of the more general type

>   **array**[subrange of members] **of** (type of set names);

Suppose we declare *components* to be of type MFSET with the intention that *components*[*x*] holds the· name of the set currently containing *x*. Then the three MFSET operations are easy to write. For example, the operation MERGE is shown in Fig. 5.22. INITIAL(A, x) simply sets *components*[*x*] to A, and FIND(x) returns *components*[*x*].

The time performance of this implementation of MFSET is easy to

```
procedure MERGE ( A, B: integer; var C: MFSET );
    var
        x: 1..n;
    begin
        for x := 1 to n do
            if C[x] = B then
                C[x] := A;
    end; { MERGE }
```

Fig. 5.22. The procedure MERGE.

analyze. Each execution of the procedure MERGE takes $O(n)$ time. On the other hand, the obvious implementations of INITIAL(A, x) and FIND(x) have constant running times.

A Faster Implementation of MFSET

Using the algorithm in Fig. 5.22, a sequence of $n-1$ MERGE instructions will take $O(n^2)$ time.[†] One way to speed up the MERGE operation is to link together all members of a component in a list. Then, instead of scanning all members when we merge component B into A, we need only run down the list of members of B. This arrangement saves time on the average. However, it could happen that the ith merge is of the form MERGE(A, B) where A is a component of size 1 and B is a component of size i, and that the result is named B. This merge operation would require $O(i)$ steps, and a sequence of $n-1$ such merge instructions would take on the order of $\sum_{i=1}^{n-1} i = n(n-1)/2$ time.

One way to avoid this worst case situation is to keep track of the size of each component and always merge the smaller into the larger.[‡] Thus, every time a member is merged into a bigger component, it finds itself in a component at least twice as big. Thus, if there are initially n components, each with one member, none of the n members can have its component changed more than $1+\log n$ times. As the time spent by this new version of MERGE is proportional to the number of members whose component names are changed, and the total number of such changes is at most $n(1+\log n)$, we see that $O(n \log n)$ work suffices for all merges.

Now let us consider the data structure needed for this implementation. First, we need a mapping from set names to records consisting of

---

† Note that $n-1$ is the largest number of merges that can be performed before all elements are in one set.

‡ Note that our ability to call the resulting component by the name of either of its constituents is important here, although in the simpler implementation, the name of the first argument was always picked.

1. a *count* giving the number of members in the set and
2. the index in the array of the first element of that set.

We also need another array of records, indexed by members, to indicate

1. the set of which each element is a member and
2. the next array element on the list for that set.

We use 0 to serve as NIL, the end-of-list marker. In a language that lent itself to such constructs, we would prefer to use pointers in this array, but Pascal does not permit pointers into arrays.

In the special case where set names, as well as members, are chosen from the subrange $1..n$, we can use an array for the mapping described above. That is, we define

> **type**
> > nametype = $1..n$;
> > elementtype = $1..n$;
> > MFSET = **record**
> > > *setheaders*: **array**[$1..n$] **of record**
> > > > { headers for set lists }
> > > > *count*: $0..n$;
> > > > *firstelement*: $0..n$
> > >
> > > **end**;
> > > *names*: **array**[$1..n$] **of record**
> > > > { table giving set containing each member }
> > > > *setname*: nametype;
> > > > *nextelement*: $0..n$
> > >
> > > **end**
> >
> > **end**;

The procedures INITIAL, MERGE, and FIND are shown in Fig. 5.23.

Figure 5.24 shows an example of the data structure used in Fig. 5.23, where set 1 is {1, 3, 4}, set 2 is {2}, and set 5 is {5, 6}.

A Tree Implementation of MFSET's

Another, completely different, approach to the implementation of MFSET's uses trees with pointers to parents. We shall describe this approach informally. The basic idea is that nodes of trees correspond to set members, with an array or other implementation of a mapping leading from set members to their nodes. Each node, except the root of each tree, has a pointer to its parent. The roots hold the name of the set, as well as an element. A mapping from set names to roots allows access to any given set, when merges are done.

Figure 5.25 shows the sets $A = \{1, 2, 3, 4\}$, $B = \{5, 6\}$, and $C = \{7\}$ represented in this form. The rectangles are assumed to be part of the root node, not separate nodes.

procedure INITIAL (*A*: nametype; *x*: elementtype; **var** *C*: MFSET);
   { initialize *A* to a set containing *x* only }
   begin
       C.*names*[*x*].*setname* := *A*;
       C.*names*[*x*].*nextelement* := 0;
          { null pointer at end of list of members of *A* }
       C.*setheaders*[*A*].*count* := 1;
       C.*setheaders*[*A*].*firstelement* := *x*
   end; { INITIAL }

procedure MERGE (*A*, *B*: nametype; **var** *C*: MFSET);
   { merge *A* and *B*, calling the result *A* or *B*, arbitrarily }
   var
      i: 0..*n*; { used to find end of smaller list }
   begin
      if *C*.*setheaders*[*A*].*count* > *C*.*setheaders*[*B*].*count* **then begin**
         { *A* is the larger set; merge *B* into *A* }
         { find end of *B*, changing set names to *A* as we go }
         i := *C*.*setheaders*[*B*].*firstelement*;
         repeat
            C.*names*[*i*].*setname* := *A*;
            i := *C*.*names*[*i*].*nextelement*
         until *C*.*names*[*i*].*nextelement* = 0;
         { append list *A* to the end of *B* and call the result *A* }
         { now *i* is the index of the last member of *B* }
         C.*names*[*i*].*setname* := *A*;
         C.*names*[*i*].*nextelement* := *C*.*setheaders*[*A*].*firstelement*;
         C.*setheaders*[*A*].*firstelement* := *C*.*setheaders*[*B*].*firstelement*;
         C.*setheaders*[*A*].*count* := *C*.*setheaders*[*A*].*count* +
                 C.*setheaders*[*B*].*count*;
         C.*setheaders*[*B*].*count* := 0;
         C.*setheaders*[*B*].*firstelement* := 0
           { above two steps not really necessary, as set *B* no longer exists }
     end
     else { *B* is at least as large as *A* }
        { code similar to case above, but with *A* and *B* interchanged }
   end; { MERGE }

function FIND (*x*: 1..*n*; **var** *C*: MFSET);
   { return the name of the set of which *x* is a member }
   begin
      return (*C*.*names*[*x*].*setname*)
   end; { FIND }

Fig. 5.23. The operations of an MFSET.

To find the set containing an element x, we first consult a mapping (e.g., an array) not shown in Fig. 5.25, to obtain a pointer to the node for x. We then follow the path from that node to the root of its tree and read the name of the set there.

The basic merge operation is to make the root of one tree be a child of the root of the other. For example, we could merge A and B of Fig. 5.25 and call the result A, by making node 5 a child of node 1. The result is shown in Fig. 5.26. However, indiscriminate merging could result in a tree of n nodes that is a single chain. Then doing a FIND operation on each of those nodes would take $O(n^2)$ time. Observe that although a merge can be done in $O(1)$ steps, the cost of a reasonable number of FIND's will dominate the total cost, and this approach is not necessarily better than the simplest one for executing n merges and n finds.

However, a simple improvement guarantees that if n is the number of elements, then no FIND will take more than $O(\log n)$ steps. We simply keep at each root a count of the number of elements in the set, and when called upon to merge two sets, we make the root of the smaller tree be a child of the root of the larger. Thus, every time a node is moved to a new tree, two things happen: the distance from the node to its root increases by one, and the node will be in a set with at least twice as many elements as before. Thus, if n is the total number of elements, no node can be moved more than $\log n$ times; hence, the distance to its root can never exceed $\log n$. We conclude that each FIND requires at most $O(\log n)$ time.

Path Compression

Another idea that may speed up this implementation of MFSET's is *path compression*. During a FIND, when following a path from some node to the root, make each node encountered along the path be a child of the root. The easiest way to do this is in two passes. First, find the root, and then retraverse the same path, making each node a child of the root.

Example 5.7. Figure 5.27(a) shows a tree before executing a FIND operation on the node for element 7 and Fig. 5.27(b) shows the result after 5 and 7 are made children of the root. Nodes 1 and 2 on the path are not moved because 1 is the root, and 2 is already a child of the root. □

Path compression does not affect the cost of MERGE's; each MERGE still takes a constant amount of time. There is, however, a subtle speedup in FIND's since path compression tends to shorten a large number of paths from various nodes to the root with relatively little effort.

Unfortunately, it is very difficult to analyze the average cost of FIND's when path compression is used. It turns out that if we do not require that smaller trees be merged into larger ones, we require no more than $O(n \log n)$ time to do n FIND's. Of course, the first FIND may take $O(n)$ time by itself for a tree consisting of one chain. But path compression can change a tree very rapidly and no matter in what order we apply FIND to elements of any tree no more than $O(n)$ time is spent on n FIND's. However, there are

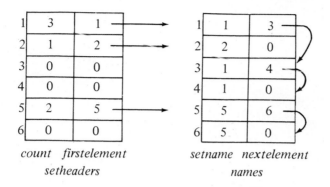

count firstelement
setheaders

setname nextelement
names

Fig. 5.24. Example of the MFSET data structure.

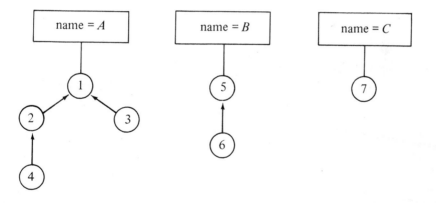

Fig. 5.25. MFSET represented by a collection of trees.

sequences of MERGE and FIND instructions that require $\Omega(n\log n)$ time.

The algorithm that both uses path compression and merges the smaller tree into the larger is asymptotically the most efficient method known for implementing MFSET's. In particular, n FIND's require no more than $O(n\alpha(n))$ time, where $\alpha(n)$ is a function that is not constant, yet grows much more slowly than $\log n$. We shall define $\alpha(n)$ below, but the analysis that leads to this bound is beyond the scope of this book.

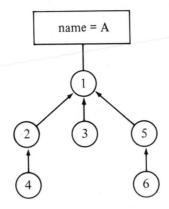

Fig. 5.26. Merging B into A.

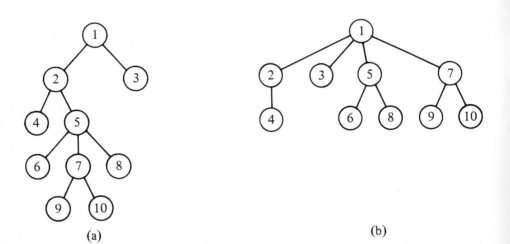

(a)   (b)

Fig. 5.27. An example of path compression.

The Function $\alpha(n)$

The function $\alpha(n)$ is closely related to a very rapidly growing function $A(x, y)$, known as *Ackermann's function*. $A(x, y)$ is defined recursively by:

$$A(0, y) = 1 \text{ for } y \geq 0$$

$$A(1, 0) = 2$$

$$A(x, 0) = x+2 \text{ for } x \geq 2$$

$$A(x, y) = A(A(x-1, y), y-1) \text{ for } x, y \geq 1$$

Each value of y defines a function of one variable. For example, the third line above tells us that for $y=0$, this function is "add 2." For $y = 1$, we have $A(x, 1) = A(A(x-1, 1), 0) = A(x-1, 1) + 2$, for $x > 1$, with $A(1, 1) = A(A(0, 1), 0) = A(1, 0) = 2$. Thus $A(x, 1) = 2x$ for all $x \geq 1$. In other words, $A(x, 1)$ is "multiply by 2." Then, $A(x, 2) = A(A(x-1, 2), 1) = 2A(x-1, 2)$ for $x > 1$. Also, $A(1, 2) = A(A(0, 2), 1) = A(1, 1) = 2$. Thus $A(x, 2) = 2^x$. Similarly, we can show that $A(x, 3) = 2^{2^{\cdots^2}}$ (stack of x 2's), while $A(x, 4)$ is so rapidly growing there is no accepted mathematical notation for such a function.

A single-variable Ackermann's function can be defined by letting $A(x) = A(x, x)$. The function $\alpha(n)$ is a pseudo-inverse of this single variable function. That is, $\alpha(n)$ is the least x such that $n \leq A(x)$. For example, $A(1) = 2$, so $\alpha(1) = \alpha(2) = 1$. $A(2) = 4$, so $\alpha(3) = \alpha(4) = 2$. $A(3) = 8$, so $\alpha(5) = \cdots = \alpha(8) = 3$. So far, $\alpha(n)$ seems to be growing rather steadily.

However, $A(4)$ is a stack of 65536 2's. Since $\log(A(4))$ is a stack of 65535 2's, we cannot hope even to write $A(4)$ explicitly, as it would take $\log(A(4))$ bits to do so. Thus $\alpha(n) \leq 4$ for all integers n one is ever likely to encounter. Nevertheless, $\alpha(n)$ eventually reaches 5, 6, 7, . . . on its unimaginably slow course toward infinity.

5.6 An ADT with MERGE and SPLIT

Let S be a set whose members are ordered by the relation $<$. The operation SPLIT(S, S_1, S_2, x) partitions S into two sets: $S_1=\{ a \mid a$ is in S and $a < x\}$ and $S_2 = \{ a \mid a$ is in S and $a \geq x\}$. The value of S after the split is undefined, unless it is one of S_1 or S_2. There are several situations where the operation of splitting sets by comparing each member with a fixed value x is essential. We shall consider one such problem here.

The Longest Common Subsequence Problem

A *subsequence* of a sequence x is obtained by removing zero or more (not necessarily contiguous) elements from x. Given two sequences x and y, a *longest common subsequence (LCS)* is a longest sequence that is a subsequence of both x and y.

For example, an LCS of 1, 2, 3, 2, 4, 1, 2 and 2, 4, 3, 1, 2, 1 is the subsequence 2, 3, 2, 1, formed as shown in Fig. 5.28. There are other LCS's as well, such as 2, 4, 1, 2, but there are no common subsequences of length 5.

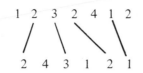

Fig. 5.28. A longest common subsequence.

There is a UNIX command called *diff* that compares files line-by-line, finding a longest common subsequence, where a line of a file is considered an element of the subsequence. That is, whole lines are analogous to the integers 1, 2, 3, and 4 in Fig. 5.28. The assumption behind the command *diff* is that the lines of each file that are not in this LCS are lines inserted, deleted or modified in going from one file to the other. For example, if the two files are versions of the same program made several days apart, *diff* will, with high probability, find the changes.

There are several general solutions to the LCS problem that work in $O(n^2)$ steps on sequences of length n. The command *diff* uses a different strategy that works well when the files do not have too many repetitions of any line. For example, programs will tend to have lines "begin" and "end" repeated many times, but other lines are not likely to repeat.

The algorithm used by *diff* for finding an LCS makes use of an efficient implementation of sets with operations MERGE and SPLIT, to work in time $O(p\log n)$, where n is the maximum number of lines in a file and p is the number of pairs of positions, one from each file, that have the same line. For example, p for the strings in Fig. 5.28 is 12. The two 1's in each string contribute four pairs, the 2's contribute six pairs, and 3 and 4 contribute one pair each. In the worst case, p could be n^2, and this algorithm would take $O(n^2\log n)$ time. However, in practice, p is usually closer to n, so we can expect an $O(n\log n)$ time complexity.

To begin the description of the algorithm let $A = a_1a_2 \cdots a_n$ and $B = b_1b_2 \cdots b_m$ be the two strings whose LCS we desire. The first step is to tabulate for each value a, the positions of the string A at which a appears. That is, we define PLACES(a) = $\{ i \mid a = a_i \}$. We can compute the sets PLACES(a) by constructing a mapping from symbols to headers of lists of positions. By using a hash table, we can create the sets PLACES(a) in $O(n)$ "steps" on the average, where a "step" is the time it takes to operate on a symbol, say to hash it or compare it with another. This time could be a constant if symbols are characters or integers, say. However, if the symbols of A

and B are really lines of text, then steps take an amount of time that depends on the average length of a line of text.

Having computed PLACES(a) for each symbol a that occurs in string A, we are ready to find an LCS. To simplify matters, we shall only show how to find the length of the LCS, leaving the actual construction of the LCS as an exercise. The algorithm considers each b_j, for $j = 1, 2, \ldots, m$, in turn. After considering b_j, we need to know, for each i between 0 and n, the length of the LCS of strings $a_1 \cdots a_i$ and $b_1 \cdots b_j$.

We shall group values of i into sets S_k, for $k = 0, 1, \ldots, n$, where S_k consists of all those integers i such that the LCS of $a_1 \cdots a_i$ and $b_1 \cdots b_j$ has length k. Note that S_k will always be a set of consecutive integers, and the integers in S_{k+1} are larger than those in S_k, for all k.

Example 5.8. Consider Fig. 5.28, with $j = 5$. If we try to match zero symbols from the first string with the first five symbols of the second (24312), we naturally have an LCS of length 0, so 0 is in S_0. If we use the first symbol from the first string, we can obtain an LCS of length 1, and if we use the first two symbols, 12, we can obtain an LCS of length 2. However, using 123, the first three symbols, still gives us an LCS of length 2 when matched against 24312. Proceeding in this manner, we discover $S_0 = \{0\}$, $S_1 = \{1\}$, $S_2 = \{2, 3\}$, $S_3 = \{4, 5, 6\}$, and $S_4 = \{7\}$. □

Suppose that we have computed the S_k's for position $j-1$ of the second string and we wish to modify them to apply to position j. We consider the set PLACES(b_j). For each r in PLACES(b_j), we consider whether we can improve some of the LCS's by adding the match between a_r and b_j to the LCS of $a_1 \cdots a_{r-1}$ and $b_1 \cdots b_j$. That is, if both $r-1$ and r are in S_k, then all $s \geq r$ in S_k really belong in S_{k+1} when b_j is considered. To see this we observe that we can obtain k matches between $a_1 \cdots a_{r-1}$ and $b_1 \cdots b_{j-1}$, to which we add a match between a_r and b_j. We can modify S_k and S_{k+1} by the following steps.

1. FIND(r) to get S_k.
2. If FIND($r-1$) is not S_k, then no benefit can be had by matching b_j with a_r. Skip the remaining steps and do not modify S_k or S_{k+1}.
3. If FIND($r-1$) $= S_k$, apply SPLIT(S_k, S_k, S_k', r) to separate from S_k those members greater than or equal to r.
4. MERGE(S_k', S_{k+1}, S_{k+1}) to move these elements into S_{k+1}.

It is important to consider the members of PLACES(b_j) largest first. To see why, suppose for example that 7 and 9 are in PLACES(b_j), and before b_j is considered, $S_3=\{6, 7, 8, 9\}$ and $S_4=\{10, 11\}$.

If we consider 7 before 9, we split S_3 into $S_3 = \{6\}$ and $S_3' = \{7, 8, 9\}$, then make $S_4 = \{7, 8, 9, 10, 11\}$. If we then consider 9, we split S_4 into $S_4=\{7, 8\}$ and $S_4' = \{9, 10, 11\}$, then merge 9, 10 and 11 into S_5. We have thus moved 9 from S_3 to S_5 by considering only one more position in the second string, representing an impossibility. Intuitively, what has happened is that we have erroneously matched b_j against both a_7 and a_9 in creating an imaginary LCS of length 5.

In Fig. 5.29, we see a sketch of the algorithm that maintains the sets S_k as we scan the second string. To determine the length of an LCS, we need only execute FIND(n) at the end.

```
        procedure LCS;
            begin
(1)             initialize S₀ = {0, 1, . . . , n} and Sᵢ = ∅ for i = 1, 2, . . . , n;
(2)             for j := 1 to n do { compute Sₖ's for position j }
(3)                 for r in PLACES(bⱼ), largest first do begin
(4)                     k := FIND(r);
(5)                     if k = FIND(r−1) then begin { r is not smallest in Sₖ }
(6)                         SPLIT(Sₖ, Sₖ, Sₖ', r);
(7)                         MERGE(Sₖ', Sₖ₊₁, Sₖ₊₁)
                        end
                    end
            end; { LCS }
```

Fig. 5.29. Sketch of longest common subsequence program.

Time Analysis of the LCS Algorithm

As we mentioned earlier, the algorithm of Fig. 5.29 is a useful approach only if there are not too many matches between symbols of the two strings. The measure of the number of matches is

$$p = \sum_{j=1}^{m} | \text{PLACES}(b_j)|$$

where $|\text{PLACES}(b_j)|$ denotes the number of elements in set $\text{PLACES}(b_j)$. In other words, p is the sum over all b_j of the number of positions in the first string that match b_j. Recall that in our discussion of file comparison, we expect p to be of the same order as m and n, the lengths of the two strings (files).

It turns out that the 2-3 tree is a good structure for the sets S_k. We can initialize these sets, as in line (1) of Fig. 5.29, in $O(n)$ steps. The FIND operation requires an array to serve as a mapping from positions r to the leaf for r and also requires pointers to parents in the 2-3 tree. The name of the set, i.e., k for S_k, can be kept at the root, so we can execute FIND in $O(\log n)$ steps by following parent pointers until we reach the root. Thus all executions of lines (4) and (5) together take $O(p \log n)$ time, since those lines are each executed exactly once for each match found.

The MERGE operation of line (5) has the special property that every member of S_k' is lower than every member of S_{k+1}, and we can take advantage of this fact when using 2-3 trees for an implementation.† To begin the

---

† Strictly speaking we should use a different name for the MERGE operation, as the implementation we propose will not work to compute the arbitrary union of disjoint sets,

MERGE, place the 2-3 tree for S_k' to the left of that for S_{k+1}. If both are of the same height, create a new root with the roots of the two trees as children. If S_k' is shorter, insert the root of that tree as the leftmost child of the leftmost node of S_{k+1} at the appropriate level. If this node now has four children, we modify the tree exactly as in the INSERT procedure of Fig. 5.20. An example is shown in Fig. 5.30. Similarly, if S_{k+1} is shorter, make its root the rightmost child of the rightmost node of S_k' at the appropriate level.

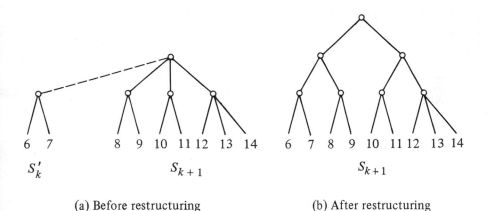

(a) Before restructuring   (b) After restructuring

Fig. 5.30. Example of MERGE.

The SPLIT operation at r requires that we travel up the tree from leaf r, duplicating every interior node along the path and giving one copy to each of the two resulting trees. Nodes with no children are eliminated, and nodes with one child are removed and have that child inserted into the proper tree at the proper level.

Example 5.9. Suppose we split the tree of Fig. 5.30(b) at node 9. The two trees, with duplicated nodes, are shown in Fig. 5.31(a). On the left, the parent of 8 has only one child, so 8 becomes a child of the parent of 6 and 7. This parent now has three children, so all is as it should be; if it had four children, a new node would have been created and inserted into the tree. We need only eliminate nodes with zero children (the old parent of 8) and the chain of nodes with one child leading to the root. The parent of 6, 7, and 8 becomes the new root, as shown in Fig. 5.31(b). Similarly, in the right-hand tree, 9 becomes a sibling of 10 and 11, and unnecessary nodes are eliminated, as is also shown in Fig. 5.31(b). □

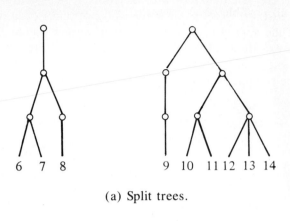

(a) Split trees.

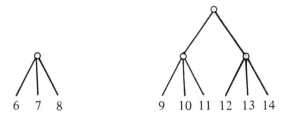

(b) Result of repairs.

Fig. 5.31. An example of SPLIT.

If we do the splitting and reorganization of the 2-3 tree bottom up, it can be shown by consideration of a large number of cases that $O(\log n)$ steps suffices. Thus, the total time spent in lines (6) and (7) of Fig. 5.29 is $O(p \log n)$, and hence the entire algorithm takes $O(p \log n)$ steps. We must add in the preprocessing time needed to compute and sort PLACES(a) for symbols a. As we mentioned, if the symbols a are "large" objects, this time can be much greater than any other part of the algorithm. As we shall see in Chapter 8, if the symbols can be manipulated and compared in single "steps," then $O(n \log n)$ time suffices to sort the first string $a_1 a_2 \cdots a_n$ (actually, to sort objects (i, a_i) on the second field), whereupon PLACES(a) can be read off from this list in $O(n)$ time. Thus, the length of the LCS can be computed in $O(\max(n, p) \log n)$ time which, since $p \geq n$ is normal, can be taken as $O(p \log n)$.

Exercises

5.1 Draw all possible binary search trees containing the four elements 1, 2, 3, 4.

5.2 Insert the integers 7, 2, 9, 0, 5, 6, 8, 1 into a binary search tree by repeated application of the procedure INSERT of Fig. 5.3.

5.3 Show the result of deleting 7, then 2 from the final tree of Exercise 5.2.

*5.4 When deleting two elements from a binary search tree using the procedure of Fig. 5.5, does the final tree ever depend on the order in which you delete them?

5.5 We wish to keep track of all 5-character substrings that occur in a given string, using a trie. Show the trie that results when we insert the 14 substrings of length five of the string ABCDABACDEBA-CADEBA.

*5.6 To implement Exercise 5.5, we could keep a pointer at each leaf, which, say, represents string *abcde*, to the interior node representing the suffix *bcde*. That way, if the next symbol, say *f*, is received, we don't have to insert all of *bcdef*, starting at the root. Furthermore, having seen *abcde*, we may as well create nodes for *bcde*, *cde*, *de*, and *e*, since we shall, unless the sequence ends abruptly, need those nodes eventually. Modify the trie data structure to maintain such pointers, and modify the trie insertion algorithm to take advantage of this data structure.

5.7 Show the 2-3 tree that results if we insert into an empty set, represented as a 2-3 tree, the elements 5, 2, 7, 0, 3, 4, 6, 1, 8, 9.

5.8 Show the result of deleting 3 from the 2-3 tree that results from Exercise 5.7.

5.9 Show the successive values of the various S_i's when implementing the LCS algorithm of Fig. 5.29 with first string *abacabada*, and second string *bdbacbad*.

5.10 Suppose we use 2-3 trees to implement the MERGE and SPLIT operations as in Section 5.8.
    a) Show the result of splitting the tree of Exercise 5.7 at 6.
    b) Merge the tree of Exercise 5.7 with the tree consisting of leaves for elements 10 and 11.

5.11 Some of the structures discussed in this chapter can be modified easily to support the MAPPING ADT. Write procedures MAKENULL, ASSIGN, and COMPUTE to operate on the following data structures.
    a) Binary search trees. The "<" ordering applies to domain elements.

h) 2-3 trees. At interior nodes, place only the key field of domain elements.

5.12 Show that in any subtree of a binary search tree, the minimum element is at a node without a left child.

5.13 Use Exercise 5.12 to produce a nonrecursive version of DELETE-MIN.

5.14 Write procedures ASSIGN, VALUEOF, MAKENULL and GETNEW for trie nodes represented as lists of cells.

*5.15 How do the trie (list of cells implementation), the open hash table, and the binary search tree compare for speed and for space utilization when elements are strings of up to ten characters?

*5.16 If elements of a set are ordered by a "$<$" relation, then we can keep one or two elements (not just their keys) at interior nodes of a 2-3 tree, and we then do not have to keep these elements at the leaves. Write INSERT and DELETE procedures for 2-3 trees of this type.

5.17 Another modification we could make to 2-3 trees is to keep only keys at interior nodes, but do not require that the keys k_1 and k_2 at a node truly be the minimum keys of the second and third subtrees, just that all keys k of the third subtree satisfy $k \geq k_2$, all keys k of the second satisfy $k_1 \leq k < k_2$, and all keys k of the first satisfy $k < k_1$.
   a) How does this convention simplify the DELETE operation?
   b) Which of the dictionary and mapping operations are made more complicated or less efficient?

*5.18 Another data structure that supports dictionaries with the MIN operation is the *AVL tree* (named for the inventors' initials) or *height − balanced tree*. These trees are binary search trees in which the heights of two siblings are not permitted to differ by more than one. Write procedures to implement INSERT and DELETE, while maintaining the AVL-tree property.

5.19 Write the Pascal program for procedure *delete*1 of Fig. 5.21.

*5.20 A *finite automaton* consists of a set of states, which we shall take to be the integers 1..n and a table *transitions*[*state*, *input*] giving a *next state* for each *state* and each *input* character. For our purposes, we shall assume that the input is always either 0 or 1. Further, certain of the states are designated *accepting states*. For our purposes, we shall assume that all and only the even numbered states are accepting. Two states p and q are *equivalent* if either they are the same state, or (i) they are both accepting or both nonaccepting, (ii) on input 0 they transfer to equivalent states, and (iii) on input 1 they transfer to equivalent states. Intuitively, equivalent states behave the same on all sequences of inputs; either both or neither lead to accepting states. Write a program using the MFSET operations that computes the sets of equivalent states of a given finite automaton.

5.21 In the tree implementation of MFSET:

    a) Show that $\Omega(n \log n)$ time is needed for certain lists of n operations if path compression is used but larger trees are permitted to be merged into smaller ones.

    b) Show that $O(n\alpha(n))$ is the worst case running time for n operations if path compression is used, and the smaller tree is always merged into the larger.

5.22 Select a data structure and write a program to compute PLACES (defined in Section 5.6) in average time $O(n)$ for strings of length n.

*5.23** Modify the LCS procedure of Fig. 5.29 to compute the LCS, not just its length.

*5.24** Write a detailed SPLIT procedure to work on 2-3 trees.

*5.25** If elements of a set represented by a 2-3 tree consist only of a key field, an element whose key appears at an interior node need not appear at a leaf. Rewrite the dictionary operations to take advantage of this fact and avoid storing any element at two different nodes.

Bibliographic Notes

Tries were first proposed by Fredkin [1960]. Bayer and McCreight [1972] introduced B-trees, which, as we shall see in Chapter 11, are a generalization of 2-3 trees. The first uses of 2-3 trees were by J. E. Hopcroft in 1970 (unpublished) for insertion, deletion, concatenation, and splitting, and by Ullman [1974] for a code optimization problem.

The tree structure of Section 5.5, using path compression and merging smaller into larger, was first used by M. D. McIlroy and R. Morris to construct minimum-cost spanning trees. The performance of the tree implementation of MFSET's was analyzed by Fischer [1972] and by Hopcroft and Ullman [1973]. Exercise 5.21(b) is from Tarjan [1974].

The solution to the LCS problem of Section 5.6 is from Hunt and Szymanski [1975]. An efficient data structure for FIND, SPLIT, and the restricted MERGE (where all elements of one set are less than those of the other) is described in van Emde Boas, Kaas, and Zijlstra [1975].

Exercise 5.6 is based on an efficient algorithm for matching patterns developed by Weiner [1973]. The 2-3 tree variant of Exercise 5.16 is discussed in detail in Wirth [1976]. The AVL tree structure in Exercise 5.18 is from Adel'son-Vel'skii and Landis [1962].

CHAPTER 6

Directed Graphs

In problems arising in computer science, mathematics, engineering, and many other disciplines we often need to represent arbitrary relationships among data objects. Directed and undirected graphs are natural models of such relationships. This chapter presents the basic data structures that can be used to represent directed graphs. Some basic algorithms for determining the connectivity of directed graphs and for finding shortest paths are also presented.

6.1 Basic Definitions

A *directed graph* (*digraph* for short) G consists of a set of vertices V and a set of arcs E. The vertices are also called *nodes* or *points*; the arcs could be called *directed edges* or *directed lines*. An arc is an ordered pair of vertices (v, w); v is called the *tail* and w the head of the arc. The arc (v, w) is often expressed by $v \rightarrow w$ and drawn as

Notice that the "arrowhead" is at the vertex called the "head" and the tail of the arrow is at the vertex called the "tail." We say that arc $v \rightarrow w$ is *from v to w*, and that w is *adjacent* to v.

Example 6.1. Figure 6.1 shows a digraph with four vertices and five arcs. □

The vertices of a digraph can be used to represent objects, and the arcs relationships between the objects. For example, the vertices might represent cities and the arcs airplane flights from one city to another. As another example, which we introduced in Section 4.2, a digraph can be used to represent the flow of control in a computer program. The vertices represent basic blocks and the arcs possible transfers of flow of control.

A *path* in a digraph is a sequence of vertices $v_1, v_2, \ldots, v_n$, such that $v_1 \rightarrow v_2, v_2 \rightarrow v_3, \ldots, v_{n-1} \rightarrow v_n$ are arcs. This path is *from* vertex v_1 to vertex v_n, and *passes through* vertices $v_2, v_3, \ldots, v_{n-1}$, and ends at vertex v_n. The *length* of a path is the number of arcs on the path, in this case, $n-1$. As a

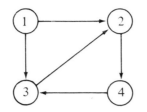

Fig. 6.1. Directed graph.

special case, a single vertex v by itself denotes a path of length zero from v to v. In Fig. 6.1, the sequence 1, 2, 4 is a path of length 2 from vertex 1 to vertex 4.

A path is *simple* if all vertices on the path, except possibly the first and last, are distinct. A simple *cycle* is a simple path of length at least one that begins and ends at the same vertex. In Fig. 6.1, the path 3, 2, 4, 3 is a cycle of length three.

In many applications it is useful to attach information to the vertices and arcs of a digraph. For this purpose we can use a *labeled digraph,* a digraph in which each arc and/or each vertex can have an associated label. A label can be a name, a cost, or a value of any given data type.

Example 6.2. Figure 6.2 shows a labeled digraph in which each arc is labeled by a letter that causes a transition from one vertex to another. This labeled digraph has the interesting property that the arc labels on every cycle from vertex 1 back to vertex 1 spell out a string of a's and b's in which both the number of a's and b's is even. □

In a labeled digraph a vertex can have both a name and a label. Quite frequently, we shall use the vertex label as the name of the vertex. Thus, the numbers in Fig. 6.2 could be interpreted as vertex names or vertex labels.

6.2 Representations for Directed Graphs

Several data structures can be used to represent a directed graph. The appropriate choice of data structure depends on the operations that will be applied to the vertices and arcs of the digraph. One common representation for a digraph $G = (V, E)$ is the *adjacency matrix.* Suppose $V = \{1, 2, \ldots, n\}$. The adjacency matrix for G is an $n \times n$ matrix A of booleans, where $A[i, j]$ is true if and only if there is an arc from vertex i to j. Often, we shall exhibit adjacency matrices with 1 for true and 0 for false; adjacency matrices may even be implemented that way. In the adjacency matrix representation the time required to access an element of an adjacency matrix is independent of the size of V and E. Thus the adjacency matrix

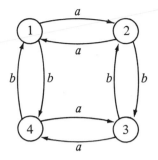

Fig. 6.2. Transition digraph.

representation is useful in those graph algorithms in which we frequently need
to know whether a given arc is present.

Closely related is the *labeled adjacency matrix* representation of a digraph,
where $A[i, j]$ is the label on the arc going from vertex i to vertex j. If there
is no arc from i to j, then a value that cannot be a legitimate label must be
used as the entry for $A[i, j]$.

Example 6.3. Figure 6.3 shows the labeled adjacency matrix for the digraph
of Fig. 6.2. Here, the label type is char, and a blank represents the absence
of an arc. □

	1	2	3	4
1		a		b
2	a		b	
3		b		a
4	b		a	

Fig. 6.3. Labeled adjacency matrix for digraph of Fig. 6.2.

The main disadvantage of using an adjacency matrix to represent a
digraph is that the matrix requires $\Omega(n^2)$ storage even if the digraph has many
fewer than n^2 arcs. Simply to read in or examine the matrix would require
$O(n^2)$ time, which would preclude $O(n)$ algorithms for manipulating digraphs
with $O(n)$ arcs.

To avoid this disadvantage we can use another common representation for
a digraph $G = (V, E)$ called the *adjacency list* representation. The adjacency
list for a vertex i is a list, in some order, of all vertices adjacent to i. We can
represent G by an array *HEAD*, where *HEAD*[i] is a pointer to the adjacency

list for vertex i. The adjacency list representation of a digraph requires storage proportional to sum of the number of vertices plus the number of arcs; it is often used when the number of arcs is much less than n^2. However, a potential disadvantage of the adjacency list representation is that it may take $O(n)$ time to determine whether there is an arc from vertex i to vertex j, since there can be $O(n)$ vertices on the adjacency list for vertex i.

Example 6.4. Figure 6.4 shows an adjacency list representation for the digraph of Fig. 6.1, where singly linked lists are used. If arcs had labels, these could be included in the cells of the linked list.

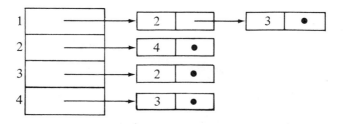

Fig. 6.4. Adjacency list representation for digraph of Fig. 6.1.

If we did insertions and deletions from the adjacency lists, we might prefer to have the *HEAD* array point to header cells that did not contain adjacent vertices.† Alternatively, if the graph were expected to remain fixed, with no (or very few) changes to be made to the adjacency lists, we might prefer *HEAD*[i] to be a cursor to an array *ADJ*, where *ADJ*[*HEAD*[i]], *ADJ*[*HEAD*[i]+1], . . . , and so on, contained the vertices adjacent to vertex i, up to that point in *ADJ* where we first encounter a 0, which marks the end of the list of adjacent vertices for i. For example, Fig. 6.1 could be represented as in Fig. 6.5. □

Directed Graph ADT's

We could define an ADT corresponding to the directed graph formally and study implementations of its operations. We shall not pursue this direction extensively, because there is little of a surprising nature, and the principal data structures for graphs have already been covered. The most common operations on directed graphs include operations to read the label of a vertex or arc, to insert or delete vertices and arcs, and to navigate by following arcs

---

† This is another manifestation of the old Pascal problem of doing insertion and deletion at arbitrary positions of singly linked lists.

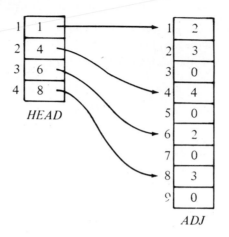

Fig. 6.5. Another adjacency list representation of Fig. 6.1.

from tail to head.

The latter operations require a little thought. Most frequently, we shall encounter in informal programs statements like

>   **for** each vertex w adjacent to vertex v **do**   (6.1)
>   { some action on w }

To implement such a step, we need the notion of an *index* type for the set of vertices adjacent to some one vertex v. For example, if adjacency lists are used to represent the graph, then an index is really a position on the adjacency list for v. If an adjacency matrix is used, an index is an integer representing an adjacent vertex. We need the following three operations on directed graphs.

1. FIRST(v) returns the index for the first vertex adjacent to v. A null vertex Λ is returned if there is no vertex adjacent to v.
2. NEXT(v, i) returns the index after index i for the vertices adjacent to v. Λ is returned if i is the last index for vertices adjacent to v.
3. VERTEX(v, i) returns the vertex with index i among the vertices adjacent to v.

Example 6.5.   If the adjacency matrix representation is chosen, VERTEX(v, i) returns i. FIRST(v) and NEXT(v, i) can be written as in Fig. 6.6 to operate on an externally defined $n \times n$ boolean matrix A. We assume A

is declared

  array [1..*n*, 1..*n*] **of** boolean

and that 0 is used for Λ. We can then implement the statement (6.1) as in Fig. 6.7. □

```
function FIRST ( v: integer ) : integer;
    var
        i: integer;
    begin
        for i := 1 to n do
            if A[v, i] then
                return (i);
        return (0)  { if we reach here, v has no adjacent vertex }
    end;  { FIRST }

function NEXT ( v: integer; i: integer ) : integer;
    var
        j : integer;
    begin
        for j := i+1 to n do
            if A[v, j] then
                return (j);
        return (0)
    end;  { NEXT }
```

Fig. 6.6. Operations to scan adjacent vertices.

```
i := FIRST(v);
while i <> Λ do begin
    w := VERTEX(v, i);
    { some action on w }
    i := NEXT(v, i)
end
```

Fig. 6.7. Iteration over vertices adjacent to *v*.

6.3 The Single Source Shortest Paths Problem

In this section we consider a common path-finding problem on directed graphs. We are given a directed graph $G = (V, E)$ in which each arc has a nonnegative label, and one vertex is specified as the *source*. Our problem is to determine the cost of the shortest path from the source to every other vertex in V, where the *length of a path* is just the sum of the costs of the arcs on the path. This problem is often called the *single-source shortest paths* prob-lem.† Note that we shall talk of paths as having "length" even if costs represent something different, like time.

We might think of G as a map of airline flights, in which each vertex represents a city and each arc $v \to w$ an airline route from city v to city w. The label on arc $v \to w$ is the time to fly from v to w.‡ Solving the single-source shortest paths problem for this directed graph would determine the minimum travel time from a given city to every other city on the map.

To solve this problem we shall use a "greedy" technique, often known as *Dijkstra's algorithm*. The algorithm works by maintaining a set S of vertices whose shortest distance from the source is already known. Initially, S con-tains only the source vertex. At each step, we add to S a remaining vertex v whose distance from the source is as short as possible. Assuming all arcs have nonnegative costs, we can always find a shortest path from the source to v that passes only through vertices in S. Call such a path *special*. At each step of the algorithm, we use an array D to record the length of the shortest spe-cial path to each vertex. Once S includes all vertices, all paths are "special," so D will hold the shortest distance from the source to each vertex.

The algorithm itself is given in Fig. 6.8. It assumes that we are given a directed graph $G = (V, E)$ where $V = \{1, 2, \ldots, n\}$ and vertex 1 is the source. C is a two-dimensional array of costs, where $C[i, j]$ is the cost of going from vertex i to vertex j on arc $i \to j$. If there is no arc $i \to j$, then we assume $C[i, j]$ is ∞, some value much larger than any actual cost. At each step $D[i]$ contains the length of the current shortest special path to vertex i.

Example 6.6. Let us apply *Dijkstra* to the directed graph of Fig. 6.9. Ini-tially, $S = \{1\}$, $D[2] = 10$, $D[3] = \infty$, $D[4] = 30$ and $D[5] = 100$. In the first iteration of the for-loop of lines (4)−(8), $w = 2$ is selected as the vertex with the minimum D value. Then we set $D[3] = \min(\infty, 10+50) = 60$. $D(4)$ and $D(5)$ do not change, because reaching them from 1 directly is shorter than going through vertex 2. The sequence of D-values after each iteration of the for-loop is shown in Fig. 6.10. □

If we wish to reconstruct the shortest path from the source to each vertex, then we can maintain another array P of vertices, such that $P[v]$ contains the vertex immediately before vertex v in the shortest path. Initialize $P[v]$ to 1 for all $v \neq 1$. The P-array can be updated right after line (8) of *Dijkstra*. If $D[w] + C[w, v] < D[v]$ at line (8), then we set $P[v] := w$. Upon termination

---

† One might expect that a more natural problem is to find the shortest path from the source to one particular *destination* vertex. However, that problem appears just as hard in general as the single-source shortest paths problem (expect that by luck we may find the path to the destination before some of the other vertices and thereby terminate the algo-rithm a bit earlier than if we wanted paths to all the vertices).

‡ We might assume that an undirected graph would be used here, since the label on arcs $v \to w$ and $w \to v$ would be the same. However, in fact, travel times are different in dif-ferent directions because of prevailing winds. Anyway, assuming that labels on $v \to w$ and $w \to v$ are identical doesn't seem to help solve the problem.

procedure *Dijkstra*;
    { Dijkstra computes the cost of the shortest paths
      from vertex 1 to every vertex of a directed graph }
  begin

(1) $S := \{1\}$;
(2)   **for** $i := 2$ **to** n **do**
(3) $D[i] := C[1, i]$; { initialize D }
(4)   **for** $i := 1$ **to** $n-1$ **do begin**
(5)   choose a vertex w in $V-S$ such that
                $D[w]$ is a minimum;
(6)   add w to S;
(7)   **for** each vertex v in $V-S$ **do**
(8) $D[v] := \min(D[v], D[w] + C[w, v])$
    end
end; { *Dijkstra* }

Fig. 6.8. Dijkstra's algorithm.

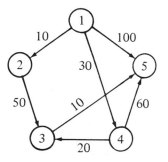

Fig. 6.9. Digraph with labeled arcs.

of *Dijkstra* the path to each vertex can be found by tracing backward the predecessor vertices in the P-array.

Example 6.7. For the digraph in Example 6.6 the P-array would have the values $P[2] = 1$, $P[3] = 4$, $P[4] = 1$, and $P[5] = 3$. To find the shortest path from vertex 1 to vertex 5, for example, we would trace the predecessors in reverse order beginning at vertex 5. From the P-array we determine 3 is the predecessor of 5, 4 the predecessor of 3, and 1 the predecessor of 4. Thus the shortest path from vertex 1 to vertex 5 is 1, 4, 3, 5. □

Iteration	S	w	$D[2]$	$D[3]$	$D[4]$	$D[5]$
initial	{1}	—	10	∞	30	100
1	{1, 2}	2	10	60	30	100
2	{1,2,4}	4	10	50	30	90
3	{1,2,4,3}	3	10	50	30	60
4	{1,2,4,3,5}	5	10	50	30	60

Fig. 6.10. Computation of *Dijkstra* on digraph of Fig. 6.9.

Why Dijkstra's Algorithm Works

Dijkstra's algorithm is an example where "greed" pays off, in the sense that what appears locally as the best thing to do turns out to be the best over all. In this case, the locally "best" thing to do is to find the distance to the vertex w that is outside S but has the shortest special path. To see why in this case there cannot be a shorter nonspecial path from the source to w, observe Fig. 6.11. There we show a hypothetical shorter path to w that first leaves S to go to vertex x, then (perhaps) wanders into and out of S several times before ultimately arriving at w.

But if this path is shorter than the shortest special path to w, then the initial segment of the path from the source to x is a special path to x shorter than the shortest special path to w. (Notice how important the fact that costs are nonnegative is here; without it our argument wouldn't work, and in fact Dijkstra's algorithm would not work correctly.) In that case, when we selected w at line (5) of Fig. 6.8, we should have selected x instead, because $D[x]$ was less than $D[w]$.

To complete a proof that Fig. 6.8 works, we should verify that at all times $D[v]$ is truly the shortest distance of a special path to vertex v. The crux of this argument is in observing that when we add a new vertex w to S at line (6), lines (7) and (8) adjust D to take account of the possibility that there is now a shorter special path to v going through w. If that path goes through the old S to w and then immediately to v, its cost, $D[w]+C[w, v]$, will be compared with $D[v]$ at line (8), and $D[v]$ will be reduced if the new special path is shorter. The only other possibility for a shorter special path is shown in Fig. 6.12, where the path travels to w, then back into the old S, to some member x of the old S, then to v.

But there really cannot be such a path. Since x was placed in S before w, the shortest of all paths from the source to x runs through the old S alone. Therefore, the path to x through w shown in Fig. 6.12 is no shorter than the path directly to x through S. As a result, the length of the path in Fig. 6.12 from the source to w, x, and v is no less from the old value of $D[v]$, since $D[v]$ was no greater than the length of the shortest path to x through S and then directly to w. Thus $D[v]$ cannot be reduced at line (8) by a path through w and x as in Fig. 6.12, and we need not consider the length of such paths.

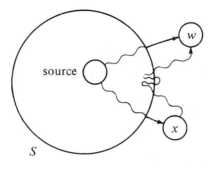

Fig. 6.11. Hypothetical shorter path to w.

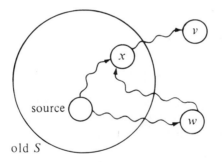

Fig. 6.12. Impossible shortest special path.

Running Time of Dijkstra's Algorithm

Suppose Fig. 6.8 operates on a digraph with n vertices and e edges. If we use an adjacency matrix to represent the digraph, then the loop of lines (7) and (8) takes $O(n)$ time, and it is executed $n-1$ times for a total time of $O(n^2)$. The rest of the algorithm is easily seen to require no more time than this.

If e is much less than n^2, we might do better by using an adjacency list representation of the digraph and using a priority queue implemented as a partially ordered tree to organize the vertices in $V-S$. The loop of lines (7) and (8) can then be implemented by going down the adjacency list for w and updating the distances in the priority queue. A total of e updates will be made, each at a cost of $O(\log n)$ time, so the total time spent in lines (7) and (8) is now $O(e \log n)$, rather than $O(n^2)$.

Lines (1)−(3) clearly take $O(n)$ time, as do lines (4) and (6). Using the priority queue to represent $V-S$, lines (5)−(6) implement exactly the DELETEMIN operation, and each of the $n-1$ iterations of these lines requires $O(\log n)$ time.

As a result, the total time spent on this version of Dijkstra's algorithm is bounded by $O(e \log n)$. This running time is considerably better than $O(n^2)$ if e is very small compared with n^2.

6.4 The All-Pairs Shortest Paths Problem

Suppose we have a labeled digraph that gives the flying time on certain routes connecting cities, and we wish to construct a table that gives the shortest time required to fly from any one city to any other. We now have an instance of the *all-pairs shortest paths* (APSP) problem. To state the problem precisely, we are given a directed graph $G = (V, E)$ in which each arc $v \rightarrow w$ has a nonnegative cost $C[v, w]$. The APSP problem is to find for each ordered pair of vertices (v, w) the smallest length of any path from v to w.

We could solve this problem using Dijkstra's algorithm with each vertex in turn as the source. A more direct way of solving the problem is to use the following algorithm due to R. W. Floyd. For convenience, let us again assume the vertices in V are numbered $1, 2, \ldots, n$. Floyd's algorithm uses an $n \times n$ matrix A in which to compute the lengths of the shortest paths. We initially set $A[i, j] = C[i, j]$ for all $i \neq j$. If there is no arc from i to j, we assume $C[i, j] = \infty$. Each diagonal element is set to 0.

We then make n iterations over the A matrix. After the kth iteration, $A[i, j]$ will have for its value the smallest length of any path from vertex i to vertex j that does not pass through a vertex numbered higher than k. That is to say, i and j, the end vertices on the path, may be any vertex, but any intermediate vertex on the path must be less than or equal to k.

In the kth iteration we use the following formula to compute A.

$$A_k[i, j] = \min \begin{cases} A_{k-1}[i, j] \\ A_{k-1}[i, k] + A_{k-1}[k, j] \end{cases}$$

The subscript k denotes the value of the A matrix after the kth iteration, and it should not be assumed that there are n different matrices. We shall eliminate these subscripts shortly. This formula has the simple interpretation shown in Fig. 6.13.

To compute $A_k[i, j]$ we compare $A_{k-1}[i, j]$, the cost of going from i to j without going through k or any higher-numbered vertex, with $A_{k-1}[i, k] + A_{k-1}[k, j]$, the cost of going first from i to k and then from k to j, without passing through a vertex numbered higher than k. If passing through vertex k produces a cheaper path than what we had for $A_{k-1}[i, j]$, then we choose that cheaper cost for $A_k[i, j]$.

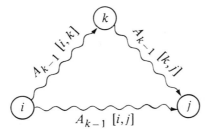

Fig. 6.13. Including k among the vertices to go from i to j.

Example 6.8. Consider the weighted digraph shown in Fig. 6.14. The values of the A matrix initially and after the three iterations are shown in Fig. 6.15.

□

Fig. 6.14. Weighted digraph.

Since $A_k[i, k] = A_{k-1}[i, k]$ and $A_k[k, j] = A_{k-1}[k, j]$, no entry with either subscript equal to k changes during the kth iteration. Therefore, we can perform the computation with only one copy of the A matrix. A program to perform this computation on $n \times n$ matrices is shown in Fig. 6.16.

The running time of this program is clearly $O(n^3)$, since the program is basically nothing more than a triply nested for-loop. To verify that this program works, it is easy to prove by induction on k that after k passes through the triple for-loop, $A[i, j]$ holds the length of the shortest path from vertex i to vertex j that does not pass through a vertex numbered higher than k.

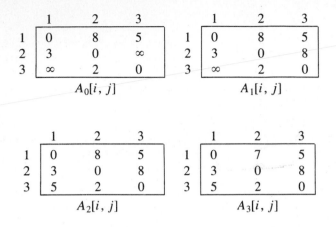

Fig. 6.15. Values of successive A matrices.

```
procedure Floyd ( var A: array[1..n, 1..n] of real;
        C: array[1..n, 1..n] of real );
    { Floyd computes shortest path matrix A given arc cost matrix C }
    var
        i, j, k: integer;
    begin
        for i := 1 to n do
            for j := 1 to n do
                A[i, j] := C[i, j];
        for i := 1 to n do
            A[i, i] := 0;
        for k := 1 to n do
            for i := 1 to n do
                for j := 1 to n do
                    if A[i, k] + A[k, j] < A[i, j] then
                        A[i, j] := A[i, k] + A[k, j]
    end; { Floyd }
```

Fig. 6.16. Floyd's algorithm.

Comparison Between Floyd's and Dijkstra's Algorithms

Since the adjacency-matrix version of *Dijkstra* finds shortest paths from one vertex in $O(n^2)$ time, it, like Floyd's algorithm, can find all shortest paths in $O(n^3)$ time. The compiler, machine, and implementation details will determine the constants of proportionality. Experimentation and measurement are the easiest way to ascertain the best algorithm for the application at hand.

If e, the number of edges, is very much less than n^2, then despite the relatively low constant factor in the $O(n^3)$ running time of *Floyd*, we would expect the adjacency list version of *Dijkstra*, taking $O(ne \log n)$ time to solve the APSP, to be superior, at least for large sparse graphs.

Recovering the Paths

In many situations we may want to print out the cheapest path from one vertex to another. One way to accomplish this is to use another matrix P, where $P[i, j]$ holds that vertex k that led *Floyd* to find the smallest value of $A[i, j]$. If $P[i, j]=0$, then the shortest path from i to j is direct, following the arc from i to j. The modified version of *Floyd* in Fig. 6.17 stores the appropriate intermediate vertices into P.

```
procedure shortest ( var A: array[1..n, 1..n] of real;
        C: array[1..n, 1..n] of real; P: array[1..n, 1..n] of integer );
        { shortest takes an n×n matrix C of arc costs and produces an
            n×n matrix A of lengths of shortest paths and an n×n matrix
            P giving a point in the "middle" of each shortest path }
    var
        i, j, k: integer;
    begin
        for i := 1 to n do
            for j := 1 to n do begin
                A[i, j] := C[i, j];
                P[i, j] := 0
            end;
        for i := 1 to n do
            A[i, i] := 0;
        for k := 1 to n do
            for i := 1 to n do
                for j := 1 to n do
                    if A[i, k] + A[k, j] < A[i, j] then begin
                        A[i, j] := A[i, k] + A[k, j];
                        P[i, j] := k
                    end
    end; { shortest }
```

Fig. 6.17. Shortest paths program.

To print out the intermediate vertices on the shortest path from vertex i to vertex j, we invoke the procedure $path(i, j)$ where *path* is given in Fig. 6.18. While on an arbitrary matrix P, *path* could loop forever, if P comes from *shortest*, we could not, say, have k on the shortest path from i to j and also have j on the shortest path from i to k. Note how our assumption of nonnegative weights is again crucial.

```
procedure path ( i, j: integer );
    var
        k: integer;
    begin
        k := P[i, j];
        if k = 0 then
            return;
        path(i, k);
        writeln(k);
        path(k, j)
    end;  { path }
```

Fig. 6.18. Procedure to print shortest path.

Example 6.9. Figure 6.19 shows the final P matrix for the digraph of Fig. 6.14. □

$$
\begin{array}{c|ccc}
 & 1 & 2 & 3 \\
\hline
1 & 0 & 3 & 0 \\
2 & 0 & 0 & 1 \\
3 & 2 & 0 & 0 \\
\end{array}
$$
$$P$$

Fig. 6.19. P matrix for digraph of Fig. 6.14.

Transitive Closure

In some problems we may be interested in determining only whether there exists a path of length one or more from vertex i to vertex j. Floyd's algorithm can be specialized readily to this problem; the resulting algorithm, which predates Floyd's, is called Warshall's algorithm.

Suppose our cost matrix C is just the adjacency matrix for the given digraph. That is, $C[i, j] = 1$ if there is an arc from i to j, and 0 otherwise. We wish to compute the matrix A such that $A[i, j] = 1$ if there is a path of length one or more from i to j, and 0 otherwise. A is often called the *transitive closure* of the adjacency matrix.

Example 6.10. Figure 6.20 shows the transitive closure for the adjacency matrix of the digraph of Fig. 6.14. □

The transitive closure can be computed using a procedure similar to *Floyd* by applying the following formula in the kth pass over the boolean A matrix.

$$A_k[i, j] = A_{k-1}[i, j] \text{ or } (A_{k-1}[i, k] \text{ and } A_{k-1}[k, j])$$

This formula states that there is a path from i to j not passing through a

	1	2	3
1	0	1	1
2	1	0	1
3	1	1	0

Fig. 6.20. Transitive closure.

vertex numbered higher than k if

1. there is already a path from i to j not passing through a vertex numbered higher than $k-1$ or

2. there is a path from i to k not passing through a vertex numbered higher than $k-1$ and a path from k to j not passing through a vertex numbered higher than $k-1$.

As before $A_k[i, k] = A_{k-1}[i, k]$ and $A_k[k, j] = A_{k-1}[k, j]$ so we can perform the computation with only one copy of the A matrix. The resulting Pascal program, named *Warshall* after its discoverer, is shown in Fig. 6.21.

```
procedure Warshall ( var A: array[1..n, 1..n] of boolean;
        C: array[1..n, 1..n] of boolean );
    { Warshall makes A the transitive closure of C }
    var
        i, j, k: integer;
    begin
        for i := 1 to n do
            for j := 1 to n do
                A[i, j] := C[i, j];
        for k := 1 to n do
            for i := 1 to n do
                for j := 1 to n do
                    if A[i, j] = false then
                        A[i, j] := A[i, k] and A[k, j]
    end;  { Warshall }
```

Fig. 6.21. Warshall's algorithm for transitive closure.

An Example: Finding the Center of a Digraph

Suppose we wish to determine the most central vertex in a digraph. This problem can be readily solved using Floyd's algorithm. First, let us make more precise the term "most central vertex." Let v be a vertex in a digraph $G = (V, E)$. The *eccentricity* of v is

$$\max_{w \text{ in } V} \{ \text{ minimum length of a path from } w \text{ to } v \}$$

The *center* of G is a vertex of minimum eccentricity. Thus, the center of a digraph is a vertex that is closest to the vertex most distant from it.

Example 6.11. Consider the weighted digraph shown in Fig. 6.22.

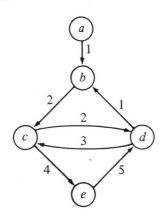

Fig. 6.22. Weighted digraph.

The eccentricities of the vertices are

vertex	eccentricity
a	∞
b	6
c	8
d	5
e	7

Thus the center is vertex d. □

Finding the center of a digraph G is easy. Suppose C is the cost matrix for G.

1. First apply the procedure *Floyd* of Fig. 6.16 to C to compute the all-pairs shortest paths matrix A.

2. Find the maximum cost in each column i. This gives us the eccentricity of vertex i.

3. Find a vertex with minimum eccentricity. This is the center of G.

This running time of this process is dominated by the first step, which takes

$O(n^3)$ time. Step (2) takes $O(n^2)$ time and step (3) $O(n)$ time.

Example 6.12. The APSP cost matrix for Fig. 6.22 is shown in Fig. 6.23. The maximum value in each column is shown below.

	a	b	c	d	e
a	0	1	3	5	7
b	∞	0	2	4	6
c	∞	3	0	2	4
d	∞	1	3	0	7
e	∞	6	8	5	0
max	∞	6	8	5	7

Fig. 6.23. APSP cost matrix.

6.5 Traversals of Directed Graphs

To solve many problems dealing with directed graphs efficiently we need to visit the vertices and arcs of a directed graph in a systematic fashion. Depth-first search, a generalization of the preorder traversal of a tree, is one important technique for doing so. Depth-first search can serve as a skeleton around which many other efficient graph algorithms can be built. The last two sections of this chapter contain several algorithms that use depth-first search as a foundation.

Suppose we have a directed graph G in which all vertices are initially marked *unvisited*. Depth-first search works by selecting one vertex v of G as a start vertex; v is marked *visited*. Then each unvisited vertex adjacent to v is searched in turn, using depth-first search recursively. Once all vertices that can be reached from v have been visited, the search of v is complete. If some vertices remain unvisited, we select an unvisited vertex as a new start vertex. We repeat this process until all vertices of G have been visited.

This technique is called depth-first search because it continues searching in the forward (deeper) direction as long as possible. For example, suppose x is the most recently visited vertex. Depth-first search selects some unexplored arc $x \rightarrow y$ emanating from x. If y has been visited, the procedure looks for another unexplored arc emanating from x. If y has not been visited, then the procedure marks y visited and initiates a new search at y. After completing the search through all paths beginning at y, the search returns to x, the vertex from which y was first visited. The process of selecting unexplored arcs emanating from x is then continued until all arcs from x have been explored.

An adjacency list $L[v]$ can be used to represent the vertices adjacent to vertex v, and an array *mark*, whose elements are chosen from (*visited*, *unvisited*), can be used to determine whether a vertex has been previously visited. The recursive procedure *dfs* is outlined in Fig. 6.24. To use it on an n-vertex graph, we initialize *mark* to *unvisited* and then commence a depth-first search from each vertex that is still unvisited when its turn comes, by

```
for v := 1 to n do
    mark[v] := unvisited;
for v := 1 to n do
    if mark[v] = unvisited then
        dfs(v)
```

Note that Fig. 6.24 is a template to which we shall attach other actions later, as we apply depth-first search. The code in Fig. 6.24 doesn't do anything but set the *mark* array.

Analysis of Depth-First Search

All the calls to *dfs* in the depth-first search of a graph with e arcs and $n \le e$ vertices take $O(e)$ time. To see why, observe that on no vertex is *dfs* called more than once, because as soon as we call *dfs*(v) we set *mark*[v] to *visited* at line (1), and we never call *dfs* on a vertex that previously had its *mark* set to *visited*. Thus, the total time spent at lines (2)$-$(3) going down the adjacency lists is proportional to the sum of the lengths of those lists, that is, $O(e)$. Thus, assuming $n \le e$, the total time spent on the depth-first search of an entire graph is $O(e)$, which is, to within a constant factor, the time needed merely to "look at" each arc.

```
        procedure dfs ( v: vertex );
            var
                w: vertex;
            begin
(1)             mark[v] := visited;
(2)             for each vertex w on L[v] do
(3)                 if mark[w] = unvisited then
(4)                     dfs(w)
            end; { dfs }
```

Fig. 6.24. Depth-first search.

Example 6.13. Assume the procedure *dfs*(v) is applied to the directed graph of Fig. 6.25 with $v = A$. The algorithm marks A visited and selects vertex B from the adjacency list of vertex A. Since B is unvisited, the search continues by calling *dfs*(B). The algorithm now marks B visited and selects the first vertex from the adjacency list for vertex B. Depending on the order of the vertices on the adjacency list of B the search will go to C or D next.

Assuming that C appears ahead of D, *dfs*(C) is invoked. Vertex A is on the adjacency list of C. However, A is already visited at this point so the

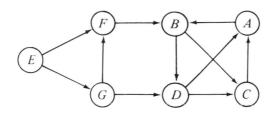

Fig. 6.25. Directed graph.

search remains at C. Since all vertices on the adjacency list at C have now been exhausted, the search returns to B, from which the search proceeds to D. Vertices A and C on the adjacency list of D were already visited, so the search returns to B and then to A.

At this point the original call of *dfs(A)* is complete. However, the digraph has not been entirely searched; vertices E, F and G are still unvisited. To complete the search, we can call *dfs(E)*.

The Depth-first Spanning Forest

During a depth-first traversal of a directed graph, certain arcs, when traversed, lead to unvisited vertices. The arcs leading to new vertices are called *tree arcs* and they form a *depth-first spanning forest* for the given digraph. The solid arcs in Fig. 6.26 form the depth-first spanning forest for the digraph of Fig. 6.25. Note that the tree arcs must indeed form a forest, since a vertex cannot be unvisited when two different arcs to it are traversed.

In addition to the tree arcs, there are three other types of arcs defined by a depth-first search of a directed graph. These are called back arcs, forward arcs, and cross arcs. An arc such as $C \rightarrow A$ is called a *back arc,* since it goes from a vertex to one of its ancestors in the spanning forest. Note that an arc from a vertex to itself is a back arc. A nonspanning arc that goes from a vertex to a proper descendant is called a *forward* arc. There are no forward arcs in Fig. 6.25.

Arcs such as $D \rightarrow C$ or $G \rightarrow D$, which go from a vertex to another vertex that is neither an ancestor nor a descendant, are called *cross* arcs. Observe that all cross arcs in Fig. 6.26 go from right to left, on the assumption that we add children to the tree in the order they were visited, from left to right, and that we add new trees to the forest from left to right. This pattern is not accidental. Had the arc $G \rightarrow D$ been the arc $D \rightarrow G$, then G would have been unvisited when the search at D was in progress, and thus on encountering arc $D \rightarrow G$, vertex G would have been made a descendant of D, and $D \rightarrow G$ would have become a tree arc.

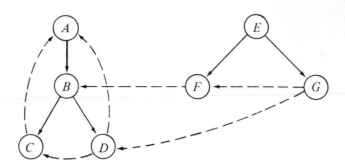

Fig. 6.26. Depth-first spanning forest for Fig. 6.25.

How do we distinguish among the four types of arcs? Clearly tree arcs are special since they lead to unvisited vertices during the depth-first search. Suppose we number the vertices of a directed graph in the order in which we first mark them visited during a depth-first search. That is, we may assign to an array

$$dfnumber[v] := count;$$
$$count := count + 1;$$

after line (1) of Fig. 6.24. Let us call this the *depth-first numbering* of a directed graph; notice how depth-first numbering generalizes the preorder numbering introduced in Section 3.1.

All descendents of a vertex v are assigned depth-first search numbers greater than or equal to the number assigned v. In fact, w is a descendant of v if and only if $dfnumber(v) \le dfnumber(w) \le dfnumber(v)$ + number of descendents of v. Thus, forward arcs go from low-numbered to high-numbered vertices and back arcs go from high-numbered to low-numbered vertices.

All cross arcs go from high-numbered vertices to low-numbered vertices. To see this, suppose that $x \to y$ is an arc and $dfnumber(x) \le dfnumber(y)$. Thus x is visited before y. Every vertex visited between the time $dfs(x)$ is first invoked and the time $dfs(x)$ is complete becomes a descendant of x in the depth-first spanning forest. If y is unvisited at the time arc $x \to y$ is explored, $x \to y$ becomes a tree arc. Otherwise, $x \to y$ is a forward arc. Thus there can be no cross arc $x \to y$ with $dfnumber(x) \le dfnumber(y)$.

In the next two sections we shall show how depth-first search can be used in solving various graph problems.

6.6 Directed Acyclic Graphs

A *directed acyclic graph*, or *dag* for short, is a directed graph with no cycles. Measured in terms of the relationships they can represent, dags are more general than trees but less general than arbitrary directed graphs. Figure 6.27 gives an example of a tree, a dag, and a directed graph with a cycle.

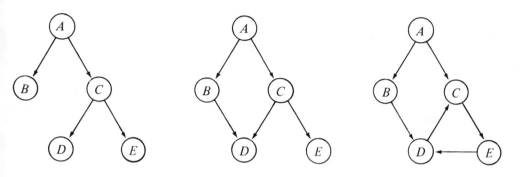

Fig. 6.27. Three directed graphs.

Among other things, dags are useful in representing the syntactic structure of arithmetic expressions with common subexpressions. For example, Fig. 6.28 shows a dag for the expression

$$((a+b)*c + ((a+b)+e) * (e+f)) * ((a+b)*c)$$

The terms $a+b$ and $(a+b)*c$ are shared common subexpressions that are represented by vertices with more than one incoming arc.

Dags are also useful in representing partial orders. A *partial order R* on a set S is a binary relation such that

1. for all a in S, $a R a$ is false (R is irreflexive)
2. for all a, b, c in S, if $a R b$ and $b R c$, then $a R c$ (R is transitive)

Two natural examples of partial orders are the "less than" ($<$) relation on integers, and the relation of proper containment ($\subset$) on sets.

Example 6.14. Let $S = \{1, 2, 3\}$ and let $P(S)$ be the power set of S, that is, the set of all subsets of S. $P(S) = \{\varnothing, \{1\}, \{2\}, \{3\}, \{1,2\}, \{1,3\}, \{2,3\}, \{1,2,3\}\}$. $\subset$ is a partial order on $P(S)$. Certainly, $A \subset A$ is false for any set A (irreflexivity), and if $A \subset B$ and $B \subset C$, then $A \subset C$ (transitivity).

Dags can be used to portray partial orders graphically. To begin, we can view a relation R as a set of pairs (arcs) suc that (a, b) is in the set if and only if $a R b$ is true. If R is a partial order on a set S, then the directed graph $G = (S, R)$ is a dag. Conversely, suppose $G = (S, R)$ is a dag and R^+ is the relation defined by $a R^+ b$ if and only if there is a path of length one or

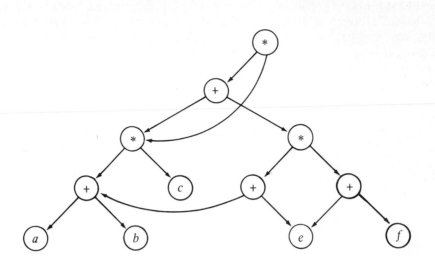

Fig. 6.28. Dag for arithmetic expression.

more from a to b. (R^+ is the transitive closure of the relation R.) Then, R^+ is a partial order on S.

Example 6.15. Figure 6.29 shows a dag ($P(S)$, R), where $S = \{1, 2, 3\}$. The relation R^+ is proper containment on the power set $P(S)$. □

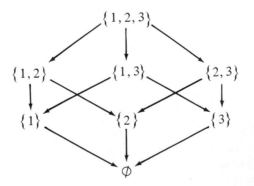

Fig. 6.29. Dag of proper containments.

Test for Acyclicity

Suppose we are given a directed graph $G = (V, E)$, and we wish to determine whether G is acyclic, that is, whether G has no cycles. Depth-first search can be used to answer this question. If a back arc is encountered during a depth-first search of G, then clearly the graph has a cycle. Conversely, if a directed graph has a cycle, then a back arc will always be encountered in any depth-first search of the graph.

To see this fact, suppose G is cyclic. If we do a depth-first search of G, there will be one vertex v having the lowest depth-first search number of any vertex on a cycle. Consider an arc $u \rightarrow v$ on some cycle containing v. Since u is on the cycle, u must be a descendant of v in the depth-first spanning forest. Thus, $u \rightarrow v$ cannot be a cross arc. Since the depth-first number of u is greater than the depth-first number of v, $u \rightarrow v$ cannot be a tree arc or a forward arc. Thus, $u \rightarrow v$ must be a back arc, as illustrated in Fig. 6.30.

Fig. 6.30. Every cycle contains a back arc.

Topological Sort

A large project is often divided into a collection of smaller tasks, some of which have to be performed in certain specified orders so that we may complete the entire project. For example, a university curriculum may have courses that require other courses as prerequisites. Dags can be used to model such situations naturally. For example, we could have an arc from course C to course D if C is a prerequisite for D.

Example 6.16. Figure 6.31 shows a dag giving the prerequisite structure on five courses. Course $C3$, for example, requires courses $C1$ and $C2$ as prerequisites. □

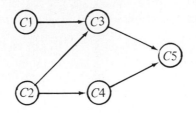

Fig. 6.31. Dag of prerequisites.

Topological sort is a process of assigning a linear ordering to the vertices of a dag so that if there is an arc from vertex i to vertex j, then i appears before j in the linear ordering. For example, $C1, C2, C3, C4, C5$ is a topological sort of the dag in Fig. 6.31. Taking the courses in this sequence would satisfy the prerequisite structure given in Fig. 6.31.

Topological sort can be easily accomplished by adding a print statement after line (4) to the depth-first search procedure in Fig. 6.24:

```
procedure topsort ( v: vertex );
    { print vertices accessible from v in reverse topological order }
    var
        w: vertex;
    begin
        mark[v] := visited;
        for each vertex w on L[v] do
            if mark[w] = unvisited then
                topsort(w);
        writeln(v)
    end;  { topsort }
```

When *topsort* finishes searching all vertices adjacent to a given vertex x, it prints x. The effect of calling *topsort*(v) is to print in a reverse topological order all vertices of a dag accessible from v by a path in the dag.

This technique works because there are no back arcs in a dag. Consider what happens when depth-first search leaves a vertex x for the last time. The only arcs emanating from v are tree, forward, and cross arcs. But all these arcs are directed towards vertices that have already been completely visited by the search and therefore precede x in the order being constructed.

6.7 Strong Components

A strongly connected component of a directed graph is a maximal set of vertices in which there is a path from any one vertex in the set to any other vertex in the set. Depth-first search can be used to determine strongly connected components of a directed graph efficiently.

Let $G = (V, E)$ be a directed graph. We can partition V into equivalence classes V_i, $1 \leq i \leq r$, such that vertices v and w are equivalent if and only if there is a path from v to w and a path from w to v. Let E_i, $1 \leq i \leq r$, be the set of arcs with head and tail in V_i. The graphs $G_i = (V_i, E_i)$ are called the *strongly connected components* (or just *strong components*) of G. A directed graph with only one strong component is said to be *strongly connected*.

Example 6.17. Figure 6.32 illustrates a directed graph with the two strong components shown in Fig. 6.33. □

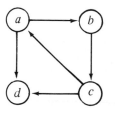

Fig. 6.32. Directed graph.

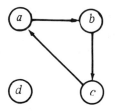

Fig. 6.33. The strong components of the digraph of Fig. 6.32.

Note that every vertex of a directed graph G is in some strong component, but that certain arcs may not be in any component. Such arcs, called *cross-component* arcs, go from a vertex in one component to a vertex in another. We can represent the interconnections among the components by constructing a *reduced graph* for G. The vertices of the reduced graph are the strongly connected components of G. There is an arc from vertex C to a different vertex C' of the reduced graph if there is an arc in G from some vertex in the component C to some vertex in component C'. The reduced graph is always a dag, because if there were a cycle, then all the components in the cycle would really be one strong component, meaning that we didn't compute strong components properly. Figure 6.34 shows the reduced graph for the digraph in Figure 6.32.

We shall now present an algorithm to find the strongly connected

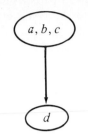

Fig. 6.34. Reduced graph.

components of a given directed graph G.

1. Perform a depth-first search of G and number the vertices in order of completion of the recursive calls; i.e., assign a number to vertex v after line (4) of Fig. 6.24.

2. Construct a new directed graph G_r by reversing the direction of every arc in G.

3. Perform a depth-first search on G_r starting the search from the highest-numbered vertex according to the numbering assigned at step (1). If the depth-first search does not reach all vertices, start the next depth-first search from the highest-numbered remaining vertex.

4. Each tree in the resulting spanning forest is a strongly connected component of G.

Example 6.18. Let us apply this algorithm to the directed graph of Fig. 6.32, starting at a and progressing first to b. After step (1) we number the vertices as shown in Fig. 6.35. Reversing the direction of the arcs we obtain the directed graph G_r in Fig. 6.36.

Performing the depth-first search on G_r, we obtain the depth-first spanning forest shown in Fig. 6.37. We begin with a as a root, because a has the highest number. From a we can only reach c and then b. The next tree has root d, since that is the highest numbered (and only) remaining vertex. Each tree in this forest forms a strongly connected component of the original directed graph. □

We have claimed that the vertices of a strongly connected component correspond precisely to the vertices of a tree in the spanning forest of the second depth-first search. To see why, observe that if v and w are vertices in the same strongly connected component, then there are paths in G from v to w and from w to v. Thus there are also paths from v to w and from w to v in G_r.

Suppose that in the depth-first search of G_r, we begin a search at some root x and reach either v or w. Since v and w are reachable from each other,

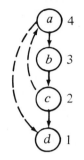

Fig. 6.35. After step 1.

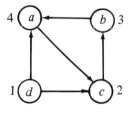

Fig. 6.36. G_r.

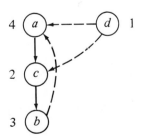

Fig. 6.37. Depth-first spanning forest for G_r.

both v and w will end up in the spanning tree with root x.

Now suppose v and w are in the same spanning tree of the depth-first spanning forest of G_r. We must show that v and w are in the same strongly connected component. Let x be the root of the spanning tree containing v and w. Since v is a descendant of x, there exists a path in G_r from x to v. Thus there exists a path in G from v to x.

In the construction of the depth-first spanning forest of G_r, vertex v was still unvisited when the depth-first search at x was initiated. Thus x has a higher number than v, so in the depth-first search of G, the recursive call at v terminated before the recursive call at x did. But in the depth-first search of G, the search at v could not have started before x, since the path in G from v to x would then imply that the search at x would start and end before the search at v ended.

We conclude that in the search of G, v is visited during the search of x and hence v is a descendant of x in the first depth-first spanning forest for G. Thus there exists a path from x to v in G. Therefore x and v are in the same strongly connected component. An identical argument shows that x and w are in the same strongly connected component and hence v and w are in the same strongly connected component, as shown by the path from v to x to w and the path from w to x to v.

Exercises

6.1 Represent the digraph of Fig. 6.38

    a) by an adjacency matrix giving arc costs.

    b) by a linked adjacency list with arc costs indicated.

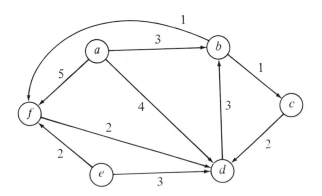

Fig. 6.38. Directed graph with arc costs.

6.2 Describe a mathematical model for the following scheduling problem. Given tasks $T_1, T_2, \ldots, T_n$, which require times $t_1, t_2, \ldots, t_n$ to complete, and a set of constraints, each of the form "T_i must be completed prior to the start of T_j," find the minimum time necessary to complete all tasks.

6.3 Implement the operations FIRST, NEXT, and VERTEX for digraphs represented by
   a) adjacency matrices,
   b) linked adjacency lists, and
   c) adjacency lists represented as in Fig. 6.5.

6.4 In the digraph of Fig. 6.38
   a) Use the algorithm *Dijkstra* to find the shortest paths from a to the other vertices.
   b) Use the algorithm *Floyd* to find the shortest distances between all pairs of points. Also construct the matrix P that lets us recover the shortest paths.

6.5 Write a complete program for Dijkstra's algorithm using a partially ordered tree as a priority queue and linked adjacency lists.

*6.6 Show that the program *Dijkstra* does not work correctly if arc costs can be negative.

**6.7 Show that the program *Floyd* still works if some of the arcs have negative cost but no cycle has negative cost.

6.8 Assuming the order of the vertices is $a, b, \ldots, f$ in Fig. 6.38, construct a depth-first spanning forest; indicate the tree, forward, back and cross arcs, and indicate the depth-first numbering of the vertices.

*6.9 Suppose we are given a depth-first spanning forest, and we list in postorder each of the spanning trees (trees composed of spanning edges), from the leftmost to the rightmost. Show that this order is the same as the order in which the calls of *dfs* ended when the spanning forest was constructed.

6.10 A *root* of a dag is a vertex r such that every vertex of the dag can be reached by a directed path from r. Write a program to determine whether a dag is rooted.

*6.11 Consider a dag with e arcs and with two distinguished vertices s and t. Construct an $O(e)$ algorithm to find a maximal set of vertex disjoint paths from s to t. By maximal we mean no additional path may be added, but it does not mean that it is the largest size such set.

6.12 Construct an algorithm to convert an expression tree with operators $+$ and $*$ into a dag by sharing common subexpressions. What is the time complexity of your algorithm?

6.13 Construct an algorithm for evaluating an arithmetic expression represented as a dag.

6.14 Write a program to find the longest path in a dag. What is the time complexity of your program?

6.15 Find the strong components of Fig. 6.38.

*6.16 Prove that the reduced graph of the strong components of Section 6.7 must be a dag.

6.17 Draw the first spanning forest, the reverse graph, and the second spanning forest developed by the strong components algorithm applied to the digraph of Fig. 6.38.

6.18 Implement the strong components algorithm discussed in Section 6.7.

*6.19 Show that the strong components algorithm takes time $O(e)$ on a directed graph of e arcs and n vertices, assuming $n \leq e$.

*6.20 Write a program that takes as input a digraph and two of its vertices. The program is to print all simple paths from one vertex to the other. What is the time complexity of your program?

*6.21 A *transitive reduction* of a directed graph $G = (V, E)$ is any graph G' with the same vertices but with as few arcs as possible, such that the transitive closure of G' is the same as the transitive closure of G. Show that if G is a dag, then the transitive reduction of G is unique.

*6.22 Write a program to compute the transitive reduction of a digraph. What is the time complexity of your program?

*6.23 $G' = (V, E')$ is called a *minimal equivalent digraph* for a digraph $G = (V, E)$ if E' is a smallest subset of E and the transitive closure of both G and G' are the same. Show that if G is acyclic, then there is only one minimal equivalent digraph, namely, the transitive reduction.

*6.24 Write a program to find a minimal equivalent digraph for a given digraph. What is the time complexity of your program?

*6.25 Write a program to find the longest simple path from a given vertex of a digraph. What is the time complexity of your program?

Bibliographic Notes

Berge [1958] and Harary [1969] are two good sources for additional material on graph theory. Some books treating graph algorithms are Deo [1975], Even [1980], and Tarjan [1983].

The single-source shortest paths algorithm in Section 6.3 is from Dijkstra [1959]. The all-pairs shortest paths algorithm is from Floyd [1962] and the transitive closure algorithm is from Warshall [1962]. Johnson [1977] discusses efficient algorithms for finding shortest paths in sparse graphs. Knuth [1968] contains additional material on topological sort.

The strong components algorithm of Section 6.7 is similar to one suggested by R. Kosaraju in 1978 (unpublished), and to one published by Sharir [1981]. Tarjan [1972] contains another strong components algorithm that needs only one depth-first search traversal.

Coffman [1976] contains many examples of how digraphs can be used to model scheduling problems as in Exercise 6.2. Aho, Garey, and Ullman [1972] show that the transitive reduction of a dag is unique, and that computing the transitive reduction of a digraph is computationally equivalent to computing the transitive closure (Exercises 6.21 and 6.22). Finding the minimal equivalent digraph (Exercises 6.23 and 6.24), on the other hand, appears to be computationally much harder; this problem is NP-complete [Sahni (1974)].

CHAPTER 7

Undirected Graphs

An undirected graph $G = (V, E)$ consists of a finite set of vertices V and a set of edges E. It differs from a directed graph in that each edge in E is an unordered pair of vertices.† If (v, w) is an undirected edge, then $(v, w) = (w, v)$. Hereafter we refer to an undirected graph as simply a graph.

Graphs are used in many different disciplines to model a symmetric relationship between objects. The objects are represented by the vertices of the graph and two objects are connected by an edge if the objects are related. In this chapter we present several data structures that can be used to represent graphs. We then present algorithms for three typical problems involving undirected graphs: constructing minimal spanning trees, biconnected components, and maximal matchings.

7.1 Definitions

Much of the terminology for directed graphs is also applicable to undirected graphs. For example, vertices v and w are *adjacent* if (v, w) is an edge [or, equivalently, if (w, v) is an edge]. We say the edge (v, w) is *incident* upon vertices v and w.

A *path* is a sequence of vertices $v_1, v_2, \ldots, v_n$ such that (v_i, v_{i+1}) is an edge for $1 \le i < n$. A path is *simple* if all vertices on the path are distinct, with the exception that v_1 and v_n may be the same. The length of the path is $n-1$, the number of edges along the path. We say the path $v_1, v_2, \ldots, v_n$ *connects* v_1 and v_n. A graph is *connected* if every pair of its vertices is connected.

Let $G = (V, E)$ be a graph with vertex set V and edge set E. A *subgraph* of G is a graph $G' = (V', E')$ where

1. V' is a subset of V.

2. E' consists of edges (v, w) in E such that both v and w are in V'.

If E' consists of all edges (v, w) in E, such that both v and w are in V', then G' is called an *induced subgraph* of G.

Example 7.1. In Fig. 7.1(a) we see a graph $G = (V, E)$ with $V = \{a, b, c, d\}$ and $E = \{(a, b), (a, d), (b, c), (b, d), (c, d)\}$. In Fig. 7.1(b) we see one

---

† Unless otherwise stated, we shall assume an edge is always a pair of distinct vertices.

of its induced subgraphs, the one defined by the set of vertices $\{a, b, c\}$ and all the edges in Fig. 7.1(a) that are not incident upon vertex d. □

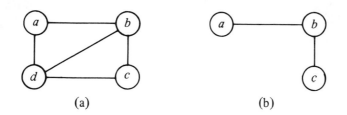

(a) (b)

Fig. 7.1. A graph and one of its subgraphs.

A *connected component* of a graph G is a maximal connected induced subgraph, that is, a connected induced subgraph that is not itself a proper subgraph of any other connected subgraph of G.

Example 7.2 Figure 7.1 is a connected graph. It has only one connected component, namely itself. Figure 7.2 is a graph with two connected components. □

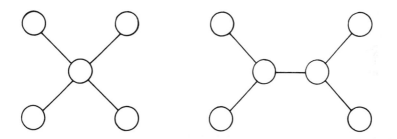

Fig. 7.2. An unconnected graph.

A (simple) *cycle* in a graph is a (simple) path of length three or more that connects a vertex to itself. We do not consider paths of the form v (path of length 0), v, v (path of length 1), or v, w, v (path of length 2) to be cycles. A graph is *cyclic* if it contains at least one cycle. A connected, acyclic graph is sometimes called a *free tree*. Figure 7.2 shows a graph consisting of two connected components where each connected component is a free tree. A free tree can be made into an ordinary tree if we pick any vertex we wish as the root and orient each edge from the root.

Free trees have two important properties, which we shall use in the next section.

1. Every free tree with $n \geq 1$ vertices contains exactly $n-1$ edges.

2. If we add any edge to a free tree, we get a cycle.

We can prove (1) by induction on n, or what is equivalent, by an argument concerning the "smallest counterexample." Suppose $G = (V, E)$ is a counterexample to (1) with the fewest vertices, say n vertices. Now n cannot be 1, because the only free tree on one vertex has zero edges, and (1) is satisfied. Therefore, n must be greater than 1.

We now claim that in the free tree there must be some vertex with exactly one incident edge. In proof, no vertex can have zero incident edges, or G would not be connected. Suppose every vertex has at least two edges incident. Then, start at some vertex v_1, and follow any edge from v_1. At each step, leave a vertex by a different edge from the one used to enter it, thereby forming a path $v_1, v_2, v_3, \ldots$.

Since there are only a finite number of vertices in V, all vertices on this path cannot be distinct; eventually, we find $v_i = v_j$ for some $i < j$. We cannot have $i = j-1$ because there are no loops from a vertex to itself, and we cannot have $i = j-2$ or else we entered and left vertex v_{i+1} on the same edge. Thus, $i \leq j-3$, and we have a cycle $v_i, v_{i+1}, \ldots, v_j = v_i$. Thus, we have contradicted the hypothesis that G had no vertex with only one edge incident, and therefore conclude that such a vertex v with edge (v, w) exists.

Now consider the graph G' formed by deleting vertex v and edge (v, w) from G. G' cannot contradict (1), because if it did, it would be a smaller counterexample than G. Therefore, G' has $n-1$ vertices and $n-2$ edges. But G has one more edge and one more vertex than G', so G has $n-1$ edges, proving that G does indeed satisfy (1). Since there is no smallest counterexample to (1), we conclude there can be no counterexample at all, so (1) is true.

Now we can easily prove statement (2), that adding an edge to a free tree forms a cycle. If not, the result of adding the edge to a free tree of n vertices would be a graph with n vertices and n edges. This graph would still be connected, and we supposed that adding the edge left the graph acyclic. Thus we would have a free tree whose vertex and edge count did not satisfy condition (1).

Methods of Representation

The methods of representing directed graphs can be used to represent undirected graphs. One simply represents an undirected edge between v and w by two directed edges, one from v to w and the other from w to v.

Example 7.3. The adjacency matrix and adjacency list representations for the graph of Fig. 7.1(a) are shown in Fig. 7.3. □

Clearly, the adjacency matrix for a graph is symmetric. In the adjacency list representation if (i, j) is an edge, then vertex j is on the list for vertex i and vertex i is on the list for vertex j.

	a	b	c	d
a	0	1	0	1
b	1	0	1	1
c	0	1	0	1
d	1	1	1	0

(a) Adjacency matrix

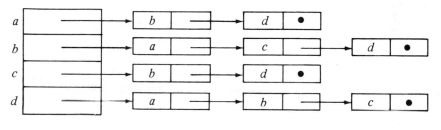

(b) Adjacency list

Fig. 7.3. Representations.

7.2 Minimum-Cost Spanning Trees

Suppose $G = (V, E)$ is a connected graph in which each edge (u, v) in E has a cost $c(u, v)$ attached to it. A *spanning tree* for G is a free tree that connects all the vertices in V. The *cost* of a spanning tree is the sum of the costs of the edges in the tree. In this section we shall show how to find a minimum-cost spanning tree for G.

Example 7.4. Figure 7.4 shows a weighted graph and its minimum-cost spanning tree. □

A typical application for minimum-cost spanning trees occurs in the design of communications networks. The vertices of a graph represent cities and the edges possible communications links between the cities. The cost associated with an edge represents the cost of selecting that link for the network. A minimum-cost spanning tree represents a communications network that connects all the cities at minimal cost.

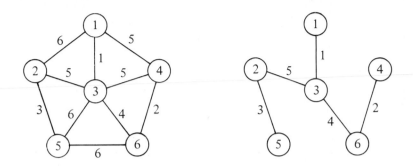

Fig. 7.4. A graph and spanning tree.

The MST Property

There are several different ways to construct a minimum-cost spanning tree. Many of these methods use the following property of minimum-cost spanning trees, which we call the *MST property*. Let $G = (V, E)$ be a connected graph with a cost function defined on the edges. Let U be some proper subset of the set of vertices V. If (u, v) is an edge of lowest cost such that $u \in U$ and $v \in V-U$, then there is a minimum-cost spanning tree that includes (u, v) as an edge.

The proof that every minimum-cost spanning tree satisfies the MST property is not hard. Suppose to the contrary that there is no minimum-cost spanning tree for G that includes (u, v). Let T be any minimum-cost spanning tree for G. Adding (u, v) to T must introduce a cycle, since T is a free tree and therefore satisfies property (2) for free trees. This cycle involves edge (u, v). Thus, there must be another edge (u', v') in T such that $u' \in U$ and $v' \in V-U$, as illustrated in Fig. 7.5. If not, there would be no way for the cycle to get from u to v without following the edge (u, v) a second time.

Deleting the edge (u', v') breaks the cycle and yields a spanning tree T' whose cost is certainly no higher than the cost of T since by assumption $c(u, v) \le c(u', v')$. Thus, T' contradicts our assumption that there is no minimum-cost spanning tree that includes (u, v).

Prim's Algorithm

There are two popular techniques that exploit the MST property to construct a minimum-cost spanning tree from a weighted graph $G = (V, E)$. One such method is known as Prim's algorithm. Suppose $V = \{1, 2, \ldots, n\}$. Prim's algorithm begins with a set U initialized to $\{1\}$. It then "grows" a spanning tree, one edge at a time. At each step, it finds a shortest edge (u, v) that connects U and $V-U$ and then adds v, the vertex in $V-U$, to U. It repeats this

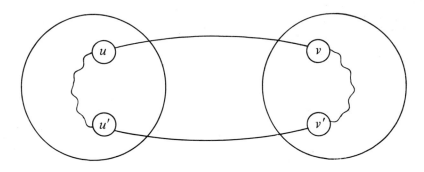

Fig. 7.5. Resulting cycle.

step until $U = V$. The algorithm is summarized in Fig. 7.6 and the sequence of edges added to T for the graph of Fig. 7.4(a) is shown in Fig. 7.7.

> **procedure** *Prim* (G: graph; **var** T: set of edges);
>   { *Prim* constructs a minimum-cost spanning tree T for G }
>   **var**
> U: set of vertices;
> u, v: vertex;
>   **begin**
> $T := \varnothing$;
> $U := \{1\}$;
>   **while** $U \neq V$ **do begin**
>   let (u, v) be a lowest cost edge such that
> u is in U and v is in $V-U$;
> $T := T \cup \{(u, v)\}$;
> $U := U \cup \{v\}$
>   **end**
>   **end**; { *Prim* }

Fig. 7.6. Sketch of Prim's algorithm.

One simple way to find the lowest-cost edge between U and $V-U$ at each step is to maintain two arrays. One array *CLOSEST*[i] gives the vertex in U that is currently closest to vertex i in $V-U$. The other array *LOWCOST*[i] gives the cost of the edge (i, *CLOSEST*[i]).

At each step we can scan *LOWCOST* to find the vertex, say k, in $V-U$ that is closest to U. We print the edge (k, *CLOSEST*[k]). We then update the *LOWCOST* and *CLOSEST* arrays, taking into account the fact that k has been

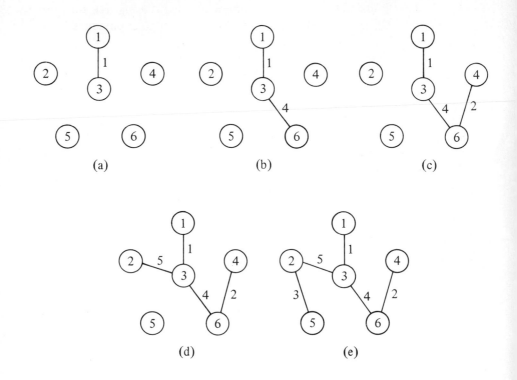

Fig. 7.7. Sequences of edges added by Prim's algorithm.

added to U. A Pascal version of this algorithm is given in Fig. 7.8. We assume C is an $n \times n$ array such that $C[i, j]$ is the cost of edge (i, j). If edge (i, j) does not exist, we assume $C[i, j]$ is some appropriate large value.

Whenever we find another vertex k for the spanning tree, we make $LOWCOST[k]$ be *infinity*, a very large value, so this vertex will no longer be considered in subsequent passes for inclusion in U. The value *infinity* is greater than the cost of any edge or the cost associated with a missing edge.

The time complexity of Prim's algorithm is $O(n^2)$, since we make $n-1$ iterations of the loop of lines (4)–(16) and each iteration of the loop takes $O(n)$ time, due to the inner loops of lines (7)–(10) and (13)–(16). As n gets large the performance of this algorithm may become unsatisfactory. We now give another algorithm due to Kruskal for finding minimum-cost spanning trees whose performance is at most $O(e \log e)$, where e is the number of edges in the given graph. If e is much less than n^2, Kruskal's algorithm is superior, although if e is about n^2, we would prefer Prim's algorithm.

```
        procedure Prim ( C: array[1..n, 1..n] of real );
            { Prim prints the edges of a minimum-cost spanning tree for a graph
                with vertices {1, 2, . . . ,n} and cost matrix C on edges }
        var
            LOWCOST: array[1..n] of real;
            CLOSEST: array[1..n] of integer;
            i, j, k, min: integer;
            { i and j are indices.  During a scan of the LOWCOST array,
                k is the index of the closest vertex found so far, and
                min = LOWCOST[k] }

        begin
(1)         for i := 2 to n do begin
                { initialize with only vertex 1 in the set U }
(2)             LOWCOST[i] := C[1, i];
(3)             CLOSEST[i] := 1
            end;
(4)         for i := 2 to n do begin
                { find the closest vertex k outside of U to
                    some vertex in U }
(5)             min := LOWCOST[2];
(6)             k := 2;
(7)             for j := 3 to n do
(8)                 if LOWCOST[j] < min then begin
(9)                     min := LOWCOST[j];
(10)                    k := j
                    end;
(11)            writeln(k, CLOSEST[k]);  { print edge }
(12)            LOWCOST[k] := infinity;  { k is added to U }
(13)            for j := 2 to n do  { adjust costs to U }
(14)                if (C[k, j] < LOWCOST[j]) and
                            (LOWCOST[j] < infinity) then begin
(15)                    LOWCOST[j] := C[k, j];
(16)                    CLOSEST[j] := k
                    end
            end
        end;  { Prim }
```

Fig. 7.8. Prim's algorithm.

Kruskal's Algorithm

Suppose again we are given a connected graph $G = (V, E)$, with $V = \{1, 2, \ldots ,n\}$ and a cost function c defined on the edges of E. Another way to construct a minimum-cost spanning tree for G is to start with a graph

$T = (V, \varnothing)$ consisting only of the n vertices of G and having no edges. Each vertex is therefore in a connected component by itself. As the algorithm proceeds, we shall always have a collection of connected components, and for each component we shall have selected edges that form a spanning tree.

To build progressively larger components, we examine edges from E, in order of increasing cost. If the edge connects two vertices in two different connected components, then we add the edge to T. If the edge connects two vertices in the same component, then we discard the edge, since it would cause a cycle if we added it to the spanning tree for that connected component. When all vertices are in one component, T is a minimum-cost spanning tree for G.

Example 7.5. Consider the weighted graph of Fig. 7.4(a). The sequence of edges added to T is shown in Fig. 7.9. The edges of cost 1, 2, 3, and 4 are considered first, and all are accepted, since none of them causes a cycle. The edges (1, 4) and (3, 4) of cost 5 cannot be accepted, because they connect vertices in the same component in Fig. 7.9(d), and therefore would complete a cycle. However, the remaining edge of cost 5, namely (2, 3), does not create a cycle. Once it is accepted, we are done. □

We can implement this algorithm using sets and set operations discussed in Chapters 4 and 5. First, we need a set consisting of the edges in E. We then apply the DELETEMIN operator repeatedly to this set to select edges in order of increasing cost. The set of edges therefore forms a priority queue, and a partially ordered tree is an appropriate data structure to use here.

We also need to maintain a set of connected components C. The operations we apply to it are:

1. MERGE(A, B, C) to merge the components A and B in C and to call the result either A or B arbitrarily.†

2. FIND(v, C) to return the name of the component of C of which vertex v is a member. This operation will be used to determine whether the two vertices of an edge are in the same or different components.

3. INITIAL(A, v, C) to make A the name of a component in C containing only vertex v initially.

These are the operations of the MERGE-FIND ADT called MFSET, which we encountered in Section 5.5. A sketch of a program called *Kruskal* to find a minimum-cost spanning tree using these operations is shown in Fig. 7.10.

We can use the techniques of Section 5.5 to implement the operations used in this program. The running time of this program is dependent on two factors. If there are e edges, it takes $O(e \log e)$ time to insert the edges into the priority queue.‡ In each iteration of the while-loop, finding the least cost

---

† Note that MERGE and FIND are defined slightly differently from Section 5.5, since C is a parameter telling where A and B can be found.

‡ We can initialize a partially ordered tree of e elements in $O(e)$ time if we do it all at once. We discuss this technique in Section 8.4, and we should probably use it here, since

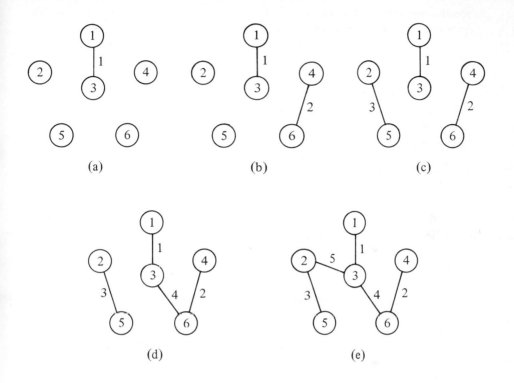

Fig. 7.9. Sequence of edges added by Kruskal's algorithm.

edge in *edges* takes $O(\log e)$ time. Thus, the priority queue operations take $O(e \log e)$ time in the worst case. The total time required to perform the MERGE and FIND operations depends on the method used to implement the MFSET. As shown in Section 5.5, there are $O(e \log e)$ and $O(e \alpha(e))$ methods. either case, Kruskal's algorithm can be implemented to run in $O(e \log e)$ time.

7.3 Traversals

In a number of graph problems, we need to visit the vertices of a graph systematically. Depth-first search and breadth-first search, the subjects of this section, are two important techniques for doing this. Both techniques can be used to determine efficiently all vertices that are connected to a given vertex.

---

if many fewer than e edges are examined before the minimum-cost spanning tree is found, we may save significant time.

```
procedure Kruskal ( V: SET of vertex;
        E: SET of edges;
        var T: SET of edges );
    var
        ncomp: integer; { current number of components }
        edges: PRIORITYQUEUE; { the set of edges }
        components: MFSET; { the set V grouped into
            a MERGE-FIND set of components }
        u, v: vertex;
        e: edge;
        nextcomp: integer; { name for new component }
        ucomp, vcomp; { component names }

    begin
        MAKENULL(T);
        MAKENULL(edges);
        nextcomp := 0;
        ncomp := number of members of V;
        for v in V do begin { initialize a
                component to contain one vertex of V }
            nextcomp := nextcomp + 1;
            INITIAL(nextcomp, v, components)
        end;
        for e in E do  { initialize priority queue of edges }
            INSERT(e, edges);
        while ncomp > 1 do begin  { consider next edge }
            e := DELETEMIN(edges)
            let e = (u, v);
            ucomp := FIND(u, components);
            vcomp := FIND(v, components);
            if ucomp <> vcomp then begin
                    { e connects two different components }
                    MERGE(ucomp, vcomp, components);
                    ncomp := ncomp - 1;
                    INSERT(e, T)
            end
        end
    end; { Kruskal }
```

Fig. 7.10. Kruskal's algorithm.

Depth-First Search

Recall from Section 6.5 the algorithm *dfs* for searching a directed graph. The same algorithm can be used to search undirected graphs, since the undirected edge (v, w) may be thought of as the pair of directed edges $v \to w$ and $w \to v$.

In fact, the depth-first spanning forests constructed for undirected graphs are even simpler than for digraphs. We should first note that each tree in the forest is one connected component of the graph, so if a graph is connected, it has only one tree in its depth-first spanning forest. Second, for digraphs we identified four kinds of arcs: tree, forward, back, and cross. For undirected graphs there are only two kinds: tree edges and back edges.

Because there is no distinction between forward edges and backward edges for undirected graphs, we elect to refer to them all as *back* arcs. In an undirected graph, there can be no cross edges, that is, edges (v, w) where v is neither an ancestor nor descendant of w in the spanning tree. Suppose there were. Then let v be a vertex that is reached before w in the search. The call to $dfs(v)$ cannot end until w has been searched, so w is entered into the tree as some descendant of v. Similarly, if $dfs(w)$ is called before $dfs(v)$, then v becomes a descendant of w.

As a result, during a depth-first search of an undirected graph G, all edges become either

1. *tree edges*, those edges (v, w) such that $dfs(v)$ directly calls $dfs(w)$ or vice versa, or

2. *back edges*, those edges (v, w) such that neither $dfs(v)$ nor $dfs(w)$ called the other directly, but one called the other indirectly (i.e., $dfs(w)$ calls $dfs(x)$, which calls $dfs(v)$, so w is an ancestor of v).

Example 7.6. Consider the connected graph G in Fig. 7.11(a). A depth-first spanning tree T resulting from a depth-first search of G is shown in Fig. 7.11(b). We assume the search began at vertex a, and we have adopted the convention of showing tree edges solid and back edges dashed. The tree has been drawn with the root on the top and the children of each vertex have been drawn in the left-to-right order in which they were first visited in the procedure of *dfs*.

To follow a few steps of the search, the procedure $dfs(a)$ calls $dfs(b)$ and adds edge (a, b) to T since b is not yet visited. At b, *dfs* calls $dfs(d)$ and adds edge (b, d) to T. At d, *dfs* calls $dfs(e)$ and adds edge (d, e) to T. At e, vertices a, b, and d are marked visited so $dfs(e)$ returns without adding any edges to T. At d, *dfs* now sees that vertices a and b are marked visited so $dfs(d)$ returns without adding any more edges to T. At b, *dfs* now sees that the remaining adjacent vertices a and e are marked visited, so $dfs(b)$ returns. The search then continues to c, f, and g. □

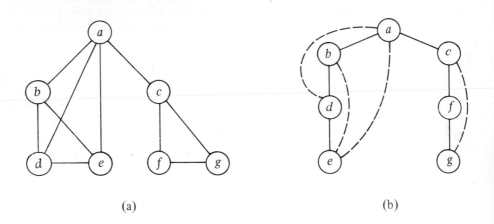

(a)   (b)

Fig. 7.11. A graph and its depth-first search.

Breadth-First Search

Another systematic way of visiting the vertices is called *breadth-first search*. The approach is called "breadth-first" because from each vertex v that we visit we search as broadly as possible by next visiting all the vertices adjacent to v. We can also apply this strategy of search to directed graphs.

As for depth-first search, we can build a spanning forest when we perform a breadth-first search. In this case, we consider edge (x, y) a tree edge if vertex y is first visited from vertex x in the inner loop of the search procedure *bfs* of Fig. 7.12.

It turns out that for the breadth-first search of an undirected graph, every edge that is not a tree edge is a cross edge, that is, it connects two vertices neither of which is an ancestor of the other.

The breadth-first search algorithm given in Fig. 7.12 inserts the tree edges into a set T, which we assume is initially empty. Every entry in the array *mark* is assumed to be initialized to the value *unvisited*; Figure 7.12 works on one connected component. If the graph is not connected, *bfs* must be called on a vertex of each component. Note that in a breadth-first search we must mark a vertex visited before enqueuing it, to avoid placing it on the queue more than once.

Example 7.7. The breadth-first spanning tree for the graph G in Fig. 7.11(a) is shown in Fig. 7.13. We assume the search began at vertex a. As before, we have shown tree edges solid and other edges dashed. We have also drawn the tree with the root at the top and the children in the left-to-right order in which they were first visited. □

The time complexity of breadth-first search is the same as that of depth-

```
procedure bfs ( v );
    { bfs visits all vertices connected to v using breadth-first search }
    var
        Q: QUEUE of vertex;
        x, y: vertex;
    begin
        mark[v] := visited;
        ENQUEUE(v, Q);
        while not EMPTY(Q) do begin
            x := FRONT(Q);
            DEQUEUE(Q);
            for each vertex y adjacent to x do
                if mark[y] = unvisited then begin
                    mark[y] := visited;
                    ENQUEUE(y, Q);
                    INSERT((x, y), T)
                end
        end
    end; { bfs }
```

Fig. 7.12. Breadth-first search.

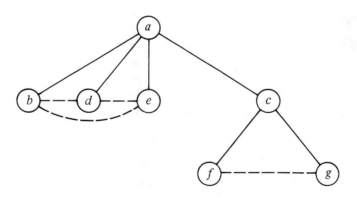

Fig. 7.13. Breadth-first search of G.

first search. Each vertex visited is placed in the queue once, so the while loop is executed once for each vertex. Each edge (x, y) is examined twice, once from x and once from y. Thus, if a graph has n vertices and e edges, the running time of *bfs* is $O(\max(n, e))$ if we use an adjacency list representation for the edges. Since $e \geq n$ is typical, we shall usually refer to the running time of breadth-first search as $O(e)$, just as we did for depth-first search.

Depth-first search and breadth-first search can be used as frameworks around which to design efficient graph algorithms. For example, either method can be used to find the connected components of a graph, since the connected components are the trees of either spanning forest.

We can test for cycles using breadth-first search in $O(n)$ time, where n is the number of vertices, independent of the number of edges. As we discussed in Section 7.1, any graph with n vertices and n or more edges must have a cycle. However, a graph could have $n-1$ or fewer edges and still have a cycle, if it had two or more connected components. One sure way to find the cycles is to build a breadth-first spanning forest. Then, every cross edge (v, w) must complete a simple cycle with the tree edges leading to v and w from their closest common ancestor, as shown in Fig. 7.14.

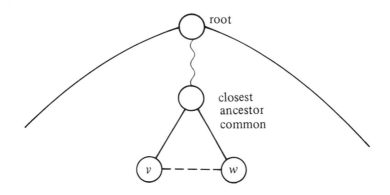

Fig. 7.14 A cycle found by breadth-first search.

7.4 Articulation Points and Biconnected Components

An *articulation point* of a graph is a vertex v such that when we remove v and all edges incident upon v, we break a connected component of the graph into two or more pieces. For example, the articulation points of Fig. 7.11(a) are a and c. If we delete a, the graph, which is one connected component, is divided into two triangles: $\{b, d, e\}$ and $\{c, f, g\}$. If we delete c, we divide the graph into $\{a, b, d, e\}$ and $\{f, g\}$. However, if we delete any one of the other vertices from the graph of Fig. 7.11(a), we do not split the connected component. A graph with no articulation points is said to be *biconnected*. Depth-first search is particularly useful in finding the biconnected components of a graph.

The problem of finding articulation points is the simplest of many important problems concerning the connectivity of graphs. As an example of applications of connectivity algorithms, we may represent a communication network as a graph in which the vertices are sites to be kept in communication with one another. A graph has *connectivity k* if the deletion of any $k-1$ vertices fails to disconnect the graph. For example, a graph has connectivity two or more if and only if it has no articulation points, that is, if and only if it is biconnected. The higher the connectivity of a graph, the more likely the graph is to survive the failure of some of its vertices, whether by failure of the processing units at the vertices or external attack.

We shall here give a simple depth-first search algorithm to find all the articulation points of a graph, and thereby test by their absence whether the graph is biconnected.

1. Perform a depth-first search of the graph, computing *dfnumber*[v] for each vertex v as discussed in Section 6.5. In essence, *dfnumber* orders the vertices as in a preorder traversal of the depth-first spanning tree.

2. For each vertex v, compute *low*[v], which is the smallest *dfnumber* of v or of any vertex w reachable from v by following down zero or more tree edges to a descendant x of v (x may be v) and then following a back edge (x, w). We compute *low*[v] for all vertices v by visiting the vertices in a postorder traversal. When we process v, we have computed *low*[y] for every child y of v. We take *low*[v] to be the minimum of

   a) *dfnumber*[v],

   b) *dfnumber*[z] for any vertex z for which there is a back edge (v, z) and

   c) *low*[y] for any child y of v.

3. Now we find the articulation points as follows.

   a) The root is an articulation point if and only if it has two or more children. Since there are no cross edges, deletion of the root must disconnect the subtrees rooted at its children, as a disconnects $\{b, d, e\}$ from $\{c, f, g\}$ in Fig. 7.11(b).

   b) A vertex v other than the root is an articulation point if and only if there is some child w of v such that *low*[w] $\geq$ *dfnumber*[v]. In this case, v disconnects w and its descendants from the rest of the graph. Conversely, if *low*[w] $<$ *dfnumber*[v], then there must be a way to get from w down the tree and back to a proper ancestor of v (the vertex whose *dfnumber* is *low*[w]), and therefore deletion of v does not disconnect w or its descendants from the rest of the graph.

Example 7.8. *dfnumber* and *low* are computed for the graph of Fig. 7.11(a) in Fig. 7.15. As an example of the calculation of *low*, our postorder traversal visits e first. At e, there are back edges (e, a) and (e, b), so *low*[e] is set to min(*dfnumber*[e], *dfnumber*[a], *dfnumber*[b]) = 1. Then d is visited, and *low*[d] is set to the minimum of *dfnumber*[d], *low*[e], and *dfnumber*[a]. The second of these arises because e is a child of d and the third because of the back edge (d, a).

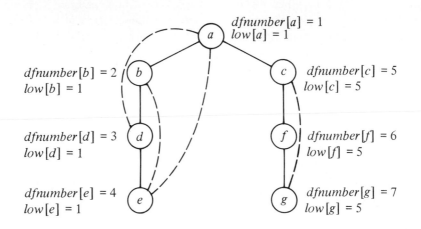

Fig. 7.15. Depth-first and *low* numberings.

After computing *low* we consider each vertex. The root, a, is an articulation point because it has two children. Vertex c is an articulation point because it has a child f with $low[f] \geq dfnumber[c]$. The other vertices are not articulation points. □

The time taken by the above algorithm on a graph of e edges and $n \leq e$ vertices is $O(e)$. The reader should check that the time spent in each of the three phases can be attributed either to the vertex visited or to an edge emanating from that vertex, there being only a constant amount of time attributable to any vertex or edge in any pass. Thus, the total time is $O(n+e)$, which is $O(e)$ under the assumption $n \leq e$.

7.5 Graph Matching

In this section we outline an algorithm to solve "matching problems" on graphs. A simple example of a matching problem occurs when we have a set of teachers to assign to a set of courses. Each teacher is qualified to teach certain courses but not others. We wish to assign a course to a qualified teacher so that no two teachers are assigned the same course. With certain distributions of teachers and courses it is impossible to assign every teacher a course; in those situations we wish to assign as many teachers as possible.

We can represent this situation by a graph as in Fig. 7.16 where the vertices are divided into two sets V_1 and V_2, such that vertices in the set V_1 represent teachers and vertices in the set V_2 courses. That teacher v is qualified to teach course w is represented by an edge (v, w). A graph such as this whose vertices can be divided into two disjoint groups with each edge having one end in each group is called *bipartite*. Assigning a teacher a course is equivalent to selecting an edge between a teacher vertex and a course vertex.

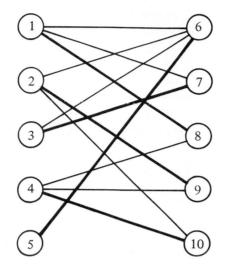

Fig. 7.16. A bipartite graph.

The matching problem can be formulated in general terms as follows. Given a graph $G = (V, E)$, a subset of the edges in E with no two edges incident upon the same vertex in V is called a *matching*. The task of selecting a maximum subset of such edges is called the *maximal matching problem*. The heavy edges in Fig. 7.16 are an example of one maximal matching in that graph. A *complete matching* is a matching in which every vertex is an end-point of some edge in the matching. Clearly, every complete matching is a maximal matching.

There is a straightforward way to find maximal matchings. We can sys-tematically generate all matchings and then pick one that has the largest number of edges. The difficulty with this method is that it has a running time that is an exponential function of the number of edges.

There are more efficient algorithms for finding maximal matchings. These algorithms generally use a technique known as "augmenting paths." Let M be a matching in a graph G. A vertex v is *matched* if it is the endpoint of an edge in M. A path connecting two unmatched vertices in which alternate edges in the path are in M is called an *augmenting path relative to M*. Observe that an augmenting path must be of odd length, and must begin and end with edges not in M. Also observe that given an augmenting path P we can always find a bigger matching by removing from M those edges that are in P, and then adding to M the edges of P that were initially not in M. This new matching is $M \oplus P$ where $\oplus$ denotes "exclusive or" on sets. That is, the new matching consists of those edges that are in M or P, but not in both.

Example 7.9. Figure 7.17(a) shows a graph and a matching M consisting of the heavy edges (1, 6), (3, 7), and (4, 8). The path 2, 6, 1, 8, 4, 9 in Fig. 7.17(b) is an augmenting path relative to M. Figure 7.18 shows the matching (1, 8), (2, 6), (3, 7), (4, 9) obtained by removing from M those edges that are in the path, and then adding to M the other edges in the path. □

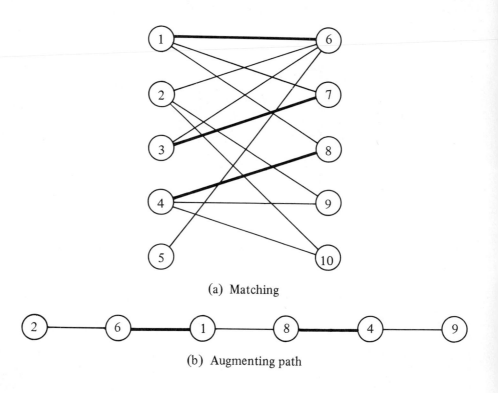

(a) Matching

(b) Augmenting path

Fig. 7.17. A matching and an augmenting path.

The key observation is that M is a maximal matching if and only if there is no augmenting path relative to M. This observation is the basis of our maximal matching algorithm.

Suppose M and N are matchings with $|M| < |N|$. ($|M|$ denotes the number of edges in M.) To see that $M \oplus N$ contains an augmenting path relative to M consider the graph $G' = (V, M \oplus N)$. Since M and N are both matchings, each vertex of V is an endpoint of at most one edge from M and an endpoint of at most one edge from N. Thus each connected component of G' forms a simple path (possibly a cycle) with edges alternating between M and N. Each path that is not a cycle is either an augmenting path relative to M or an augmenting path relative to N depending on whether it has more edges from N or

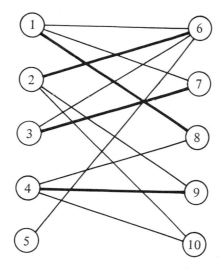

Fig. 7.18. The larger matching.

from M. Each cycle has an equal number of edges from M and N. Since $|M| < |N|$, $M \oplus N$ has more edges from N than M, and hence has at least one augmenting path relative to M.

We can now outline our procedure to find a maximal matching M for a graph $G = (V, E)$.

1. Start with $M = \varnothing$.
2. Find an augmenting path P relative to M and replace M by $M \oplus P$.
3. Repeat step (2) until no further augmenting paths exist, at which point M is a maximal matching.

It remains only to show how to find an augmenting path relative to a matching M. We shall do this for the simpler case where G is a bipartite graph. We shall build an *augmenting path graph* for G for levels $i = 0, 1, 2, \ldots$ using a process similar to breadth-first search. At level $i = 0$ we begin with some new unmatched vertex. At odd level i, we add new vertices that are adjacent to a vertex at level $i-1$, by a non-matching edge, and we also add that edge. At even level i, we add new vertices that are adjacent to a vertex at level $i-1$ because of an edge in the matching M, together with that edge.

We continue building the augmenting path graph level-by-level until an unmatched vertex is added at an odd level, or until no more vertices can be added. If an unmatched vertex v is added at an odd level, the path from v to the beginning vertex at level 0 is an augmenting path relative to M. If no more vertices can be added to the current augmenting path graph, we build another augmenting path graph beginning at a new unmatched starting vertex. If there are no more new unmatched vertices, then there is no augmenting path relative to M. If there is an augmenting path, an augmenting path graph terminating in an unmatched vertex at an odd level will eventually be constructed.

Example 7.10. Figure 7.19 illustrates the augmenting path graph for the graph in Fig. 7.17(a) relative to the matching in Fig. 7.18, in which we have chosen vertex 5 as the unmatched vertex at level 0. At level 1 we add the non-matching edge (5, 6). At level 2 we add the matching edge (6, 2). At level 3 we can add either of the non-matching edges (2, 9) or (2, 10). Since both vertices 9 and 10 are currently unmatched, we can terminate the construction of the augmenting path graph after the addition of either one of these vertices. Both paths 9, 2, 6, 5 and 10, 2, 6, 5 are augmenting paths relative to the matching in Fig. 7.18. □

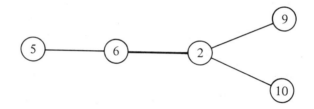

Fig. 7.19. Augmenting path graph.

Suppose G has n vertices and e edges. Constructing the augmenting path graphs for a given matching takes $O(e)$ time if we use an adjacency list representation for edges. Thus, to find each new augmenting path takes $O(e)$ time. To find a maximal matching, we construct at most $n/2$ augmenting paths, since each enlarges the current matching by at least one edge. Therefore, a maximal matching may be found in $O(ne)$ time for a bipartite graph G.

Exercises

7.1 Describe an algorithm to insert and delete edges in the adjacency list representation for an undirected graph. Remember that an edge (i, j) appears on the adjacency list for both vertex i and vertex j.

7.2 Modify the adjacency list representation for an undirected graph so that the first edge on the adjacency list for a vertex can be deleted in constant time. Write an algorithm to delete the first edge at a vertex using your new representation. *Hint.* How do you arrange that the two cells representing edge (i, j) can be found quickly from one another?

7.3 Consider the graph of Fig. 7.20.
   a) Find a minimum-cost spanning tree by Prim's algorithm.
   b) Find a minimum-cost spanning tree by Kruskal's algorithm.
   c) Find a depth-first spanning tree starting at a and at d.
   d) Find a breadth-first spanning tree starting at a and at d.

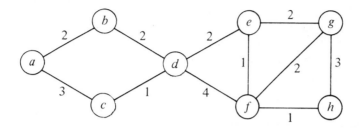

Fig. 7.20. A graph.

7.4 Let T be a depth-first spanning tree and B be the back edges for a connected undirected graph $G = (V, E)$.

   *a) Show that when each back edge from B is added to T a unique cycle results. Call such a cycle a *basic* cycle.

   **b) The *linear combination* of cycles C_1, $C_2, \ldots, C_n$ is $C_1 \oplus C_2 \oplus \cdots \oplus C_n$. Prove that the linear combination of two distinct nondisjoint cycles is a cycle.

   **c) Show that every cycle in G can be expressed as a linear combination of basic cycles.

*7.5 Let $G = (V, E)$ be a graph. Let R be a relation on V such that $u\ R\ v$ if and only if u and v lie on a common (not necessarily simple) cycle. Prove that R is an equivalence relation on V.

7.6 Implement both Prim's algorithm and Kruskal's algorithm. Compare the running times of your programs on a set of "random" graphs.

7.7 Write a program to find all the connected components of a graph.

7.8 Write an $O(n)$ program to determine whether a graph of n vertices has a cycle.

7.9 Write a program to enumerate all simple cycles of a graph. How many such cycles can there be? What is the time complexity of your program?

7.10 Show that all edges in a breadth-first search are either tree edges or cross edges.

7.11 Implement the algorithm for finding articulation points discussed in Section 7.4.

*__7.12__ Let $G = (V, E)$ be a *complete* graph, that is, a graph in which there is an edge between every pair of distinct vertices. Let $G' = (V, E')$ be a directed graph in which E' is E with each edge given an arbitrary orientation. Show that G' has a directed path that includes every vertex exactly once.

**__7.13__ Show that an n-vertex complete graph has n^{n-2} spanning trees.

7.14 Find all maximal matchings for the graph of Fig. 7.16.

7.15 Write a program to find a maximal matching for a bipartite graph.

7.16 Let M be a matching and let m be the number of edges in a maximal matching.

   a) Prove there exists an augmenting path relative to M of length at most $2(|M|/m - |M|) + 1$.

   b) Prove that if P is a shortest augmenting path relative to M and if P' is an augmenting path relative to $M \oplus P$, then $|P'| \geq |P| + |P \cap P'|$.

*__7.17__ Prove that a graph is bipartite if and only if it has no odd length cycles. Give an example of a non-bipartite graph for which the augmenting path graph technique of Section 7.5 no longer works.

7.18 Let M and N be matchings in a bipartite graph. Prove that $M \oplus N$ has at least $|M| - |N|$ vertex disjoint augmenting paths relative to M.

Bibliographic Notes

Methods for constructing minimal spanning trees have been studied since at least Boruvka [1926]. The two algorithms given in this chapter are based on Kruskal [1956] and Prim [1957]. Johnson [1975] shows how k-ary partially ordered trees can be used to implement Prim's algorithm. Cheriton and Tarjan [1978] and Yao [1975] present $O(e \log \log n)$ spanning tree algorithms. Tarjan [1981] provides a comprehensive survey and history of spanning tree algorithms.

    Hopcroft and Tarjan [1973] and Tarjan [1972] popularized the use of depth-first search in graph algorithms. The biconnected components algorithm is from there.

    Graph matching was studied by Hall [1948] and augmenting paths by Berge [1957] and by Edmonds [1965]. Hopcroft and Karp [1973] give an $O(n^{2.5})$ algorithm for maximal matching in bipartite graphs, and Micali and Vazirani [1980] give an $O(\sqrt{|V|} \cdot E)$ algorithm for maximal matching in general graphs. Papadimitriou and Steiglitz [1982] contains a good discussion of general matching.

CHAPTER 8

Sorting

The process of sorting, or ordering, a list of objects according to some linear order such as $\leq$ for numbers is so fundamental, and is done so frequently, that a careful examination of the subject is warranted. We shall divide the subject into two parts — internal and external sorting. Internal sorting takes place in the main memory of a computer, where we can use the random access capability of the main memory to advantage in various ways. External sorting is necessary when the number of objects to be sorted is too large to fit in main memory. There, the bottleneck is usually the movement of data between main and secondary storage, and we must move data in large blocks to be efficient. The fact that physically contiguous data is most conveniently moved in one block constrains the kinds of external sorting algorithms we can use. External sorting will be covered in Chapter 11.

8.1 The Internal Sorting Model

In this chapter, we shall present the principal internal sorting algorithms. The simplest algorithms usually take $O(n^2)$ time to sort n objects and are only useful for sorting short lists. One of the most popular sorting algorithms is quicksort, which takes $O(n \log n)$ time on average. Quicksort works well for most common applications, although, in the worst case, it can take $O(n^2)$ time. There are other methods, such as heapsort and mergesort, that take $O(n \log n)$ time in the worst case, although their average case behavior may not be quite as good as that of quicksort. Mergesort, however, is well-suited for external sorting. We shall consider several other algorithms called "bin" or "bucket" sorts. These algorithms work only on special kinds of data, such as integers chosen from a limited range, but when applicable, they are remarkably fast, taking only $O(n)$ time in the worst case.

Throughout this chapter we assume that the objects to be sorted are records consisting of one or more fields. One of the fields, called the *key*, is of a type for which a linear-ordering relationship $\leq$ is defined. Integers, reals, and arrays of characters are common examples of such types, although we may generally use any key type for which a "less than" or "less than or equal to" relation is defined.

The *sorting* problem is to arrange a sequence of records so that the values of their key fields form a nondecreasing sequence. That is, given records r_1, $r_2, \ldots, r_n$, with key values $k_1, k_2, \ldots, k_n$, respectively, we must produce the same records in an order $r_{i_1}, r_{i_2}, \ldots, r_{i_n}$, such that $k_{i_1} \leq k_{i_2} \leq \cdots \leq k_{i_n}$. The records all need not have distinct values, nor do we require that records with the same key value appear in any particular order.

We shall use several criteria to evaluate the running time of an internal sorting algorithm. The first and most common measure is the number of algorithm steps required to sort n records. Another common measure is the number of comparisons between keys that must be made to sort n records. This measure is often useful when a comparison between a pair of keys is a relatively expensive operation, as when keys are long strings of characters. If the size of records is large, we may also want to count the number of times a record must be moved. The application at hand usually makes the appropriate cost measure evident.

8.2 Some Simple Sorting Schemes

Perhaps the simplest sorting method one can devise is an algorithm called "bubblesort." The basic idea behind bubblesort is to imagine that the records to be sorted are kept in an array held vertically. The records with low key values are "light" and bubble up to the top. We make repeated passes over the array, from bottom to top. As we go, if two adjacent elements are out of order, that is, if the "lighter" one is below, we reverse them. The effect of this operation is that on the first pass the "lightest" record, that is, the record with the lowest key value, rises all the way to the top. On the second pass, the second lowest key rises to the second position, and so on. We need not, on pass two, try to bubble up to position one, because we know the lowest key already resides there. In general, pass i need not try to bubble up past position i. We sketch the algorithm in Fig. 8.1, assuming A is an **array**$[1..n]$ **of** recordtype, and n is the number of records. We assume here, and throughout the chapter, that one field called *key* holds the key value for each record.

```
(1)      for i := 1 to n−1 do
(2)          for j := n downto i+1 do
(3)              if A[j].key < A[j−1].key then
(4)                  swap(A[j], A[j−1])
```

Fig. 8.1. The bubblesort algorithm.

The procedure *swap* is used in many sorting algorithms and is defined in Fig. 8.2.

Example 8.1. Figure 8.3 shows a list of famous volcanoes, and a year in which each erupted.

For this example we shall use the following type definitions in our sorting programs:

```
procedure swap ( var x, y: recordtype )
     { swap exchanges the values of x and y }
     var
        temp: recordtype;
     begin
        temp := x;
        x := y;
        y := temp
     end;  { swap }
```

Fig. 8.2. The procedure *swap*.

NAME	YEAR
Pelee	1902
Etna	1669
Krakatoa	1883
Agung	1963
Vesuvius	79
St. Helens	1980

Fig. 8.3. Famous volcanoes.

```
type
     keytype = array[1..10] of char;
     recordtype = record
        key: keytype;  { the volcano name }
        year: integer
     end;
```

The bubblesort algorithm of Fig. 8.1 applied to the list of Fig. 8.3 sorts the list in alphabetical order of names, if the relation $\leq$ on objects of this key type is the ordinary lexicographic order. In Fig. 8.4, we see the five passes made by the algorithm when $n = 6$. The lines indicate the point above which the names are known to be the smallest in the list and in the correct order. However, after $i = 5$, when all but the last record is in its place, the last must also be in its correct place, and the algorithm stops.

At the start of the first pass, St. Helens bubbles past Vesuvius, but not past Agung. On the remainder of that pass, Agung bubbles all the way up to the top. In the second pass, Etna bubbles as far as it can go, to position 2. In the third pass, Krakatoa bubbles past Pelee, and the list is now in lexicographically sorted order although according to the algorithm of Fig. 8.1, two additional passes are made. □

Pelee	Agung	Agung	Agung	Agung	Agung
Etna	Pelee	Etna	Etna	Etna	Etna
Krakatoa	Etna	Pelee	Krakatoa	Krakatoa	Krakatoa
Agung	Krakatoa	Krakatoa	Pelee	Pelee	Pelee
Vesuvius	St. Helens	St. Helens	St. Helens	St. Helens	St. Helens
St. Helens	Vesuvius	Vesuvius	Vesuvius	Vesuvius	Vesuvius
initial	after $i=1$	after $i=2$	after $i=3$	after $i=4$	after $i=5$

Fig. 8.4. The passes of bubblesort.

Insertion Sorting

The second sorting method we shall consider is called "insertion sort," because on the ith pass we "insert" the ith element $A[i]$ into its rightful place among $A[1]$, $A[2]$, . . . ,$A[i-1]$, which were previously placed in sorted order. After doing this insertion, the records occupying $A[1]$, . . . ,$A[i]$ are in sorted order. That is, we execute

> **for** $i := 2$ **to** n **do**
>   move $A[i]$ forward to the position $j \leq i$ such that
> $A[i] < A[k]$ for $j \leq k < i$, and
>   either $A[i] \geq A[j-1]$ or $j = 1$

To make the process of moving $A[i]$ easier, it helps to introduce an element $A[0]$, whose key has a value smaller than that of any key among $A[1]$, . . . ,$A[n]$. We shall postulate the existence of a constant $-\infty$ of type keytype that is smaller than the key of any record that could appear in practice. If no constant $-\infty$ can be used safely, we must, when deciding whether to push $A[i]$ up before position j, check first whether $j = 1$, and if not, compare $A[i]$ (which is now in position j) with $A[j-1]$. The complete program is shown in Fig. 8.5.

```
(1)        A[0].key := −∞;
(2)        for i := 2 to n do begin
(3)            j := i;
(4)            while A[j] < A[j−1] do begin
(5)                swap(A[j], A[j−1]);
(6)                j := j−1
               end
        end
```

Fig. 8.5. Insertion sort.

Example 8.2. We show in Fig. 8.6 the initial list from Fig. 8.3, and the result of the passes of insertion sort for $i = 2, 3, \ldots, 6$. After each pass, the elements above the line are guaranteed to be sorted relative to each other, although their order bears no relation to the records below the line, which will be inserted later. □

$-\infty$	$-\infty$	$-\infty$	$-\infty$	$-\infty$	$-\infty$
Pelee	Etna	Etna	Agung	Agung	Agung
Etna	Pelee	Krakatoa	Etna	Etna	Etna
Krakatoa	Krakatoa	Pelee	Krakatoa	Krakatoa	Krakatoa
Agung	Agung	Agung	Pelee	Pelee	Pelee
Vesuvius	Vesuvius	Vesuvius	Vesuvius	Vesuvius	St. Helens
St. Helens	St. Helens	St. Helens	St. Helens	St. Helens	Vesuvius
initial	after $i=2$	after $i=3$	after $i=4$	after $i=5$	after $i=6$

Fig. 8.6. The passes of insertion sort.

Selection Sort

The idea behind "selection sorting" is also elementary. In the ith pass, we select the record with the lowest key, among $A[i], \ldots, A[n]$, and we swap it with $A[i]$. As a result, after i passes, the i lowest records will occupy $A[1], \ldots, A[i]$, in sorted order. That is, selection sort can be described by

> **for** $i := 1$ **to** $n-1$ **do**
>   select the smallest among $A[i], \ldots, A[n]$ and
>   swap it with $A[i]$;

A more complete program is shown in Fig. 8.7.

Example 8.3. The passes of selection sort on the list of Fig. 8.3 are shown in Fig. 8.8. For example, on pass 1, the ultimate value of lowindex is 4, the position of Agung, which is swapped with Vesuvius in $A[1]$.

The lines in Fig. 8.8 indicate the point above which elements known to be smallest appear in sorted order. After $n-1$ passes, record $A[n]$, Vesuvius in Fig. 8.8, is also in its rightful place, since it is the element known not to be among the $n-1$ smallest. □

Time Complexity of the Methods

Bubblesort, insertion sort, and selection sort each take $O(n^2)$ time, and will take $\Omega(n^2)$ time on some, in fact, on most input sequences of n elements. Consider bubblesort in Fig. 8.1. No matter what recordtype is, *swap* takes a constant time. Thus lines (3) and (4) of Fig. 8.1 take at most c_1 time units for some constant c_1. Hence for a fixed value of i, the loop of lines (2–4)

```
        var
                lowkey: keytype;  { the currently smallest key found
                        on a pass through A[i], . . . ,A[n] }
                lowindex : integer;  { the position of lowkey }
        begin
(1)             for i := 1 to n−1  do begin
                        { select the lowest among A[i], . . . ,A[n] and swap it with A[i] }
(2)                     lowindex := i;
(3)                     lowkey := A[i].key;
(4)                     for j := i + 1 to n do
                                { compare each key with current lowkey }
(5)                             if A[j].key < lowkey then begin
(6)                                     lowkey := A[j].key;
(7)                                     lowindex := j
                                end;
(8)                     swap(A[i], A[lowindex]);
                end
        end;
```

Fig. 8.7. Selection sort.

Pelee	Agung	Agung	Agung	Agung	Agung
Etna	Etna	Etna	Etna	Etna	Etna
Krakatoa	Krakatoa	Krakatoa	Krakatoa	Krakatoa	Krakatoa
Agung	Pelee	Pelee	Pelee	Pelee	Pelee
Vesuvius	Vesuvius	Vesuvius	Vesuvius	Vesuvius	St. Helens
St. Helens	St. Helens	St. Helens	St. Helens	St. Helens	Vesuvius
initial	after $i=1$	after $i=2$	after $i=3$	after $i=4$	after $i=5$

Fig. 8.8. Passes of selection sort.

takes at most $c_2(n-1)$ steps, for some constant c_2; the latter constant is somewhat larger than c_1 to account for decrementation and testing of j. Consequently, the entire program takes

$$c_3n + \sum_{i=1}^{n-1} c_2(n-1) = \frac{1}{2}c_2n^2+(c_3-\frac{1}{2}c_2)n$$

steps, where the term c_3n accounts for incrementing and testing i. As the latter formula does not exceed $(c_2/2+c_3)n^2$, for $n \geq 1$, we see that the time complexity of bubblesort is $O(n^2)$. The algorithm requires $\Omega(n^2)$ steps, since even if no swaps are ever needed (i.e., the input happens to be already sorted), the test of line (3) is executed $n(n-1)/2$ times.

Next consider the insertion sort of Fig. 8.5. The while-loop of lines (4–6) of Fig. 8.5 cannot take more than $O(i)$ steps, since j is initialized to i at line

(3), and decreases each time around the loop. The loop terminates by the time $j=1$, since $A[0]$ is $-\infty$, which forces the test of line (4) to be false when $j=1$. We may conclude that the for-loop of lines (2–6) takes at most $c\sum_{i=2}^{n} i$ steps for some constant c. This sum is $O(n^2)$.

The reader may check that if the array is initially sorted in reverse order, then we actually go around the while-loop of lines (4–6) $i-1$ times, so line (4) is executed $\sum_{i=2}^{n}(i-1) = n(n-1)/2$ times. Therefore, insertion sort requires $\Omega(n^2)$ time in the worst case. It can be shown that this smaller bound holds in the average case as well.

Finally, consider selection sort in Fig. 8.7. We may check that the inner for-loop of lines (4−7) takes $O(n-i)$ time, since j ranges from $i+1$ to n. Thus the total time taken by the algorithm is $c \sum_{i=1}^{n-1}(n-i)$, for some constant c. This sum, which is $cn(n-1)/2$, is seen easily to be $O(n^2)$. Conversely, one can show that line (5), at least, is executed $\sum_{i=1}^{n-1} \sum_{j=i+1}^{n} (1) = n(n-1)/2$ times regardless of the initial array A, so selection sort takes $\Omega(n^2)$ time in the worst case and the average case as well.

Counting Swaps

If the size of records is large, the procedure *swap*, which is the only place in the three algorithms where records are copied, will take far more time than any of the other steps, such as comparison of keys and calculations on array indices. Thus, while all three algorithms take time proportional to n^2, we might be able to compare them in more detail if we count the uses of *swap*.

To begin, bubblesort executes the swap step of line (4) of Fig. 8.3 at most

$$\sum_{i=1}^{n-1} \sum_{j=i+1}^{n} (1) = n(n-1)/2$$

times, or about $n^2/2$ times. But since the execution of line (4) depends on the outcome of the test of line (3), we might expect that the actual number of swaps will be considerably less than $n^2/2$.

In fact, bubblesort swaps exactly half the time on the average, so the expected number of swaps if all initial sequences are equally likely is about $n^2/4$. To see this, consider two initial lists of keys that are the reverse of one another, say $L_1 = k_1, k_2, \ldots, k_n$ and $L_2 = k_n, k_{n-1}, \ldots, k_1$. A swap is the only way k_i and k_j can cross each other if they are initially out of order. But k_i and k_j are out of order in exactly one of L_1 and L_2. Thus, the total number of swaps executed when bubblesort is applied to L_1 and L_2 is equal to the number of pairs of elements, that is, $\binom{n}{2}$ or $n(n-1)/2$. Therefore, the average number of swaps for L_1 and L_2 is $n(n-1)/4$ or about $n^2/4$. Since all possible orderings can be paired with their reversals, as L_1 and L_2 were, we see that the average number of swaps over all orderings will likewise be about $n^2/4$.

The number of swaps made by insertion sort on the average is exactly what it is for bubblesort. The same argument applies; each pair of elements gets swapped either in a list L or its reverse, but never in both.

However, in the case that *swap* is an expensive operation we can easily see that selection sort is superior to either bubblesort or insertion sort. Line (8) in Fig. 8.7 is outside the inner loop of the selection sort algorithm, so it is executed exactly $n-1$ times on any array of length n. Since line (8) has the only call to *swap* in selection sort, we see that the rate of growth in the number of swaps by selection sort, which is $O(n)$, is less than the growth rates of the number of swaps by the other two algorithms, which is $O(n^2)$. Intuitively, unlike bubblesort or insertion sort, selection sort allows elements to "leap" over large numbers of other elements without being swapped with each of them individually.

A useful strategy to use when records are long and swaps are expensive is to maintain an array of pointers to the records, using whatever sorting algorithm one chooses. One can then swap pointers rather than records. Once the pointers to the records have been arranged into the proper order, the records themselves can be arranged into the final sorted order in $O(n)$ time.

Limitations of the Simple Algorithms

We should not forget that each of the algorithms mentioned in this section has an $O(n^2)$ running time, in both the worst case and average case. Thus, for large n, none of these algorithms compares favorably with the $O(n\log n)$ algorithms to be discussed in the next sections. The value of n at which these more complex $O(n\log n)$ algorithms become better than the simple $O(n^2)$ algorithms depends on a variety of factors such as the quality of the object code generated by the compiler, the machine on which the programs are run, and the size of the records we must swap. Experimentation with a time profiler is a good way to determine the cutover point. A reasonable rule of thumb is that unless n is at least around one hundred, it is probably a waste of time to implement an algorithm more complicated than the simple ones discussed in this section. Shellsort, a generalization of bubblesort, is a simple, easy-to-implement $O(n^{1.5})$ sorting algorithm that is reasonably efficient for modest values of n. Shellsort is presented in Exercise 8.3.

8.3 Quicksort

The first $O(n\log n)$ algorithm† we shall discuss, and probably the most efficient for internal sorting, has been given the name "quicksort." The essence of quicksort is to sort an array $A[1], \ldots, A[n]$ by picking some key value v in the array as a *pivot* element around which to rearrange the elements in the array. We hope the pivot is near the median key value in the array, so that it

---

† Technically, quicksort is only $O(n\log n)$ in the average case; it is $O(n^2)$ in the worst case.

is preceded by about half the keys and followed by about half. We permute the elements in the array so that for some j, all the records with keys less than v appear in $A[1], \ldots, A[j]$, and all those with keys v or greater appear in $A[j+1], \ldots, A[n]$. We then apply quicksort recursively to $A[1], \ldots, A[j]$ and to $A[j+1], \ldots, A[n]$ to sort both these groups of elements. Since all keys in the first group precede all keys in the second group, the entire array will thus be sorted.

Example 8.4. In Fig. 8.9 we display the recursive steps that quicksort might take to sort the sequence of integers 3, 1, 4, 1, 5, 9, 2, 6, 5, 3. In each case, we have chosen to take as our value v the larger of the two leftmost distinct values. The recursion stops when we discover that the portion of the array we have to sort consists of identical keys. We have shown each level as consisting of two steps, one before partitioning each subarray, and the second after. The rearrangement of records that takes place during partitioning will be explained shortly. □

Let us now begin the design of a recursive procedure $quicksort(i, j)$ that operates on an array A with elements $A[1], \ldots, A[n]$, defined externally to the procedure. $quicksort(i, j)$ sorts $A[i]$ through $A[j]$, in place. A preliminary sketch of the procedure is shown in Fig. 8.10. Note that if $A[i], \ldots, A[j]$ all have the same key, the procedure does nothing to A.

We begin by developing a function $findpivot$ that implements the test of line (1) of Fig. 8.10, determining whether the keys of $A[i], \ldots, A[j]$ are all the same. If $findpivot$ never finds two different keys, it returns 0. Otherwise, it returns the index of the larger of the first two different keys. This larger key becomes the pivot element. The function $findpivot$ is written in Fig. 8.11.

Next, we implement line (3) of Fig. 8.10, where we face the problem of permuting $A[i], \ldots, A[j]$, in place‡, so that all the keys smaller than the pivot value appear to the left of the others. To do this task, we introduce two cursors, l and r, initially at the left and right ends of the portion of A being sorted, respectively. At all times, the elements to the left of l, that is, $A[i], \ldots, A[l-1]$ will have keys less than the pivot. Elements to the right of r, that is, $A[r+1], \ldots, A[j]$, will have keys equal to or greater than the pivot, and elements in the middle will be mixed, as suggested in Fig. 8.12.

Initially, $i = l$ and $j = r$ so the above statement holds since nothing is to the left of l or to the right of r. We repeatedly do the following steps, which move l right and r left, until they finally cross, whereupon $A[i], \ldots, A[l-1]$ will contain all the keys less than the pivot and $A[r+1], \ldots, A[j]$ all the keys equal to or greater than the pivot.

1. *Scan.* Move l right over any records with keys less than the pivot. Move r left over any keys greater than or equal to the pivot. Note that our

---

‡ We could copy $A[i], \ldots, A[j]$ and arrange them as we do so, finally copying the result back into $A[i], \ldots, A[j]$. We choose not to do so because that approach would waste space and take longer than the in-place arrangement method we use.

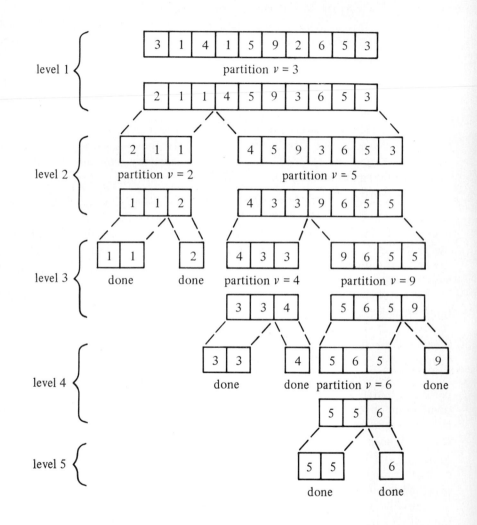

Fig. 8.9. The operation of quicksort.

selection of the pivot by *findpivot* guarantees that there is at least one key less than the pivot and at least one not less than the pivot, so l and r will surely come to rest before moving outside the range i to j.

2. *Test.* If $l > r$ (which in practice means $l = r+1$), then we have successfully partitioned $A[i], \ldots, A[j]$, and we are done.

3. *Switch.* If $l < r$ (note we cannot come to rest during the scan with $l = r$, because one or the other of these will move past any given key), then

```
(1)     if A[i] through A[j] contains at least two distinct keys then begin
(2)          let v be the larger of the first two distinct keys found;
(3)          permute A[i], . . . ,A[j] so that for some k between
                 i+1 and j, A[i], . . . ,A[k−1] all have keys less than
                 v and A[k], . . . ,A[j] all have keys ≥ v;
(4)          quicksort(i, k−1);
(5)          quicksort(k, j)
        end
```

Fig. 8.10. Sketch of quicksort.

```
function findpivot ( i, j: integer ) : integer;
     { returns 0 if A[i], . . . ,A[j] have identical keys, otherwise
     returns the index of the larger of the leftmost two different keys }
     var
          firstkey: keytype; { value of first key found, i.e., A[i].key }
          k: integer; { runs left to right looking for a different key }
     begin
          firstkey := A[i].key;
          for k := i + 1 to j do { scan for different key }
               if A[k].key > firstkey then { select larger key }
                    return (k)
               else if A[k].key < firstkey then
                    return (i);
          return (0) { different keys were never found }
     end; { findpivot }
```

Fig. 8.11. The procedure findpivot.

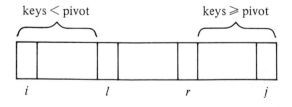

Fig. 8.12. Situation during the permutation process.

swap $A[l]$ with $A[r]$. After doing so, $A[l]$ has a key less than the pivot and $A[r]$ has a key at least equal to the pivot, so we know that in the next scan phase, l will move at least one position right, over the old $A[r]$, and r will move at least one position left.

The above loop is awkward, since the test that terminates it is in the middle. To put it in the form of a repeat-loop, we move the switch phase to the beginning. The effect is that initially, when $i = l$ and $j = r$, we shall swap $A[i]$ with $A[j]$. This could be right or wrong; it doesn't matter, as we assume no particular order for the keys among $A[i], \ldots, A[j]$ initially. The reader, however, should be aware of this "trick" and not be puzzled by it. The function *partition*, which performs the above operations and returns l, the point at which the upper half of the partitioned array begins, is shown in Fig. 8.13.

```
        function partition ( i, j: integer; pivot: keytype ) : integer;
            { partitions A[i], . . . , A[j] so keys < pivot are at the left
            and keys ≥ pivot are on the right.  Returns the beginning of the
            group on the right. }
            var
                l, r: integer;  { cursors as described above }
            begin
(1)             l := i;
(2)             r := j;
                repeat
(3)                 swap(A[l], A[r]);
                    { now the scan phase begins }
(4)                 while A[l].key < pivot do
(5)                     l := l + 1;
(6)                 while A[r].key >= pivot do
(7)                     r := r − 1
                until
(8)                 l > r;
(9)             return (l)
        end;  { partition }
```

Fig. 8.13. The procedure *partition*.

We are now ready to elaborate the quicksort sketch of Fig. 8.10. The final program is shown in Fig. 8.14. To sort an array A of type **array**[1..n] **of** recordtype we simply call *quicksort*(1, n).

The Running Time of Quicksort

We shall show that quicksort takes $O(n\log n)$ time on the average to sort n elements, and $O(n^2)$ time in the worst case. The first step in substantiating both claims is to prove that *partition* takes time proportional to the number of elements that it is called upon to separate, that is, $O(j-i+1)$ time.

```
        procedure quicksort ( i, j: integer );
            { sort elements A[i], . . . ,A[j] of external array A }
            var
                pivot: keytype; { the pivot value }
                pivotindex: integer; { the index of an element of A where
                    key is the pivot }
                k: integer; { beginning index for group of elements ≥ pivot }
            begin
(1)             pivotindex := findpivot(i, j);
(2)             if pivotindex <> 0 then begin  { do nothing if all keys are equal }
(3)                     pivot := A[pivotindex].key;
(4)                     k := partition(i, j, pivot);
(5)                     quicksort(i, k−1);
(6)                     quicksort(k, j)
                end
        end; { quicksort }
```

Fig. 8.14. The procedure quicksort.

To see why that statement is true, we must use a trick that comes up frequently in analyzing algorithms; we must find certain "items" to which time
may be "charged," and then show how to charge each step of the algorithm so
no item is charged more than some constant. Then the total time spent is no
greater than this constant times the number of "items."

In our case, the "items" are the elements from $A[i]$ to $A[j]$, and we
charge to each element all the time spent by *partition* from the time l or r first
points to that element to the time l or r leaves that element. First note that
neither l nor r ever returns to an element. Because there is at least one element in the low group and one in the high group, and because *partition* stops
as soon as l exceeds r, we know that each element will be charged at most
once.

We move off an element in the loops of lines (4) and (6) of Fig. 8.13,
either by increasing l or decreasing r. How long could it be between times
when we execute $l := l+1$ or $r := r−1$? The worst that can happen is at the
beginning. Lines (1) and (2) initialize l and r. Then we might go around the
loop without doing anything to l or r. On second and subsequent passes, the
swap at line (3) guarantees that the while-loops of lines (4) and (6) will be
successful at least once each, so the worst that can be charged to an execution
of $l := l+1$ or $r := r−1$ is the cost of line (1), line (2), twice line (3), and
the tests of lines (4), (6), (8), and (4) again. This is only a constant amount,
independent of i or j, and subsequent executions of $l := l + 1$ or $r := r − 1$
are charged less — at most one execution of lines (3) and (8) and once around
the loops of lines (4) or (6).

There are also the final two unsuccessful tests at lines (4), (6), and (8), that may not be charged to any "item," but these represent only a constant amount and may be charged to any item. After all charges have been made, we still have some constant c so that no item has been charged more than c time units. Since there are $j-i+1$ "items," that is, elements in the portion of the array to be sorted, the total time spent by *partition*$(i, j, pivot)$ is $O(j-i+1)$.

Now let us turn to the running time spent by *quicksort*(i, j). We can easily check that the time spent by the call to *findpivot* at line (1) of Fig. 8.14 is $O(j-i+1)$, and in most cases much smaller. The test of line (2) takes a constant amount of time, as does step (3) if executed. We have just argued that line (4), the call to *partition*, will take $O(j-i+1)$ time. Thus, exclusive of recursive calls it makes to *quicksort*, each individual call of *quicksort* takes time at most proportional to the number of elements it is called upon to sort.

Put another way, the total time taken by *quicksort* is the sum over all elements of the number of times that element is part of the subarray on which a call to *quicksort* is made. Let us refer back to Fig. 8.9, where we see calls to *quicksort* organized into levels. Evidently, no element can be involved in two calls at the same level, so the time taken by *quicksort* can be expressed as the sum over all elements of the *depth*, or maximum level, at which that element is found. For example, the 1's in Fig. 8.9 are of depth 3 and the 6 of depth 5.

In the worst case, we could manage to select at each call to *quicksort* a worst possible pivot, say the largest of the key values in the subarray being sorted. Then we would divide the subarray into two smaller subarrays, one with a single element (the element with the pivot as key) and the other with everything else. That sequence of partitions leads to a tree like that of Fig. 8.15, where $r_1, r_2, \ldots, r_n$ is the sequence of records in order of increasing keys.

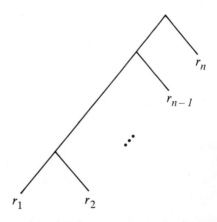

Fig. 8.15. A worst possible sequence of pivot selections.

The depth of r_i is $n-i+2$ for $2 \le i \le n$, and the depth of r_1 is n. Thus the sum of the depths is

$$n + \sum_{i=2}^{n} (n-i+2) = \frac{n^2}{2} + \frac{3n}{2} - 1$$

which is $\Omega(n^2)$. Thus in the worst case, quicksort takes time proportional to n^2 to sort n elements.

Average Case Analysis of Quicksort

Let us, as always, interpret "average case" for a sorting algorithm to mean the average over all possible initial orderings, equal probability being attributed to each possible ordering. For simplicity, we shall assume no two elements have equal keys. In general, equalities among elements make our sorting task easier, not harder, anyway.

A second assumption that makes our analysis of quicksort easier is that, when we call quicksort(i, j), then all orders for $A[i], \ldots, A[j]$ are equally likely. The justification is that prior to this call, no pivots with which $A[i], \ldots, A[j]$ were compared distinguished among them; that is, for each such pivot v, either all were less than v or all were equal to or greater then v. A careful examination of the quicksort program we developed shows that each pivot element is likely to wind up near the right end of the subarray of elements equal to or greater than this pivot, but for large subarrays, the fact that the minimum element (the previous pivot, that is) is likely to appear near the right end doesn't make a measurable difference.†

Now, let $T(n)$ be the average time taken by quicksort to sort n elements. Clearly $T(1)$ is some constant c_1, since on one element, quicksort makes no recursive calls to itself. When $n > 1$, since we assume all elements have unequal keys, we know quicksort will pick a pivot and split the subarray, taking $c_2 n$ time to do this, for some constant c_2, then call quicksort on the two subarrays. It would be nice if we could claim that the pivot was equally likely to be any of the first, second, $\ldots, n^{\text{th}}$ element in the sorted order for the subarray being sorted. However, to guarantee ourselves that quicksort would find at least one key less than each pivot and at least one equal to or greater than the pivot (so each piece would be smaller than the whole, and therefore infinite loops would not be possible), we always picked the larger of the first two elements found. It turns out that this selection doesn't affect the distribution of sizes of the subarrays, but it does tend to make the left groups (those less than the pivot) larger then the right groups.

Let us develop a formula for the probability that the left group has i of the n elements, on the assumption that all elements are unequal. For the left group to have i elements, the pivot must be the $i+1^{\text{st}}$ among the n. The

---

† If there is reason to believe nonrandom orders of elements might make *quicksort* run slower than expected, the *quicksort* program should permute the elements of the array at random before sorting.

pivot, by our method of selection, could either have been in the first position, with one of the i smaller elements second, or it could have been second, with one of the i smaller ones first. The probability that any particular element, such as the $i+1^{st}$, appears first in a random order is $1/n$. Given that it did appear first, the probability that the second element is one of the i smaller elements out of the $n-1$ remaining elements is $i/(n-1)$. Thus the probability that the pivot appears in the first position and is number $i+1$ out of n in the proper order is $i/n(n-1)$. Similarly, the probability that the pivot appears in the second position and is number $i+1$ out of n in the sorted order is $i/n(n-1)$, so the probability that the left group is of size i is $2i/n(n-1)$, for $1 \le i < n$.

Now we can write a recurrence for $T(n)$.

$$T(n) \le \sum_{i=1}^{n-1} \frac{2i}{n(n-1)} [T(i) + T(n-i)] + c_2 n \qquad (8.1)$$

Equation (8.1) says that the average time taken by quicksort is the time, $c_2 n$, spent outside of the recursive calls, plus the average time spent in recursive calls. The latter time is expressed in (8.1) as the sum, over all possible i, of the probability that the left group is of size i (and therefore the right group is of size $n-i$) times the cost of the two recursive calls: $T(i)$ and $T(n-i)$, respectively.

Our first task is to transform (8.1) so that the sum is simplified, and in fact, so (8.1) takes on the form it would have had if we had picked a truly random pivot at each step. To make the transformation, observe that for any function $f(i)$ whatsoever, by substituting i for $n-i$ we may prove

$$\sum_{i=1}^{n-1} f(i) = \sum_{i=1}^{n-1} f(n-i) \qquad (8.2)$$

By replacing half the left side of (8.2) by the right side, we see

$$\sum_{i=1}^{n-1} f(i) = \frac{1}{2} \sum_{i=1}^{n-1} f(i) + f(n-i) \qquad (8.3)$$

Applying (8.3) to (8.1), with $f(i)$ equal to the expression inside the summation of (8.1), we get

$$T(n) \le \frac{1}{2} \sum_{i=1}^{n-1} \left\{ \frac{2i}{n(n-1)}[T(i) + T(n-i)] + \frac{2(n-i)}{n(n-1)}[T(n-i) + T(i)] \right\} + c_2 n$$

$$\le \frac{1}{n-1} \sum_{i=1}^{n-1} [T(i) + T(n-i)] + c_2 n \qquad (8.4)$$

Next, we can apply (8.3) to (8.4), with $f(i) = T(i)$. This transformation yields

$$T(n) \le \frac{2}{n-1} \sum_{i=1}^{n-1} T(i) + c_2 n \qquad (8.5)$$

Note that (8.4) is the recurrence we would get if all sizes between 1 and $n-1$ for the left group were equally likely. Thus, picking the larger of two keys as the pivot really doesn't affect the size distribution. We shall study recurrences of this form in greater detail in Chapter 9. Here we shall solve recurrence (8.5) by guessing a solution and showing that it works. The solution we guess is that $T(n) \le cn\log n$ for some constant c and all $n \ge 2$.

To prove that this guess is correct, we perform an induction on n. For $n = 2$, we have only to observe that for some constant c, $T(2) \le 2c\log 2 = 2c$. To prove the induction, assume $T(i) \le ci\log i$ for $i < n$, and substitute this formula for $T(i)$ in the right side of (8.5) to show the resulting quantity is no greater than $cn\log n$. Thus, (8.5) becomes

$$T(n) \le \frac{2c}{n-1} \sum_{i=1}^{n-1} i\log i + c_2 n \tag{8.6}$$

Let us split the summation of (8.6) into low terms, where $i \le n/2$, and therefore $\log i$ is no greater than $\log(n/2)$, which is $(\log n)-1$, and high terms, where $i > n/2$, and $\log i$ may be as large as $\log n$. Then (8.6) becomes

$$T(n) \le \frac{2c}{n-1} \left[\sum_{i=1}^{n/2} i\log i + \sum_{i=n/2+1}^{n-1} i\log i \right] + c_2 n$$

$$\le \frac{2c}{n-1} \left[\sum_{i=1}^{n/2} i(\log n - 1) + \sum_{i=n/2+1}^{n-1} i\log n \right] + c_2 n$$

$$\le \frac{2c}{n-1} \left[\frac{n}{4}(\frac{n}{2}+1)\log n - \frac{n}{4}(\frac{n}{2}+1) + \frac{3}{4}n(\frac{n}{2}-1)\log n \right] + c_2 n$$

$$\le \frac{2c}{n-1} \left[\left(\frac{n^2}{2} - \frac{n}{2}\right)\log n - \left(\frac{n^2}{8} + \frac{n}{4}\right) \right] + c_2 n$$

$$\le cn\log n - \frac{cn}{4} - \frac{3cn}{4(n-1)} + c_2 n \tag{8.7}$$

If we pick $c \ge 4c_2$, then the sum of the second and fourth terms of (8.7) is no greater than zero. The third term of (8.7) makes a negative contribution, so from (8.7) we may claim that $T(n) \le cn\log n$, if we pick $c = 4c_2$. This completes the proof that quicksort requires $O(n\log n)$ time in the average case.

Improvements to Quicksort

Not only is *quicksort* fast, its average running time is less than that of all other currently known $O(n\log n)$ sorting algorithms (by a constant factor, of course). We can improve the constant factor still further if we make some effort to pick pivots that divide each subarray into close-to-equal parts. For example, if we always divide subarrays equally, then each element will be of depth exactly $\log n$, in the tree of partitions analogous to Fig. 8.9. In comparison, the average depth of an element for *quicksort* as constituted in Fig. 8.14, is about $1.4\log n$. Thus we could hope to speed up *quicksort* by choosing pivots carefully.

For example, we could choose three elements of a subarray at random, and pick the middle one as the pivot. We could pick k elements at random for any k, sort them either by a recursive call to quicksort or by one of the simpler sorts of Section 8.2, and pick the median element, that is, the $\frac{k+1}{2}^{\text{th}}$ element, as pivot.† It is an interesting exercise to determine the best value of k, as a function of the number of elements in the subarray to be sorted. If we pick k too small, it costs time because on the average the pivot will divide the elements unequally. If we pick k too large, we spend too much time finding the median of the k elements.

Another improvement to quicksort concerns what happens when we get small subarrays. Recall from Section 8.2 that the simple $O(n^2)$ methods are better than the $O(n\log n)$ methods such as *quicksort* for small n. How small n is depends on many factors, such as the time spent making a recursive call, which is a property of the machine architecture and the strategy for implementing procedure calls used by the compiler of the language in which the sort is written. Knuth [1973] suggests 9 as the size subarray on which *quicksort* should call a simpler sorting algorithm.

There is another "speedup" to *quicksort* that is really a way to trade space for time. The same idea will work for almost any sorting algorithm. If we have the space available, create an array of pointers to the records in array A. We make the comparisons between the keys of records pointed to. But we don't move the records; rather, we move the pointers to records in the same way quicksort moves the records themselves. At the end, the pointers, read in left-to-right order, point to the records in the desired order, and it is then a relatively simple matter to rearrange the records of A into the correct order.

In this manner, we make only n swaps of records, instead of $O(n\log n)$ swaps, which makes a substantial difference if records are large. On the negative side, we need extra space for the array of pointers, and accessing keys for comparisons is slower than it was, since we have first to follow a pointer, then go into the record, to get the key field.

---

† Since we only want the median, not the entire sorted list of k elements, it may be better to use one of the fast median-finding algorithms of Section 8.7.

8.4 Heapsort

In this section we develop a sorting algorithm called *heapsort* whose worst case as well as average case running time is $O(n \log n)$. This algorithm can be expressed abstractly using the four set operations INSERT, DELETE, EMPTY, and MIN introduced in Chapters 4 and 5. Suppose L is a list of the elements to be sorted and S a set of elements of type recordtype, which will be used to hold the elements as they are being sorted. The MIN operator applies to the key field of records; that is, MIN(S) returns the record in S with the smallest key value. Figure 8.16 presents the abstract sorting algorithm that we shall transform into heapsort.

```
(1)      for x on list L do
(2)          INSERT(x, S);
(3)      while not EMPTY(S) do begin
(4)          y := MIN(S);
(5)          writeln(y);
(6)          DELETE(y, S)
         end
```

Fig. 8.16. An abstract sorting algorithm.

In Chapters 4 and 5 we presented several data structures, such as 2-3 trees, capable of supporting each of these set operations in $O(\log n)$ time per operation, if sets never grow beyond n elements. If we assume list L is of length n, then the number of operations performed is n INSERT's, n MIN's, n DELETE's, and $n+1$ EMPTY tests. The total time spent by the algorithm of Fig. 8.16 is thus $O(n \log n)$, if a suitable data structure is used.

The partially ordered tree data structure introduced in Section 4.11 is well suited for the implementation of this algorithm. Recall that a partially ordered tree can be represented by a *heap*, an array $A[1], \ldots, A[n]$, whose elements have the partially ordered tree property: $A[i].key \le A[2*i].key$ and $A[i].key \le A[2*i+1].key$. If we think of the elements at $2i$ and $2i+1$ as the "children" of the element at i, then the array forms a balanced binary tree in which the key of the parent never exceeds the keys of the children.

We showed in Section 4.11 that the partially ordered tree could support the operations INSERT and DELETEMIN in $O(\log n)$ time per operation. While the partially ordered tree cannot support the general DELETE operation in $O(\log n)$ time (just finding an arbitrary element takes linear time in the worst case), we should notice that in Fig. 8.16 the only elements we delete are the ones found to be minimal. Thus lines (4) and (6) of Fig. 8.16 can be combined into a DELETEMIN function that returns the element y. We can thus implement Fig. 8.16 using the partially ordered tree data structure of Section 4.11.

We shall make one more modification to the algorithm of Fig. 8.16 to avoid having to print the elements as we delete them. The set S will always be stored as a heap in the top of array A, as $A[1], \ldots, A[i]$, if S has i elements. By the partially ordered tree property, the smallest element is always in $A[1]$. The elements already deleted from S can be stored in $A[i+1], \ldots, A[n]$, sorted in reverse order, that is, with $A[i+1] \geq A[i+2] \geq \cdots \geq A[n]$.† Since $A[1]$ must be smallest among $A[1], \ldots, A[i]$, we may effect the DELETEMIN operation by simply swapping $A[1]$ with $A[i]$. Since the new $A[i]$ (the old $A[1]$) is no smaller than $A[i+1]$ (or else the former would have been deleted from S before the latter), we now have $A[i], \ldots, A[n]$ sorted in decreasing order. We may now regard S as occupying $A[1], \ldots, A[i-1]$.

Since the new $A[1]$ (old $A[i]$) violates the partially ordered tree property, we must push it down the tree as in the procedure DELETEMIN of Fig. 4.23. Here, we use the procedure *pushdown* shown in Fig. 8.17 that operates on the externally defined array A. By a sequence of swaps, *pushdown* pushes element $A[first]$ down through its descendants to its proper position in the tree. To restore the partially ordered tree property to the heap, we call *pushdown* with $first = 1$.

We now see how lines (4−6) of Fig. 8.16 are to be done. Selecting the minimum at line (4) is easy; it is always in $A[1]$ because of the partially ordered tree property. Instead of printing at line (5), we swap $A[1]$ with $A[i]$, the last element in the current heap. Deleting the minimum element from the partially ordered tree is now easy; we just decrement i, the cursor that indicates the end of the current heap. We then invoke $pushdown(1, i)$ to restore the partially ordered tree property to the heap $A[1], \ldots, A[i]$.

The test for emptiness of S at line (3) of Fig. 8.16 is done by testing the value of i, the cursor marking the end of the current heap. Now we have only to consider how to perform lines (1) and (2). We may assume L is originally present in $A[1], \ldots, A[n]$ in some order. To establish the partially ordered tree property initially, we call $pushdown(j, n)$ for all $j = n/2, n/2-1, \ldots, 1$. We observe that after calling $pushdown(j, n)$, no violations of the partially ordered tree property occur in $A[j], \ldots, A[n]$ because pushing a record down the tree does not introduce new violations, since we only swap a violating record with its smaller child. The complete procedure, called *heapsort*, is shown in Fig. 8.18.

Analysis of Heapsort

Let us examine the procedure *pushdown* to see how long it takes. An inspection of Fig. 8.17 confirms that the body of the while-loop takes constant time. Also, after each iteration, r has at least twice the value it had before. Thus, since r starts off equal to *first*, after i iterations we have $r \geq first * 2^i$.

---

† We could, at the end, reverse array A, but if we wish A to end up sorted lowest first, then simply apply a DELETEMAX operator in place of DELETEMIN, and partially order A in such a way that a parent has a key no smaller (rather than no larger) than its children.

```
procedure pushdown ( first, last: integer );
    { assumes A[first], . . . , A[last] obeys partially ordered tree property
      except possibly for the children of A[first]. The procedure pushes
      A[first] down until the partially ordered tree property is restored }
var
    r: integer;  { indicates the current position of A[first] }
begin
    r := first;  { initialization }
    while r <= last div 2 do
        if r = last div 2 then begin  { r has one child at 2*r }
            if A[r].key > A[2*r].key then
                swap(A[r], A[2*r]);
            r := last  { forces a break from the while-loop }
        end
        else  { r has two children, elements at 2*r and 2*r+1 }
            if A[r].key > A[2*r].key and
                A[2*r].key <= A[2*r+1].key then begin
                    { swap r with left child }
                    swap(A[r], A[2*r]);
                    r := 2*r
            end
            else if A[r].key > A[2*r+1].key and
                A[2*r+1].key < A[2*r].key then begin
                    { swap r with right child }
                    swap(A[r], A[2*r+1]);
                    r := 2*r+1
            end
            else  { r does not violate partially ordered tree property }
                r := last  { to break while-loop }
    end;  { pushdown }
```

Fig. 8.17. The procedure *pushdown*.

Surely, $r > last/2$ if $first * 2^i > last/2$, that is if

$$i > \log(last/first) - 1 \qquad (8.8)$$

Hence, the number of iterations of the while-loop of *pushdown* is at most $\log(last/first)$.

Since $first \geq 1$ and $last \leq n$ in each call to *pushdown* by the heapsort algorithm of Fig. 8.18, (8.8) says that each call to *pushdown*, at line (2) or (5) of Fig. 8.18, takes $O(\log n)$ time. Clearly the loop of lines (1) and (2) iterates $n/2$ times, so the time spent in lines (1–2) is $O(n\log n)$.[†] Also, the loop of

---

† In fact, this time is $O(n)$, by a more careful argument. For j in the range $n/2$ to $n/4+1$, (8.8) says only one iteration of *pushdown*'s while-loop is needed. For j between $n/4$ and

```
          procedure heapsort;
              { sorts array A[1], . . . ,A[n] into decreasing order }
              var
                  i: integer;  { cursor into A }
              begin
                  { establish the partially ordered tree property initially }
(1)               for i := n div 2 downto 1 do
(2)                   pushdown(i, n);
(3)               for i := n downto 2 do begin
(4)                   swap(A[1], A[i]);
                          { remove minimum from front of heap }
(5)                   pushdown(1, i−1)
                          { re-establish partially ordered tree property }
              end
          end;  { heapsort }
```

Fig. 8.18. The procedure *heapsort*.

lines $(3-5)$ iterates $n-1$ times. Thus, a total of $O(n)$ time is spent in all repetitions of *swap* at line (4), and $O(n\log n)$ time is spent during all repetitions of line (5). Hence, the total time spent in the loop of lines $(3-5)$ is $O(n\log n)$, and all of *heapsort* takes $O(n\log n)$ time.

Despite its $O(n\log n)$ worst case time, *heapsort* will, on the average take more time than *quicksort*, by a small constant factor. Heapsort is of intellectual interest, because it is the first $O(n\log n)$ worst-case sort we have covered. It is of practical utility in the situation where we want not all n elements sorted, but only the smallest k of them for some k much less than n. As we mentioned, lines $(1-2)$ really take only $O(n)$ time. If we make only k iterations of lines $(3-5)$, the time spent in that loop is $O(k\log n)$. Thus, *heapsort*, modified to produce only the first k elements takes $O(n + k\log n)$ time. If $k \le n/\log n$, that is, we want at most $(1/\log n)^{th}$ of the entire sorted list, then the required time is $O(n)$.

8.5 Bin Sorting

One might wonder whether $\Omega(n\log n)$ steps are necessary to sort n elements. In the next section, we shall show that to be the case for sorting algorithms that assume nothing about the data type of keys, except that they can be ordered by some function that tells whether one key value is "less than"

$n/8+1$, only two iterations, and so on. The total number of iterations as j ranges between $n/2$ and 1 is bounded by $\frac{n}{4}*1 + \frac{n}{8}*2 + \frac{n}{16}*3 + \ldots$. Note that the improved bound for lines $(1-2)$ does not imply an improved bound for *heapsort* as a whole; all the time is taken in lines $(3-5)$.

another. We can often sort in less than $O(n \log n)$ time, however, when we know something special about the keys being sorted.

Example 8.5. Suppose keytype is integer, and the values of the key are known to be in the range 1 to n, with no duplicates, where n is the number of elements. Then if A and B are of type **array**$[1..n]$ **of** recordtype, and the n elements to be sorted are initially in A, we can place them in the array B, in order of their keys by

$$\text{for } i := 1 \text{ to } n \text{ do}$$
$$B[A[i].key] := A[i]; \tag{8.9}$$

This code calculates where record $A[i]$ belongs and places it there. The entire loop takes $O(n)$ time. It works correctly only when there is exactly one record with key v for each value of v between 1 and n. A second record with key value v would also be placed in $B[v]$, obliterating the previous record with key v.

There are other ways to sort an array A with keys $1, 2, \ldots, n$ in place in only $O(n)$ time. Visit $A[1], \ldots, A[n]$ in turn. If the record in $A[i]$ has key $j \neq i$, swap $A[i]$ with $A[j]$. If, after swapping, the record with key k is now in $A[i]$, and $k \neq i$, swap $A[i]$ with $A[k]$, and so on. Each swap places some record where it belongs, and once a record is where it belongs, it is never moved. Thus, the following algorithm sorts A in place, in $O(n)$ time, provided there is one record with each of keys $1, 2, \ldots, n$.

```
for i := 1 to n do
    while A[i].key <> i do
        swap(A[i], A[A[i].key]);
```

□

The program (8.9) given in Example 8.5 is a simple example of a "bin-sort," a sorting process where we create a "bin" to hold all the records with a certain key value. We examine each record r to be sorted and place it in the bin for the key value of r. In the program (8.9) the bins are the array elements $B[1], \ldots, B[n]$, and $B[i]$ is the bin for key value i. We can use array elements as bins in this simple case, because we know we never put more than one record in a bin. Moreover, we need not assemble the bins into a sorted list here, because B serves as such a list.

In the general case, however, we must be prepared both to store more than one record in a bin and to string the bins together (or *concatenate*) the bins in proper order. To be specific, suppose that, as always, $A[1], \ldots, A[n]$ is an array of type recordtype, and that the keys for records are of type key-type. Suppose for the purposes of this section only, that keytype is an enumerated type, such as $1..m$ or char. Let listtype be a type that represents lists of elements of type recordtype; listtype could be any of the types for lists mentioned in Chapter 2, but a linked list is most effective, since we shall be growing lists of unpredictable size in each bin, yet the total lengths of the lists is fixed at n, and therefore an array of n cells can supply the lists for the various bins as needed.

Finally, let B be an array of type **array**[keytype] **of** listtype. Thus, B is an array of bins, which are lists (or, if the linked list representation is used, headers for lists). B is indexed by keytype, so there is one bin for each possible key value. In this manner we can effect the first generalization of (8.9) — the bins have arbitrary capacity.

Now we must consider how the bins are to be concatenated. Abstractly, we must form from lists $a_1, a_2, \ldots, a_i$ and $b_1, b_2, \ldots, b_j$ the *concatenation* of the lists, which is $a_1, a_2, \ldots, a_i, b_1, b_2, \ldots, b_j$. The implementation of this operation CONCATENATE(L_1, L_2), which replaces list L_1 by the concatenation $L_1 L_2$, can be done in any of the list representations we studied in Chapter 2.

For efficiency, however, it is useful to have, in addition to a header, a pointer to the last element on each list (or to the header if the list is empty). This modification facilitates finding the last element on list L_1 without running down the entire list. Figure 8.19 shows, by dashed lines, the revised pointers necessary to concatenate L_1 and L_2 and to have the result be called L_1. List L_2 is assumed to "disappear" after concatenation, in the sense that the header and end pointer for L_2 become null.

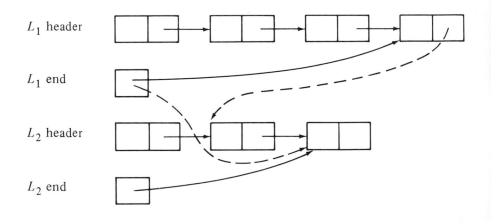

Fig. 8.19. Concatenation of linked lists.

We can now write a program to binsort arbitrary collections of records, where the key field is of an enumerated type. The program, shown in Fig. 8.20, is written in terms of list processing primitives. As we mentioned, a linked list is the preferred implementation, but options exist. Recall also that the setting for the procedure is that an array A of type **array**$[1..n]$ **of** record-type holds the elements to be sorted, and array B, of type **array** [keytype] **of** listtype represents the bins. We assume keytype is expressible as *lowkey..highkey*, as any enumerated type must be, for some quantities *lowkey* and *highkey*.

```
        procedure binsort;
            { binsort array A, leaving the sorted list in B[lowkey] }
            var
                i: integer;
                v: keytype;
            begin
                { place the records into bins }
(1)             for i := 1 to n do
                    { push A[i] onto the front of the bin for its key }
(2)                 INSERT(A[i], FIRST(B[A[i].key]), B[A[i].key]);
(3)             for v := succ(lowkey) to highkey do
                    { concatenate all the bins onto the end of B[lowkey] }
(4)                 CONCATENATE(B[lowkey], B[v])
            end;  { binsort }
```

Fig. 8.20. The abstract binsort program.

Analysis of Binsort

We claim that if there are n elements to be sorted, and there are m different key values (hence m different bins), then the program of Fig. 8.20 takes $O(n+m)$ time, if the proper data structure for bins is used. In particular, if $m \leq n$, then binsort takes $O(n)$ time. The data structure we have in mind is a linked list. Pointers to list ends, as indicated in Fig. 8.19, are useful but not required.

The loop of lines $(1-2)$ of Fig. 8.20, which places the records in bins, takes $O(n)$ time, since the INSERT operation of line (2) requires constant time, the insertion always occurring at the beginning of the list. For the loop of lines $(3-4)$, concatenating the bins, temporarily assume that pointers to the ends of lists exist. Then step (4) takes constant time, so the loop takes $O(m)$ time. Hence the entire binsort program takes $O(n+m)$ time.

If pointers to the ends of lists do not exist, then in line (4) we must spend time running down to the end of $B[v]$ before concatenating it to $B[lowkey]$. In this manner, the end of $B[lowkey]$ will always be available for the next concatenation. The extra time spent running to the end of each bin once totals $O(n)$, since the sum of the lengths of the bins is n. This extra time does not affect the order of magnitude of running time for the algorithm, since $O(n)$ is no greater than $O(n+m)$.

Sorting Large Key Sets

If m, the number of keys, is no greater than n, the number of elements, then the $O(n+m)$ running time of Fig. 8.20 is really $O(n)$. But what if $m=n^2$, say. Evidently Fig. 8.20 will take $O(n+n^2)$, which is $O(n^2)$ time. But can we still take advantage of the fact that the key set is limited and do better? The surprising answer is that even if the set of possible key values is $1,2,\ldots,n^k$, for any fixed k, then there is a generalization of the binsorting technique that takes only $O(n)$ time.

Example 8.6. Consider the specific problem of sorting n integers in the range 0 to n^2-1. We sort the n integers in two phases. The first phase appears not to be of much help at all, but it is essential. We use n bins, one for each of the integers $0, 1, \ldots, n-1$. We place each integer i on the list to be sorted, into the bin numbered i **mod** n. However, unlike Fig. 8.20, it is important that we append each integer to the end of the list for the bin, not the beginning. If we are to append efficiently, we require the linked list representation of bins, with pointers to list ends.

For example, suppose $n=10$, and the list to be sorted is the perfect squares from 0^2 to 9^2 in the random order 36, 9, 0, 25, 1, 49, 64, 16, 81, 4. In this case, where $n=10$, the bin for integer i is just the rightmost digit of i written in decimal. Fig. 8.21(a) shows the placement of our list into bins. Notice that integers appear in bins in the same order that they appear in the original list; e.g., bin 6 has contents 36, 16, not 16, 36, since 36 precedes 16 on the original list.

Now, we concatenate the bins in order, producing the list

$$0, 1, 81, 64, 4, 25, 36, 16, 9, 49 \qquad\qquad (8.10)$$

from Fig. 8.21(a). If we use the linked list data structure with pointers to list ends, then the placement of the n integers into bins and the concatenation of the bins can each be done in $O(n)$ time.

The integers on the list created by concatenating the bins are redistributed into bins, but using a different bin selection strategy. Now place integer i into bin $\lfloor i/n \rfloor$, that is, the greatest integer equal to or less than i/n. Again, append integers to the ends of the lists for the bins. When we now concatenate the bins in order, we find the list is sorted.

For our example, Fig. 8.21(b) shows the list (8.10) distributed into bins, with i going into bin $\lfloor i/10 \rfloor$.

To see why this algorithm works, we have only to notice that when several integers are placed in one bin, as 0, 1, 4, and 9 were placed in bin 0, they must be in increasing order, since the list (8.10) resulting from the first pass ordered them by rightmost digit. Hence, in any bin the rightmost digits must form an increasing sequence. Of course, any integer placed in bin i must precede an integer placed in a bin higher than i, so concatenating the bins in order produces the sorted list.

More generally, we can think of the integers between 0 and n^2-1 as two-

Bin	Contents	Bin	Contents
0	0	0	0, 1, 4, 9
1	1, 81	1	16
2		2	25
3		3	36
4	64, 4	4	49
5	25	5	
6	36, 16	6	64
7		7	
8		8	81
9	9, 49	9	
	(a)		(b)

Fig. 8.21. Two-pass binsorting.

digit numbers in base n and use the same argument to see that the sorting strategy works. Consider the integers $i = an + b$ and $j = cn + d$, where a, b, c, and d are each in the range 0 to $n-1$; i.e., they are base-n digits. Suppose $i<j$. Then, $a>c$ is not possible, so we may assume $a \leq c$. If $a<c$, then i appears in a lower bin than j after the second pass, so i will precede j in the final order. If $a=c$, then b must be less than d. Then after the first pass, i precedes j, since i was placed in bin b and j in bin d. Thus, while both i and j are placed in bin a (the same as bin c), i is inserted first, and j must follow it in the bin. □

General Radix Sorting

Suppose keytype is a sequence of fields, as in

```
type
    keytype = record
        day: 1..31;
        month: (jan, . . . ,dec);                          (8.11)
        year: 1900..1999;
    end;
```

or an array of elements of the same type, as in

```
type
    keytype = array[1..10] of char;                        (8.12)
```

We shall assume from here that keytype consists of k components,

$f_1, f_2, \ldots, f_k$ of types $t_1, t_2, \ldots, t_k$. For example, in (8.11) $t_1 = 1..31$, $t_2 = (\text{jan}, \ldots, \text{dec})$, and $t_3 = 1900..1999$. In (8.12), $k = 10$, and $t_1 = t_2 = \cdots = t_k = \text{char}$.

Let us also assume that we desire to sort records in the *lexicographic* order of their keys. That is, key value $(a_1, a_2, \ldots, a_k)$ is less than key value $(b_1, b_2, \ldots, b_k)$, where a_i and b_i are the values in field f_i, for $i = 1, 2, \ldots, k$, if either

1. $a_1 < b_1$, or
2. $a_1 = b_1$ and $a_2 < b_2$, or

.

.

.

k. $a_1 = b_1, a_2 = b_2, \ldots, a_{k-1} = b_{k-1}$, and $a_k < b_k$.

That is, for some j between 0 and $k-1$, $a_1 = b_1, \ldots, a_j = b_j$,† and $a_{j+1} < b_{j+1}$.

We can view keys of the type defined above as if key values were integers expressed in some strange radix notation. For example, (8.12) where each field is a character, can be viewed as expressing integers in base 128, or however many characters there are in a character set for the machine at hand. Type definition (8.11) can be viewed as if the rightmost place were in base 100 (corresponding to the values between 1900 and 1999), the next place in base 12, and the third in base 31. Because of this view, generalized binsort has become known as *radix* sorting. In the extreme, we can even use it to sort integers up to any fixed limit by seeing them as arrays of digits in base 2, or another base.

The key idea behind radix sorting is to binsort all the records, first on f_k, the "least significant digit," then concatenate the bins, lowest value first, binsort on f_{k-1}, and so on. As in Example 8.6, when inserting into bins, make sure that each record is appended to the end of the list, not the beginning. The radix sorting algorithm is sketched in Fig. 8.22; the reason it works was illustrated in Example 8.6. In general, after binsorting on $f_k, f_{k-1}, \ldots, f_i$, the records will appear in lexicographic order if the key consisted only of fields $f_i, \ldots, f_k$.

Analysis of Radix Sort

First, we must use the proper data structures to make radix sort efficient. Notice that we assume the list to be sorted is already in the form of a linked list, rather than an array. In practice, we need only to add one additional field, the link field, to the type recordtype, and we can link $A[i]$ to $A[i+1]$ for $i = 1, 2, \ldots, n-1$ and thus make a linked list out of the array A in $O(n)$ time. Notice also that if we present the elements to be sorted in this way, we

---

† Note that a sequence ranging from 1 to 0 (or more generally, from x to y, where $y < x$) is deemed to be an empty sequence.

never copy a record. We just move records from one list to another.

```
        procedure radixsort;
            { sorts list A of n records with keys consisting of fields f₁, . . . , f_k
              of types t₁, . . . , t_k, respectively.  The procedure uses
              arrays B_i of type array[t_i] of listtype for 1 ≤ i ≤ k,
              where listtype is a linked list of records. }
            begin
(1)             for i := k downto 1 do begin
(2)                 for each value v of type t_i do  { clear bins }
(3)                     make B_i[v] empty;
(4)                 for each record r on list A do
(5)                     move r from A onto the end of bin B_i[v],
                        where v is the value of field f_i of the key of r
(6)                 for each value v of type t_i, from lowest to highest do
(7)                     concatenate B_i[v] onto the end of A
            end
        end;  { radixsort }
```

Fig. 8.22. Radix sorting.

As before, for concatenation to be done quickly, we need pointers to the end of each list. Then, the loop of lines (2−3) in Fig. 8.22 takes $O(s_i)$ time, where s_i is the number of different values of type t_i. The loop of lines (4−5) takes $O(n)$ time, and the loop of lines (6−7) takes $O(s_i)$ time. Thus, the total time taken by radix sort is $\sum_{i=1}^{k} O(s_i+n)$, which is $O(kn+\sum_{i=1}^{k} s_i)$, or $O(n+\sum_{i=1}^{k} s_i)$ if k is assumed a constant.

Example 8.7. If keys are integers in the range 0 to n^k-1, for some constant k, we can generalize Example 8.6 and view keys as base-n integers k digits long. Then t_i is $0..(n-1)$ for all i between 1 and k, so $s_i=n$. The expression $O(n+\sum_{i=1}^{k} s_i)$ becomes $O(n+kn)$ which, since k is a constant, is $O(n)$.

As another example, if keys are character strings of length k, for constant k, then $s_i=128$ (say) for all i, and $\sum_{i=1}^{k} s_i$ is a constant. Thus radix sort on fixed length character strings is also $O(n)$. In fact, whenever k is a constant and the s_i's are constant, or even $O(n)$, then radix sort takes $O(n)$ time. Only if k grows with n can radix sort not be $O(n)$. For example, if keys are regarded as binary strings of length $\log n$, then $k = \log n$, and $s_i=2$ for $1 \le i \le k$. Thus radix sort would be $O(kn + \sum_{i=1}^{k} s_i)$, which is $O(n \log n)$.† □

_____
† But in this case, if $\log n$-bit integers can fit in one word, we are better off treating keys as consisting of one field only, of type $1..n$, and using ordinary binsort.

8.6 A Lower Bound for Sorting by Comparisons

There is a "folk theorem" to the effect that sorting n elements "requires $n \log n$ time." We saw in the last section that this statement is not always true; if the key type is such that binsort or radix sort can be used to advantage, then $O(n)$ time suffices. However, these sorting algorithms rely on keys being of a special type — a type with a limited set of values. All the other general sorting algorithms we have studied rely only on the fact that we can test whether one key value is less than another.

The reader should notice that in all the sorting algorithms prior to Section 8.5, progress toward determining the proper order for the elements is made when we compare two keys, and then the flow for control in the algorithm goes one of only two ways. In contrast, an algorithm like that of Example 8.5 causes one of n different things to happen in only one step, by storing a record with an integer key in one of n bins depending on the value of that integer. All programs in Section 8.5 use a capability of programming languages and machines that is much more powerful than a simple comparison of values, namely the ability to find in one step a location in an array, given the index of that location. But this powerful type of operation is not available if keytype were, say, real. We cannot, in Pascal or most other languages, declare an array indexed by real numbers, and even if we could, we could not then concatenate in a reasonable amount of time all the bins corresponding to the computer-representable real numbers.

Decision Trees

Let us focus our attention on sorting algorithms whose only use of the elements to be sorted is in comparisons of two keys. We can draw a binary tree in which the nodes represent the "status" of the program after making some number of comparisons of keys. We can also see a node as representing those initial orderings of that data that will bring the program to this "status." Thus a program "status" is essentially knowledge about the initial ordering gathered so far by the program.

If any node represents two or more possible initial orderings, then the program cannot yet know the correct sorted order, and so must make another comparison of keys, such as "is $k_1 < k_2$?" We may then create two children for the node. The left child represents those initial orderings that are consistent with the outcome that $k_1 < k_2$; the right child represents the orderings that are consistent with the fact that $k_1 > k_2$.† Thus each child represents a "status" consisting of the information known at the parent plus either the fact $k_1 < k_2$ or the fact $k_1 > k_2$, depending on whether the child is a left or right child, respectively.

Example 8.8. Consider the insertion sort algorithm with $n = 3$. Suppose that

---

† We may as well assume all keys are different, since if we can sort a collection of distinct keys we shall surely produce a correct order when some keys are the same.

initially $A[1]$, $A[2]$, and $A[3]$ have key values a, b, and c, respectively. Any of the six orderings of a, b, and c could be the correct one, so we begin construction of our decision tree with the node labeled (1) in Fig. 8.23, which represents all possible orderings. The insertion sort algorithm first compares $A[2]$ with $A[1]$, that is, b with a. If b turns out to be the smaller, then the correct ordering can only be bac, bca, or cba, the three orderings in which b precedes a. These three orderings are represented by node (2) in Fig. 8.23. The other three orderings are represented by node (3), the right child of (1). They are the orderings for which a precedes b.

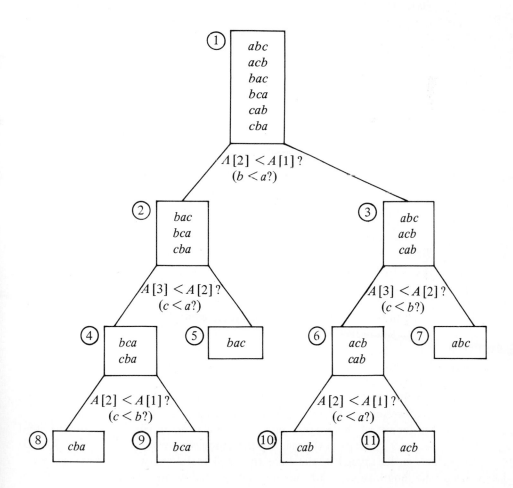

Fig. 8.23. Decision tree for insertion sort with $n=3$.

Now consider what happens if the initial data is such that we reach node

(2). We have just swapped $A[1]$ with $A[2]$, and find $A[2]$ can rise no higher since it is at the "top" already. The current ordering of the elements is *bac*. Insertion sort next begins inserting $A[3]$ into its rightful place, by comparing $A[3]$ with $A[2]$. Since $A[2]$ now holds a and $A[3]$ holds c, we are comparing c with a; node (4) represents the two orderings from node (2) in which c precedes a, while (5) represents the one ordering where it does not.

If the algorithm reaches the status of node (5), it will end, since it has already moved $A[3]$ as high as it can go. On the other hand in the status of node (4), we found $A[3]<A[2]$, and so swapped them, leaving b in $A[1]$ and c in $A[2]$. Insertion sort next compares these two elements and swaps them if $c<b$. Nodes (8) and (9) represents the orderings consistent with $c<b$ and its opposite, respectively, as well as the information gathered going from nodes (1) to (2) to (4), namely $b<a$ and $c<a$. Insertion sort ends, whether or not it has swapped $A[2]$ with $A[1]$, making no more comparisons. Fortunately, each of the leaves (5), (8), and (9) represent a single ordering, so we have gathered enough information to determine the correct sorted order.

The description of the tree descending from node (3) is quite symmetric with what we have seen, and we omit it. Figure 8.23, since all its leaves are associated with a single ordering, is seen to determine the correct sorted order of three key values a, b, and c. □

The Size of Decision Trees

Figure 8.23 has six leaves, corresponding to the six possible orderings of the initial list a, b, c. In general, if we sort a list of n elements, there are $n!=n(n-1)(n-2) \cdots (2)(1)$ possible *outcomes*, which are the correct sorted orders for the initial list $a_1, a_2, \ldots, a_n$. That is, any of the n elements could be first, any of the remaining $n-1$ second, any of the remaining $n-2$ third, and so on. Thus any decision tree describing a correct sorting algorithm working on a list of n elements must have at least $n!$ leaves, since each possible ordering must be alone at a leaf. In fact, if we delete nodes corresponding to unnecessary comparisons, and if we delete leaves that correspond to no ordering at all (since the leaves can only be reached by an inconsistent series of comparison results), there will be exactly $n!$ leaves.

Binary trees that have many leaves must have long paths. The length of a path from the root to a leaf gives the number of comparisons made when the ordering represented by that leaf is the sorted order for a certain input list L. Thus, the length of the longest path from the root to a leaf is a lower bound on the number of steps performed by the algorithm in the worst case. Specifically, if L is the input list, the algorithm will make at least as many comparisons as the length of the path, in addition, probably, to other steps that are not comparisons of keys.

We should therefore ask how short can all paths be in a binary tree with k leaves. A binary tree in which all paths are of length p or less can have 1 root, 2 nodes at level 1, 4 nodes at level 2, and in general, 2^i nodes at level i. Thus, the largest number of leaves of a tree with no nodes at levels higher than p is 2^p. Put another way, a binary tree with k leaves must have a path of length at least $\log k$. If we let $k = n!$, then we know that any sorting algorithm that only uses comparisons to determine the sorted order must take $\Omega(\log(n!))$ time in the worst case.

But how fast does $\log(n!)$ grow? A close approximation to $n!$ is $(n/e)^n$, where $e = 2.7183 \cdots$ is the base of natural logarithms. Since $\log((n/e)^n) = n\log n - n\log e$, we see that $\log(n!)$ is on the order of $n\log n$. We can get a precise lower bound by noting that $n(n-1) \cdots (2)(1)$ is the product of at least $n/2$ factors that are each at least $n/2$. Thus $n! \geq (n/2)^{n/2}$. Hence $\log(n!) \geq (n/2)\log(n/2) = (n/2)\log n - n/2$. Thus sorting by comparisons requires $\Omega(n\log n)$ time in the worst case.

The Average Case Analysis

One might wonder if there could be an algorithm that used only comparisons to sort, and took $\Omega(n\log n)$ time in the worst case, as all such algorithms must, but on the average took time that was $O(n)$ or something less than $O(n\log n)$. The answer is no, and we shall only mention how to prove the statement, leaving the details to the reader.

What we want to prove is that in any binary tree with k leaves, the average depth of a leaf is at least $\log k$. Suppose that were not the case, and let tree T be the counterexample with the fewest nodes. T cannot be a single node, because our statement says of one-leaf trees only that they have average depth of at least 0. Now suppose T has k leaves. A binary tree with $k \geq 2$ leaves looks either like the tree in Fig. 8.24(a) or that in (b).

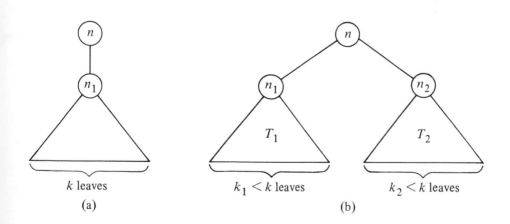

Fig. 8.24. Possible forms of binary tree T.

Figure 8.24(a) cannot be the smallest counterexample, because the tree rooted at n_1 has as many leaves as T, but even smaller average depth. If Fig. 8.24(b) were T, then the trees rooted at n_1 and n_2, being smaller then T, would not violate our claim. That is, the average depth of leaves in T_1 is at least $\log(k_1)$, and the average depth in T_2 is at least $\log(k_2)$. Then the average depth in T is

$$\left(\frac{k_1}{k_1+k_2}\right) \log(k_1) + \left(\frac{k_2}{k_1+k_2}\right) \log(k_2)+1.$$

Since $k_1+k_2=k$, we can express the average depth as

$$\frac{1}{k}(k_1\log(2k_1) + k_2\log(2k_2)) \tag{8.13}$$

The reader can check that when $k_1 = k_2 = k/2$, (8.13) has value $k\log k$. The reader must show that (8.13) has a minimum when $k_1 = k_2$, given the constraint that $k_1+k_2 = k$. We leave this proof as an exercise to the reader skilled in differential calculus. Granted that (8.13) has a minimum value of $k\log k$, we see that T was not a counterexample at all.

8.7 Order Statistics

The problem of computing *order statistics* is, given a list of n records and an integer k, to find the key of the record that is k^{th} in the sorted order of the records. In general, we refer to this problem as "finding the kth out of n." Special cases occur when $k = 1$ (finding the minimum), $k = n$ (finding the maximum), and the case where n is odd and $k = (n+1)/2$, called finding the *median*.

Certain cases of the problem are quite easy to solve in linear time. For example, finding the minimum of n elements in $O(n)$ time requires no special insight. As we mentioned in connection with heapsort, if $k \leq n/\log n$ then we can find the kth out of n by building a heap, which takes $O(n)$ time, and then selecting the k smallest elements in $O(n+k\log n) = O(n)$ time. Symmetrically, we can find the kth of n in $O(n)$ time when $k \geq n-n/\log n$.

A Quicksort Variation

Probably the fastest way, on the average, to find the kth out of n is to use a recursive procedure similar to quicksort, call it *select*(i, j, k), that finds the kth element among $A[i], \ldots, A[j]$ within some larger array $A[1], \ldots, A[n]$. The basic steps of *select* are:

1. Pick a pivot element, say v.
2. Use the procedure *partition* from Fig. 8.13 to divide $A[i], \ldots, A[j]$ into two groups: $A[i], \ldots, A[m-1]$ with keys less than v, and $A[m], \ldots, A[j]$ with keys v or greater.
3. If $k \leq m-i$, so the kth among $A[i], \ldots, A[j]$ is in the first group, then call *select*$(i, m-1, k)$. If $k > m-i$, then call *select*$(m, j, k-m+i)$.

Eventually we find that $select(i, j, k)$ is called, where all of $A[i], \ldots, A[j]$ have the same key (usually because $j=i$). Then we know the desired key; it is the key of any of these records.

As with quicksort, the function $select$ outlined above can take $\Omega(n^2)$ time in the worst case. For example, suppose we are looking for the first element, but by bad luck, the pivot is always the highest available key. However, on the average, $select$ is even faster then quicksort; it takes $O(n)$ time. The principle is that while quicksort calls itself twice, $select$ calls itself only once. We could analyze $select$ as we did quicksort, but the mathematics is again complex, and a simple intuitive argument should be convincing. On the average, $select$ calls itself on a subarray half as long as the subarray it was given. Suppose that to be conservative, we say that each call is on an array 9/10 as long as the previous call. Then if $T(n)$ is the time spent by $select$ on an array of length n, we know that for some constant c we have

$$T(n) \le T(\frac{9}{10}n) + cn \qquad (8.14)$$

Using techniques from the next chapter, we can show that the solution to (8.14) is $T(n) = O(n)$.

A Worst-Case Linear Method for Finding Order Statistics

To guarantee that a function like $select$ has worst case, rather than average case, complexity $O(n)$, it suffices to show that in linear time we can find some pivot that is guaranteed to be some positive fraction of the way from either end. For example, the solution to (8.14) shows that if the pivot of n elements is never less than the $\frac{n}{10}^{th}$ element, nor greater than the $\frac{9n}{10}^{th}$ element, so the recursive call to $select$ is on at most nine-tenths of the array, then this variant of $select$ will be $O(n)$ in the worst case.

The trick to finding a good pivot is contained in the following two steps.

1. Divide the n elements into groups of 5, leaving aside between 0 and 4 elements that cannot be placed in a group. Sort each group of 5 by any algorithm and take the middle element from each group, a total of $\lfloor n/5 \rfloor$ elements.

2. Use $select$ to find the median of these $\lfloor n/5 \rfloor$ elements, or if $\lfloor n/5 \rfloor$ is even, an element in a position as close to the middle as possible. Whether $\lfloor n/5 \rfloor$ is even or odd, the desired element is in position $\lfloor \frac{n+5}{10} \rfloor$.

This pivot is far from the extremes, provided that not too many records have the pivot for key.† To simplify matters, let us temporarily assume that all keys are different. Then we claim the chosen pivot, the $\lfloor \frac{n+5}{10} \rfloor^{th}$ element of the

---

† In the extreme case, when all keys are equal, the pivot provides no separation at all. Obviously the pivot is the kth element for any k, and another approach is needed.

$\lfloor \frac{n}{5} \rfloor$ middle elements out of groups of 5 is greater than at least $3\lfloor \frac{n-5}{10} \rfloor$ of the n elements. For it exceeds $\lfloor \frac{n-5}{10} \rfloor$ of the middle elements, and each of those exceeds two more, from the five of which it is the middle. If $n \geq 75$, then $3\lfloor \frac{n-5}{10} \rfloor$ is at least $n/4$. Similarly, we may check that the chosen pivot is less than or equal to at least $3\lfloor \frac{n-5}{10} \rfloor$ elements, so for $n \geq 75$, the pivot lies between the 1/4 and 3/4 point in the sorted order. Most importantly, when we use the pivot to partition the n elements, the kth element will be isolated to within a range of at most $3n/4$ of the elements. A sketch of the complete algorithm is given in Fig. 8.25; as for the sorting algorithms it assumes an array $A[1], \ldots , A[n]$ of recordtype, and that recordtype has a field *key* of type keytype. The algorithm to find the kth element is just a call to $select(1,n,k)$, of course.

```
        function select ( i, j, k: integer ) : keytype;
            { returns the key of the kth element in sorted order
                among A[i], . . . ,A[j] }
        var
            m: integer;  { used as index }
        begin
(1)         if j−i < 74 then begin  { too few to use select recursively }
(2)             sort A[i], . . . ,A[j] by some simple algorithm;
(3)             return (A[i+k−1].key)
            end
            else begin  { apply select recursively }
(4)             for m := 0 to (j−i−4) div 5 do
                    { get the middle elements of groups of 5
                        into A[i], A[i+1], . . . }
(5)                 find the third element among A[i+5*m] through
                        A[i+5*m+4] and swap it with A[i+m];
(6)             pivot := select(i, i+(j−i−4) div 5, (j−i−4) div 10);
                    { find median of middle elements. Note j−i−4
                        here is n−5 in the informal description above }
(7)             m := partition(i, j, pivot);
(8)             if k <= m−i then
(9)                 return (select(i, m−1, k))
                else
(10)                return (select(m, j, k−(m−i)))
            end
        end;  { select }
```

Fig. 8.25. Worst-case linear algorithm for finding k^{th} element.

To analyze the running time of *select* of Fig. 8.25, let n be $j−i+1$. Lines

(2) and (3) are only executed if n is 74 or less. Thus, even though step (2) may take $O(n^2)$ steps in general, there is some constant c_1, however large, such that for $n \leq 74$, lines (1−3) take no more than c_1 time.

Now consider lines (4−10). Line (7), the partition step, was shown in connection with quicksort to take $O(n)$ time. The loop of lines (4−5) is iterated about $n/5$ times, and each execution of line (5), requiring the sorting of 5 elements, takes some constant time, so the loop takes $O(n)$ time as a whole.

Let $T(n)$ be the time taken by a call to *select* on n elements. Then line (6) takes at most $T(n/5)$ time. Since $n \geq 75$ when line (10) is reached, and we have argued that if $n \geq 75$, at most $3n/4$ elements are less than the pivot, and at most $3n/4$ are equal to or greater than the pivot, we know that line (9) or (10) takes time at most $T(3n/4)$. Hence, for some constants c_1 and c_2 we have

$$T(n) \leq \begin{cases} c_1 \text{ if } n \leq 74 \\ \\ c_2 n + T(n/5) + T(3n/4) \text{ if } n \geq 75 \end{cases} \tag{8.15}$$

The term $c_2 n$ in (8.15) represents lines (1), (4), (5), and (7); the term $T(n/5)$ comes from line (6), and $T(3n/4)$ represents lines (9) and (10).

We shall show that $T(n)$ in (8.15) is $O(n)$. Before we proceed, the reader should now appreciate that the "magic number" 5, the size of the groups in line (5), and our selection of $n = 75$ at the breakpoint below which we did not use *select* recursively, were designed so the arguments $n/5$ and $3n/4$ of T in (8.15) would sum to something less than n. Other choices of these parameters could be made, but observe when we solve (8.15) how the fact that $1/5 + 3/4 < 1$ is needed to prove linearity.

Equation (8.15) can be solved by guessing a solution and verifying by induction that it holds for all n. We shall chose a solution of the form cn, for some constant c. If we pick $c \geq c_1$, we know $T(n) \leq cn$ for all n between 1 and 74, so consider the case $n \geq 75$. Assume by induction that $T(m) \leq cm$ for $m < n$. Then by (8.15),

$$T(n) \leq c_2 n + cn/5 + 3cn/4 \leq c_2 n + 19cn/20 \tag{8.16}$$

If we pick $c = \max(c_1, 20c_2)$, then by (8.16), we have $T(n) \leq cn/20 + cn/5 + 3cn/4 = cn$, which we needed to show. Thus, $T(n)$ is $O(n)$.

The Case Where Some Equalities Among Keys Exist

Recall that we assumed no two keys were equal in Fig. 8.25. The reason this assumption was needed is that otherwise line (7) cannot be shown to partition A into blocks of size at most $3n/4$. The modification we need to handle key equalities is to add, after step (7), another step like partition, that groups together all records with key equal to the pivot. Say there are $p \geq 1$ such keys. If $m-i \leq k \leq m-i+p$, then no recursion is necessary; simply return $A[m].key$. Otherwise, line (8) is unchanged, but line (10) calls $select(m+p, j, k-(m-i)-p)$.

Exercises

8.1 Here are eight integers: 1, 7, 3, 2, 0, 5, 0, 8. Sort them using (a) bubblesort, (b) insertion sort, and (c) selection sort.

8.2 Here are sixteen integers: 22, 36, 6, 79, 26, 45, 75, 13, 31, 62, 27, 76, 33, 16, 62, 47. Sort them using (a) quicksort, (b) insertion sort, (c) heapsort, and (d) bin sort, treating them as pairs of digits in the range 0–9.

8.3 The procedure *Shellsort* of Fig. 8.26, sometimes called *diminishing-increment sort*, sorts an array $A[1..n]$ of integers by sorting $n/2$ pairs $(A[i], A[n/2+i])$ for $1 \leq i \leq n/2$ in the first pass, $n/4$ four-tuples $(A[i], A[n/4+i], A[n/2+i], A[3n/4+i])$ for $1 \leq i \leq n/4$ in the second pass, $n/8$ eight-tuples in the third pass, and so on. In each pass the sorting is done using insertion sort in which we stop sorting once we encounter two elements in the proper order.

```
procedure Shellsort ( var A: array[1..n] of integer );
var
    i, j, incr: integer;
begin
    incr := n div 2;
    while incr > 0 do begin
        for i := incr + 1 to n do begin
            j := i - incr;
            while j > 0 do
                if A[j] > A[j+incr] then begin
                    swap(A[j], A[j+incr]);
                    j := j - incr
                end
                else
                    j := 0 { break }
        end;
        incr := incr div 2
    end
end; { Shellsort }
```

Fig. 8.26. *Shellsort*.

a) Sort the sequences of integers in Exercises 8.1 and 8.2 using *Shellsort*.

*b) Show that if $A[i]$ and $A[n/2^k+i]$ became sorted in pass k (i.e., they were swapped), then these two elements remain sorted in pass $k+1$.

c) The distances between elements compared and swapped in a pass diminish as $n/2$, $n/4$, . . . ,2, 1 in Fig. 8.26. Show that *Shellsort* will work with any sequence of distances as long as the last distance is 1.

**d) Show that *Shellsort* works in $O(n^{1.5})$ time.

*8.4 Suppose you are to sort a list L consisting of a sorted list followed by a few "random" elements. Which of the sorting methods discussed in this chapter would be especially suitable for such a task?

*8.5 A sorting algorithm is *stable* if it preserves the original order of records with equal keys. Which of the sorting methods in this chapter are stable?

*8.6 Suppose we use a variant of quicksort where we always choose as the pivot the first element in the subarray being sorted.

a) What modifications to the quicksort algorithm of Fig. 8.11 do we have to make to avoid infinite loops when there is a sequence of equal elements?

b) Show that the modified quicksort has average-case running time $O(n\log n)$.

8.7 Show that any sorting algorithm that moves elements only one position at a time must have time complexity at least $\Omega(n^2)$.

8.8 In heapsort the procedure *pushdown* of Fig. 8.17 establishes the partially ordered tree property in time $O(n)$. Instead of starting at the leaves and pushing elements down to form a heap, we could start at the root and push elements up. What is the time complexity of this method?

*8.9 Suppose we have a set of words, i.e., strings of the letters $a-z$, whose total length is n. Show how to sort these words in $O(n)$ time. Note that if the maximum length of a word is constant, binsort will work. However, you must consider the case where some of the words are very long.

*8.10 Show that the average-case running time of insertion sort is $\Omega(n^2)$.

**8.11 Consider the following algorithm *randomsort* to sort an array $A[1..n]$ of integers: If the elements $A[1]$, $A[2]$, . . . ,$A[n]$ are in sorted order, stop; otherwise, choose a random number i between 1 and n, swap $A[1]$ and $A[i]$, and repeat. What is the expected running time of *randomsort*?

*8.12 We showed that sorting by comparisons takes $\Omega(n\log n)$ comparisons in the worse case. Prove that this lower bound holds in the average case as well.

*8.13 Prove that the procedure *select*, described informally at the beginning of Section 8.7, has average case running time of $O(n)$.

8.14 Implement CONCATENATE for the data structure of Fig. 8.19.

8.15 Write a program to find the k smallest elements in an array of length n. What is the time complexity of your program? For what value of k does it become advantageous to sort the array?

8.16 Write a program to find the largest and smallest elements in an array. Can this be done in fewer than $2n-3$ comparisons?

8.17 Write a program to find the *mode* (the most frequently occurring element) of a list of elements. What is the time complexity of your program?

*8.18 Show that any algorithm to purge duplicates from a list requires at least $\Omega(n\log n)$ time under the decision tree model of computation of Section 8.6.

*8.19 Suppose we have k sets, S_1, S_2, ..., S_k, each containing n real numbers. Write a program to list all sums of the form $s_1 + s_2 + \cdots + s_k$, where s_i is in S_i, in sorted order. What is the time complexity of your program?

8.20 Suppose we have a sorted array of strings s_1, s_2, ..., s_n. Write a program to determine whether a given string x is a member of this sequence. What is the time complexity of your program as a function of n and the length of x?

Bibliographic Notes

Knuth [1973] is a comprehensive reference on sorting methods. Quicksort is due to Hoare [1962] and subsequent improvements to it were published by Singleton [1969] and Frazer and McKellar [1970]. Heapsort was discovered by Williams [1964] and improved by Floyd [1964]. The decision tree complexity of sorting was studied by Ford and Johnson [1959]. The linear selection algorithm in Section 8.7 is from Blum, Floyd, Pratt, Rivest, and Tarjan [1972].

Shellsort is due to Shell [1959] and its performance has been analyzed by Pratt [1979]. See Aho, Hopcroft, and Ullman [1974] for one solution to Exercise 8.9.

CHAPTER 9

Algorithm
Analysis
Techniques

What is a good algorithm? There is no easy answer to this question. Many of the criteria for a good algorithm involve subjective issues such as simplicity, clarity, and appropriateness for the expected data. A more objective, but not necessarily more important, issue is run-time efficiency. Section 1.5 covered the basic techniques for establishing the running time of simple programs. However, in more complex cases such as where programs are recursive, some new techniques are needed. This short chapter presents some general techniques for solving recurrence equations that arise in the analysis of the running times of recursive algorithms.

9.1 Efficiency of Algorithms

One way to determine the run-time efficiency of an algorithm is to program it and measure the execution time of the particular implementation on a specific computer for a selected set of inputs. Although popular and useful, this approach has some inherent problems. The running times depend not only on the underlying algorithm, but also on the instruction set of the computer, the quality of the compiler, and the skill of the programmer. The implementation may also be tuned to work well on the particular set of test inputs chosen. These dependencies may become strikingly evident with a different computer, a different compiler, a different programmer, or a different set of test inputs. To overcome these objections, computer scientists have adopted asymptotic time complexity as a fundamental measure of the performance of an algorithm. The term *efficiency* will refer to this measure, and particularly to the worst-case (as opposed to average) time complexity.

The reader should recall from Chapter 1 the definitions of $O(f(n))$ and $\Omega(f(n))$. The efficiency, i.e., worst-case complexity, of an algorithm is said to be $O(f(n))$, or just $f(n)$ for short, if the function of n that gives the maximum, over all inputs of length n, of the number of steps taken by the algorithm on that input, is $O(f(n))$. Put another way, there is some constant c such that for sufficiently large n, $cf(n)$ is an upper bound on the number of steps taken by the algorithm on any input of length n.

There is the implication in the assertion that "the efficiency of a given algorithm is $f(n)$" that the efficiency is also $\Omega(f(n))$, so that $f(n)$ is the slowest growing function of n that bounds the worst-case running time from above. However, this latter requirement is not part of the definition of $O(f(n))$, and sometimes it is not possible to be sure that we have the slowest growing upper bound.

Our definition of efficiency ignores constant factors in running time, and there are several pragmatic reasons for doing so. First, since most algorithms are written in a high level language, we must describe them in terms of "steps," which each take a constant amount of time when translated into the machine language of any computer. However, exactly how much time a step requires depends not only on the step itself, but on the translation process and the instruction repertoire of the machine. Thus to attempt to be more precise than to say that the running time of an algorithm is "on the order of $f(n)$", i.e., $O(f(n))$, would bog us down in the details of specific machines and would apply only to those machines.

A second important reason why we deal with asymptotic complexity and ignore constant factors is that the asymptotic complexity, far more than constant factors, determines for what size inputs the algorithm may be used to provide solutions on a computer. Chapter 1 discussed this viewpoint in detail. The reader should be alert, however, to the possibility that for some very important problems, like sorting, we may find it worthwhile to analyze algorithms in such detail that statements like "algorithm A should run twice as fast as algorithm B on a typical computer" are possible.

A second situation in which it may pay to deviate from our worst-case notion of efficiency occurs when we know the expected distribution of inputs to an algorithm. In such situations, average case analysis can be much more meaningful than worst case analysis. For example, in the previous chapter we analyzed the average running time of quicksort under the assumption that all permutations of the correct sorted order are equally likely to occur as inputs.

9.2 Analysis of Recursive Programs

In Chapter 1 we showed how to analyze the running time of a program that does not call itself recursively. The analysis for a recursive program is rather different, normally involving the solution of a difference equation. The techniques for solving difference equations are sometimes subtle, and bear considerable resemblance to the methods for solving differential equations, some of whose terminology we borrow.

Consider the sorting program sketched in Fig. 9.1. There the procedure *mergesort* takes a list of length n as input, and returns a sorted list as its output. The procedure $merge(L_1, L_2)$ takes as input two sorted lists L_1 and L_2, scans them each, element by element, from the front. At each step, the larger of the two front elements is deleted from its list and emitted as output. The result is a single sorted list containing the elements of L_1 and L_2. The details of *merge* are not important here, although we discuss this sorting algorithm in detail in Chapter 11. What is important is that the time taken by *merge* on lists of length $n/2$ is $O(n)$.

```
function mergesort ( L: LIST; n: integer ) : LIST;
    { L is a list of length n. A sorted version of L
      is returned. We assume n is a power of 2. }
    begin
        if n = 1 then
            return (L);
        else begin
            break L into two halves, L₁ and L₂, each of length n/2;
            return (merge(mergesort(L₁,n/2), mergesort(L₂, n/2)));
        end
    end; { mergesort }
```

Fig. 9.1. Recursive procedure *mergesort*.

Let $T(n)$ be the worst case running time of the procedure *mergesort* of Fig. 9.1. We can write a *recurrence* (or *difference*) equation that upper bounds $T(n)$, as follows

$$T(n) \leq \begin{cases} c_1 & \text{if } n = 1 \\ 2T(n/2) + c_2 n & \text{if } n > 1 \end{cases} \tag{9.1}$$

The term c_1 in (9.1) represents the constant number of steps taken when L has length 1. In the case that $n > 1$, the time taken by *mergesort* can be divided into two parts. The recursive calls to *mergesort* on lists of length $n/2$ each take time $T(n/2)$, hence the term $2T(n/2)$. The second part consists of the test to discover that $n \neq 1$, the breaking of list L into two equal parts and the procedure *merge*. These three operations take time that is either a constant, in the case of the test, or proportional to n for the split and the merge. Thus the constant c_2 can be chosen so the term $c_2 n$ is an upper bound on the time taken by *mergesort* to do everything except the recursive calls. We now have equation (9.1).

Observe that (9.1) applies only when n is even, and hence it will provide an upper bound in *closed form* (that is, as a formula for $T(n)$ not involving any $T(m)$ for $m < n$) only if n is a power of 2. However, even if we only know $T(n)$ when n is a power of 2, we have a good idea of $T(n)$ for all n. In particular, for essentially all algorithms, we may suppose that $T(n)$ lies between $T(2^i)$ and $T(2^{i+1})$ if n lies between 2^i and 2^{i+1}. Moreover, if we devote a little more effort to finding the solution, we could replace the term $2T(n/2)$ in (9.1) by $T((n+1)/2) + T((n-1)/2)$ for odd $n > 1$. Then we could solve the revised difference equation to get a closed form solution for all n.

9.3 Solving Recurrence Equations

There are three different approaches we might take to solving a recurrence equation.

1. Guess a solution $f(n)$ and use the recurrence to show that $T(n) \leq f(n)$. Sometimes we guess only the form of $f(n)$, leaving some parameters unspecified (e.g., guess $f(n) = an^2$ for some a) and deduce suitable values for the parameters as we try to prove $T(n) \leq f(n)$ for all n.

2. Use the recurrence itself to substitute for any $T(m)$, $m<n$, on the right until all terms $T(m)$ for $m>1$ have been replaced by formulas involving only $T(1)$. Since $T(1)$ is always a constant, we have a formula for $T(n)$ in terms of n and constants. This formula is what we have referred to as a "closed form" for $T(n)$.

3. Use the general solution to certain recurrence equations of common types found in this section or elsewhere (see the bibliographic notes).

This section examines the first two methods.

Guessing a Solution

Example 9.1. Consider method (1) applied to Equation (9.1). Suppose we guess that for some a, $T(n) = an\log n$. Substituting $n = 1$, we see that this guess will not work, because $an\log n$ has value 0, independent of the value of a. Thus, we might next try $T(n) = an\log n + b$. Now $n = 1$ requires that $b \geq c_1$.

For the induction, we assume that

$$T(k) \leq ak\log k + b \tag{9.2}$$

for all $k < n$ and try to establish that

$$T(n) \leq an\log n + b$$

To begin our proof, assume $n \geq 2$. From (9.1), we have

$$T(n) \leq 2T(n/2)+c_2n$$

From (9.2), with $k = n/2$, we obtain

$$T(n) \leq 2[a\frac{n}{2}\log\frac{n}{2} + b] + c_2n \tag{9.3}$$

$$\leq an\log n - an + c_2n + 2b$$

$$\leq an\log n + b$$

provided $a \geq c_2 + b$.

We thus see that $T(n) \leq an\log n + b$ holds provided two constraints are satisfied: $b \geq c_1$ and $a \geq c_2 + b$. Fortunately, there are values we can choose for a that satisfy these two constraints. For example, choose $b = c_1$

and $a = c_1 + c_2$. Then, by induction on n, we conclude that for all $n \geq 1$

$$T(n) \leq (c_1 + c_2)n\log n + c_1 \qquad (9.4)$$

In other words, $T(n)$ is $O(n\log n)$. $\square$

Two observations about Example 9.1 are in order. If we assume that $T(n)$ is $O(f(n))$, and our attempt to prove $T(n) \leq cf(n)$ by induction fails, it does not follow that $T(n)$ is not $O(f(n))$. In fact, an inductive hypothesis of the form $T(n) \leq cf(n) - 1$ may succeed!

Secondly we have not yet determined the exact asymptotic growth rate for $f(n)$, although we have shown that it is no worse than $O(n\log n)$. If we guess a slower growing solution, like $f(n) = an$, or $f(n) = an\log\log n$, we cannot prove the claim that $T(n) \leq f(n)$. However, the matter can only be settled conclusively by examining *mergesort* and showing that it really does take $\Omega(n\log n)$ time; in fact, it takes time proportional to $n\log n$ on all inputs, not just on the worst possible inputs. We leave this observation as an exercise.

Example 9.1 exposes a general technique for proving some function to be an upper bound on the running time of an algorithm. Suppose we are given the recurrence equation

$$T(1) = c$$

$$T(n) \leq g(T(n/2), n), \text{ for } n > 1 \qquad (9.5)$$

Note that (9.5) generalizes (9.1), where $g(x, y)$ is $2x + c_2 y$. Also observe that we could imagine equations more general than (9.5). For example, the formula g might involve all of $T(n-1)$, $T(n-2)$, ..., $T(1)$, not just $T(n/2)$. Also, we might be given values for $T(1)$, $T(2)$, ..., $T(k)$, and the recurrence would then apply only for $n > k$. The reader can, as an exercise, consider how to solve these more general recurrences by method (1) — guessing a solution and verifying it.

Let us turn our attention to (9.5) rather than its generalizations. Suppose we guess a function $f(a_1, \ldots, a_j, n)$, where $a_1, \ldots, a_j$ are parameters, and attempt to prove by induction on n that $T(n) \leq f(a_1, \ldots, a_j, n)$. For example, our guess in Example 9.1 was $f(a_1, a_2, n) = a_1 n\log n + a_2$, but we used a and b for a_1 and a_2. To check that for some values of $a_1, \ldots, a_j$ we have $T(n) \leq f(a_1, \ldots, a_j, n)$ for all $n \geq 1$, we must satisfy

$$f(a_1, \ldots, a_j, 1) \geq c$$

$$f(a_1, \ldots, a_j, n) \geq g(f(a_1, \ldots, a_j, n/2), n) \qquad (9.6)$$

That is, by the inductive hypothesis, we may substitute f for T on the right side of the recurrence (9.5) to get

$$T(n) \leq g(f(a_1, \ldots, a_j, n/2), n) \qquad (9.7)$$

When the second line of (9.6) holds, we can combine it with (9.7) to prove that $T(n) \leq f(a_1, \ldots, a_n, n)$, which is what we wanted to prove by induction on n.

For example, in Example 9.1, we have $g(x, y) = 2x + c_2y$ and $f(a_1, a_2, n) = a_1 n \log n + a_2$. Here we must try to satisfy

$$f(a_1, a_2, 1) = a_2 \geq c_1$$

$$f(a_1, a_2, n) = a_1 n \log n + a_2 \geq 2(a_1(\frac{n}{2})\log(\frac{n}{2}) + a_2) + c_2 n$$

As we discussed, $a_2 = c_1$ and $a_1 = c_1 + c_2$ is a satisfactory choice.

Expanding Recurrences

If we cannot guess a solution, or we are not sure we have the best bound on $T(n)$, we can use a method that in principle always succeeds in solving for $T(n)$ exactly, although in practice we often have trouble summing series and must resort to computing an upper bound on the sum. The general idea is to take a recurrence like (9.1), which says $T(n) \leq 2T(n/2) + c_2 n$, and use it to get a bound on $T(n/2)$ by substituting $n/2$ for n. That is

$$T(n/2) \leq 2T(n/4) + c_2 n/2 \qquad (9.8)$$

Substituting the right side of (9.8) for $T(n/2)$ in (9.1) yields

$$T(n) \leq 2(2T(n/4) + c_2 n/2) + c_2 n = 4T(n/4) + 2c_2 n \qquad (9.9)$$

Similarly, we could substitute $n/4$ for n in (9.1) and use it to get a bound on $T(n/4)$. That bound, which is $2T(n/8) + c_2 n/4$, can be substituted in the right of (9.9) to yield

$$T(n) \leq 8T(n/8) + 3c_2 n \qquad (9.10)$$

Now the reader should perceive a pattern. By induction on i we can obtain the relationship

$$T(n) \leq 2^i T(n/2^i) + i c_2 n \qquad (9.11)$$

for any i. Assuming n is a power of 2, say 2^k, this expansion process comes to a halt as soon as we get $T(1)$ on the right of (9.11). That will occur when $i = k$, whereupon (9.11) becomes

$$T(n) \leq 2^k T(1) + k c_2 n \qquad (9.12)$$

Then, since $2^k = n$, we know $k = \log n$. As $T(1) \leq c_1$, (9.12) becomes

$$T(n) \leq c_1 n + c_2 n \log n \qquad (9.13)$$

Equation (9.13) is actually as tight a bound as we can put on $T(n)$, and proves that $T(n)$ is $O(n \log n)$.

9.4 A General Solution for a Large Class of Recurrences

Consider the recurrence that one gets by dividing a problem of size n into a subproblems each of size n/b. For convenience, we assume that a problem of size 1 takes one time unit and that the time to piece together the solutions to the subproblems to make a solution for the size n problem is $d(n)$, in the same time units. For our *mergesort* example, we have $a = b = 2$, and $d(n) = c_2 n / c_1$, in units of c_1. Then if $T(n)$ is the time to solve a problem of size n, we have

$$T(1) = 1 \qquad\qquad (9.14)$$

$$T(n) = aT(n/b) + d(n)$$

Note that (9.14) only applies to n's that are an integer power of b, but if we assume $T(n)$ is smooth, getting a tight upper bound on $T(n)$ for those values of n tells us how $T(n)$ grows in general.

Also note that we use equality in (9.14) while in (9.1) we had an inequality. The reason is that here $d(n)$ can be arbitrary, and therefore exact, while in (9.1) the assumption that $c_2 n$ was the worst case time to merge, for one constant c_2 and all n, was only an upper bound; the actual worst case running time on inputs of size n may have been less than $2T(n/2) + c_2 n$. Actually, whether we use $=$ or $\le$ in the recurrence makes little difference, since we wind up with an upper bound on the worst case running time anyway.

To solve (9.14) we use the technique of repeatedly substituting for T on the right side, as we did for a specific example in our previous discussion of expanding recurrences. That is, substituting n/b^i for n in the second line of (9.14) yields

$$T(\frac{n}{b^i}) = aT(\frac{n}{b^{i+1}}) + d(\frac{n}{b^i}) \qquad\qquad (9.15)$$

Thus, starting with (9.14) and substituting (9.15) for $i = 1, 2, \ldots$, we get

$$T(n) = aT(\frac{n}{b}) + d(n)$$

$$= a[aT(\frac{n}{b^2})+d(\frac{n}{b})]+d(n) = a^2 T(\frac{n}{b^2})+ad(\frac{n}{b})+d(n)$$

$$= a^2[aT(\frac{n}{b^3})+d(\frac{n}{b^2})]+ad(\frac{n}{b})+d(n) = a^3 T(\frac{n}{b^3})+a^2 d(\frac{n}{b^2})+ad(\frac{n}{b})+d(n)$$

$$= \cdots$$

$$= a^i T(\frac{n}{b^i}) + \sum_{j=0}^{i-1} a^j d(\frac{n}{b^j})$$

Now, if we assume $n = b^k$, we can use the fact that $T(n/b^k) = T(1) = 1$ to get from the above, with $i = k$, the formula

$$T(n) = a^k + \sum_{j=0}^{k-1} a^j d(b^{k-j}) \qquad\qquad (9.16)$$

If we use the fact that $k = \log_b n$, the first term of (9.16) can be written as $a^{\log_b n}$, or equivalently $n^{\log_b a}$ (take logarithms to the base b of both expressions to see that they are the same). This expression is n to a constant power. For example, in the case of *mergesort*, where $a = b = 2$, the first term is n. In general, the larger a is, i.e., the more subproblems we need to solve, the higher the exponent will be; the higher b is, i.e., the smaller each subproblem is, the lower will be the exponent.

Homogeneous and Particular Solutions

It is interesting to see the different roles played by the two terms in (9.16). The first, a^k or $n^{\log_b a}$, is called the *homogeneous solution*, in analogy with differential equation terminology. The homogeneous solution is the exact solution when $d(n)$, called the *driving function*, is 0 for all n. In other words, the homogeneous solution represents the cost of solving all the subproblems, even if subproblems can be combined "for free."

On the other hand, the second term of (9.16) represents the cost of creating the subproblems and combining their results. We call this term the *particular solution*. The particular solution is affected by both the driving function and the number and size of subproblems. As a rule of thumb if the homogeneous solution is greater than the driving function, then the particular solution has the same growth rate as the homogeneous solution. If the driving function grows faster than the homogeneous solution by more than n^{ϵ} for some $\epsilon > 0$, then the particular solution has the same growth rate as the driving function. If the driving function has the same growth rate as the homogeneous solution, or grows faster by at most $\log^k n$ for some k, then the particular solution grows as $\log n$ times the driving function.

It is important to recognize that when searching for improvements in an algorithm, we must be alert to whether the homogeneous solution is larger than the driving function. For example, if the homogeneous solution is larger, then finding a faster way to combine subproblems will have essentially no effect on the efficiency of the overall algorithm. What we must do in that case is find a way to divide a problem into fewer or smaller subproblems. That will affect the homogeneous solution and lower the overall running time.

If the driving function exceeds the homogeneous solution, then one must try to decrease the driving function. For example, in the *mergesort* case, where $a = b = 2$, and $d(n) = cn$, we shall see that the particular solution is $O(n \log n)$. However, reducing $d(n)$ to a slightly sublinear function, say $n^{0.9}$, will, as we shall see, make the particular solution less than linear as well and reduce the overall running time to $O(n)$, which is the homogeneous solution.†

---

† But don't hold out much hope for discovering a way to merge two sorted lists of $n/2$ elements in less than linear time; we couldn't even look at all the elements on the list in that case.

Multiplicative Functions

The particular solution in (9.16) is hard to evaluate, even if we know what $d(n)$ is. However, for certain common functions $d(n)$, we can solve (9.16) exactly, and there are others for which we can get a good upper bound. We say a function f on integers is *multiplicative* if $f(xy) = f(x)f(y)$ for all positive integers x and y.

Example 9.2. The multiplicative functions that are of greatest interest to us are of the form n^α for any positive α. To prove $f(n) = n^\alpha$ is multiplicative, we have only to observe that $(xy)^\alpha = x^\alpha y^\alpha$. $\square$

Now if $d(n)$ in (9.16) is multiplicative, then $d(b^{k-j}) = (d(b))^{k-j}$, and the particular solution of (9.16) is

$$\sum_{j=0}^{k-1} a^j (d(b))^{k-j} = d(b)^k \sum_{j=0}^{k-1} \left(\frac{a}{d(b)} \right)^j$$

$$= d(b)^k \frac{\left(\dfrac{a}{d(b)} \right)^k - 1}{\dfrac{a}{d(b)} - 1}$$

$$= \frac{a^k - d(b)^k}{\dfrac{a}{d(b)} - 1} \tag{9.17}$$

There are three cases to consider, depending on whether a is greater than, less than, or equal to $d(b)$.

1. If $a > d(b)$, then the formula (9.17) is $O(a^k)$, which we recall is $n^{\log_b a}$, since $k = \log_b n$. In this case, the particular and homogeneous solutions are the same, and depend only on a and b, and not on the driving function d. Thus, improvements in the running time must come from decreasing a or increasing b; decreasing $d(n)$ is of no help.

2. If $a < d(b)$, then (9.17) is $O(d(b)^k)$, or equivalently $O(n^{\log_b d(b)})$. In this case, the particular solution exceeds the homogeneous, and we may also look to the driving function $d(n)$ as well as to a and b, for improvements. Note the important special case where $d(n) = n^\alpha$. Then $d(b) = b^\alpha$, and $\log_b(b^\alpha) = \alpha$. Thus the particular solution is $O(n^\alpha)$, or $O(d(n))$.

3. If $a = d(b)$, we must reconsider the calculation involved in (9.17), as our formula for the sum of a geometric series is now inappropriate. In this case we have

$$\sum_{j=0}^{k-1} a^j (d(b))^{k-j} = d(b)^k \sum_{j=0}^{k-1} \left(\frac{a}{d(b)} \right)^j$$

$$= d(b)^k \sum_{j=0}^{k-1} 1$$

$$= d(b)^k k$$

$$= n^{\log_b d(b)} \log_b n \tag{9.18}$$

Since $a = d(b)$, the particular solution given by (9.18) is $\log_b n$ times the homogeneous solution, and again the particular solution exceeds the homogeneous. In the special case $d(n) = n^\alpha$, (9.18) reduces to $O(n^\alpha \log n)$, by observations similar to those in Case (2).

Example 9.3. Consider the following recurrences, with $T(1) = 1$.

1. $T(n) = 4T(n/2) + n$
2. $T(n) = 4T(n/2) + n^2$
3. $T(n) = 4T(n/2) + n^3$

In each case, $a = 4$, $b = 2$, and the homogeneous solution is n^2. In Equation (1), with $d(n) = n$, we have $d(b) = 2$. Since $a = 4 > d(b)$, the particular solution is also n^2, and $T(n)$ is $O(n^2)$ in (1).

In Equation (3), $d(n) = n^3$, $d(b) = 8$, and $a < d(b)$. Thus the particular solution is $O(n^{\log_b d(b)}) = O(n^3)$, and $T(n)$ of Equation (3) is $O(n^3)$. We can deduce that the particular solution is of the same order as $d(n) = n^3$, using the observations made above about $d(n)$'s of the form n^α in analyzing the case $a < d(b)$ of (9.17). In Equation (2) we have $d(b) = 4 = a$, so (9.18) applies. As $d(n)$ is of the form n^α, the particular solution, and therefore $T(n)$ itself, is $O(n^2 \log n)$. □

Other Driving Functions

These are other functions that are not multiplicative, yet for which we can get solutions for (9.16) or even (9.17). We shall consider two examples. The first generalizes to any function that is the product of a multiplicative function and a constant greater than or equal to one. The second is typical of a case where we must examine (9.16) in detail and get a close upper bound on the particular solution.

Example 9.4. Consider

$$T(1) = 1$$

$$T(n) = 3T(n/2) + 2n^{1.5}$$

Now $2n^{1.5}$ is not multiplicative, but $n^{1.5}$ is. Let $U(n) = T(n)/2$ for all n. Then

$$U(1) = 1/2$$

$$U(n) = 3U(n/2) + n^{1.5}$$

The homogeneous solution, if $U(1)$ were 1, would be $n^{\log 3} = n^{1.59}$; since

$U(1) = 1/2$ we can easily show the homogeneous solution is $n^{1.59}/2$; certainly it is $O(n^{1.59})$. For the particular solution, we can ignore the fact that $U(1) \neq 1$, since increasing $U(1)$ will surely not lower the particular solution. Then since $a = 3$, $b = 2$, and $b^{1.5} = 2.82 < a$, the particular solution is also $O(n^{1.59})$, and that is the growth rate of $U(n)$. Since $T(n) = 2U(n)$, $T(n)$ is also $O(n^{1.59})$ or $O(n^{\log 3})$. $\square$

Example 9.5. Consider

$$T(1) = 1$$

$$T(n) = 2T(n/2) + n\log n$$

The homogeneous solution is easily seen to be n, since $a = b = 2$. However, $d(n) = n\log n$ is not multiplicative, and we must sum the particular solution formula in (9.16) by ad hoc means. That is, we want to evaluate

$$\sum_{j=0}^{k-1} 2^j 2^{k-j} \log(2^{k-j}) = 2^k \sum_{j=0}^{k-1} (k-j)$$

$$= 2^{k-1} k(k+1)$$

Since $k = \log n$ we have the particular solution $O(n\log^2 n)$, and this solution, being greater than the homogeneous, is also the value of $T(n)$ that we obtain. $\square$

Exercises

9.1 Write recurrence equations for the time and space complexity of the following algorithm, assuming n is a power of 2.

```
function path ( s, t, n: integer ) : boolean;
    begin
        if n = 1 then
            if edge(s, t) then
                return (true)
            else
                return (false);
        { if we reach here, n > 1 }
        for i := 1 to n do
            if path(s, i, n div 2) and path(i, t, n div 2) then
                return (true);
        return (false)
    end; { path }
```

The function $edge(i, j)$ returns true if vertices i and j of an n-vertex graph are connected by an edge or if $i = j$; $edge(i, j)$ returns false otherwise. What does the program do?

9.2 Solve the following recurrences, where $T(1) = 1$ and $T(n)$ for $n \geq 2$ satisfies:

a) $T(n) = 3T(n/2) + n$

b) $T(n) = 3T(n/2) + n^2$

c) $T(n) = 8T(n/2) + n^3$

9.3 Solve the following recurrences, where $T(1) = 1$ and $T(n)$ for $n \geq 2$ satisfies:

a) $T(n) = 4T(n/3) + n$

b) $T(n) = 4T(n/3) + n^2$

c) $T(n) = 9T(n/3) + n^2$

9.4 Give tight big-oh and big-omega bounds on $T(n)$ defined by the following recurrences. Assume $T(1) = 1$.

a) $T(n) = T(n/2) + 1$

b) $T(n) = 2T(n/2) + \log n$

c) $T(n) = 2T(n/2) + n$

d) $T(n) = 2T(n/2) + n^2$

*9.5 Solve the following recurrences by guessing a solution and checking your answer.

a) $T(1) = 2$

$T(n) = 2T(n-1) + 1$ for $n \geq 2$

b) $T(1) = 1$

$T(n) = 2T(n-1) + n$ for $n \geq 2$

9.6 Check your answers to Exercise 9.5 by solving the recurrences by repeated substitution.

9.7 Generalize Exercise 9.6 by solving all recurrences of the form

$T(1) = 1$

$T(n) = aT(n-1) + d(n)$ for $n \geq 1$

in terms of a and $d(n)$.

*9.8 Suppose in Exercise 9.7 that $d(n) = c^n$ for some constant $c \geq 1$. How does the solution to $T(n)$ depend on the relationship between a and c. What is $T(n)$?

**9.9 Solve for $T(n)$:

$T(1) = 1$

$T(n) = \sqrt{n}\,T(\sqrt{n}) + n$ for $n \geq 2$

9.10 Find closed form expressions for the following sums.

a) $\sum_{i=0}^{n} i$

b) $\sum_{i=0}^{n} i^k$

c) $\sum_{i=0}^{n} 2^i$

d) $\sum_{i=0}^{n} \binom{n}{i}$

*9.11 Show that the number of different orders in which to multiply a sequence of n matrices is given by the recurrence

$$T(1) = 1$$
$$T(n) = \sum_{i=1}^{n-1} T(i)T(n-i)$$

Show that $T(n+1) = \dfrac{1}{n+1}\binom{2n}{n}$. The $T(n)$'s are called *Catalan* numbers.

**9.12 Show that the number of comparisons required to sort n elements using mergesort is given by

$$T(1) = 0$$
$$T(n) = T(\lfloor n/2 \rfloor) + T(\lceil n/2 \rceil) + n - 1$$

where $\lfloor x \rfloor$ denotes the integer part of x and $\lceil x \rceil$ denotes the smallest integer $\geq x$. Show that the solution to this recurrence is

$$T(n) = n\lceil \log n \rceil - 2^{\lceil \log n \rceil} + 1$$

9.13 Show that the number of Boolean functions of n variables is given by the recurrence

$$T(1) = 4$$
$$T(n) = (T(n-1))^2$$

Solve for $T(n)$.

**9.14 Show that the number of binary trees of height $\leq n$ is given by the recurrence

$$T(1) = 1$$
$$T(n) = (T(n-1))^2 + 1$$

Show that $T(n) = \lfloor k^{2^n} \rfloor$ for some constant k. What is the value of k?

Bibliographic Notes

Bentley, Haken, and Saxe [1978], Greene and Knuth [1983], Liu [1968], and Lueker [1980] contain additional material on the solution of recurrences. Aho and Sloane [1973] show that many nonlinear recurrences of the form $T(n) = (T(n-1))^2 + g(n)$ have a solution of the form $T(n) = \lfloor k^{2^n} \rfloor$ where k is a constant, as in Exercise 9.14.

CHAPTER 10

Algorithm
Design
Techniques

Over the years computer scientists have identified a number of general techniques that often yield effective algorithms in solving large classes of problems. This chapter presents some of the more important techniques, such as divide-and-conquer, dynamic programming, greedy techniques, backtracking, and local search. In trying to devise an algorithm to solve a given problem, it is often useful to ask a question such as "What kind of solution does divide-and-conquer, dynamic programming, a greedy approach, or some other standard technique yield?"

It should be emphasized, however, that there are problems, such as the NP-complete problems, for which these or any other known techniques will not produce efficient solutions. When such a problem is encountered, it is often useful to determine if the inputs to the problem have special characteristics that could be exploited in trying to devise a solution, or if an easily found approximate solution could be used in place of the difficult-to-compute exact solution.

10.1 Divide-and-Conquer Algorithms

Perhaps the most important, and most widely applicable, technique for designing efficient algorithms is a strategy called "divide-and-conquer." It consists of breaking a problem of size n into smaller problems in such a way that from solutions to the smaller problems we can easily construct a solution to the entire problem. We have already seen a number of applications of this technique, such as *mergesort* or binary search trees.

To illustrate the method consider the familiar "towers of Hanoi" puzzle. It consists of three pegs A, B, and C. Initially peg A has on it some number of disks, starting with the largest one on the bottom and successively smaller ones on top, as shown in Fig. 10.1. The object of the puzzle is to move the disks one at a time from peg to peg, never placing a larger disk on top of a smaller one, eventually ending with all disks on peg B.

One soon learns that the puzzle can be solved by the following simple algorithm. Imagine the pegs arranged in a triangle. On odd-numbered moves, move the smallest disk one peg clockwise. On even-numbered moves

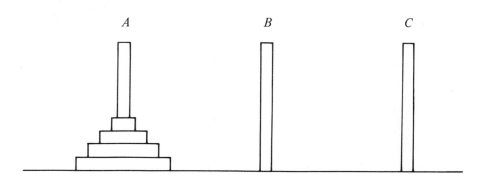

Fig. 10.1. Initial position in towers of Hanoi puzzle.

make the only legal move not involving the smallest disk.

The above algorithm is concise, and correct, but it is hard to understand why it works, and hard to invent on the spur of the moment. Consider instead the following divide-and-conquer approach. The problem of moving the n smallest disks from A to B can be thought of as consisting of two sub-problems of size $n-1$. First move the $n-1$ smallest disks from peg A to peg C, exposing the nth smallest disk on peg A. Move that disk from A to B. Then move the $n-1$ smallest disks from C to B. Moving the $n-1$ smallest disks is accomplished by a recursive application of the method. As the n disks involved in the moves are smaller than any other disks, we need not concern ourselves with what is below them on pegs A, B, or C. Although the actual movement of individual disks is not obvious, and hand simulation is hard because of the stacking of recursive calls, the algorithm is conceptually simple to understand, to prove correct and, we would like to think, to invent in the first place. It is probably the ease of discovery of divide-and-conquer algo-rithms that makes the technique so important, although in many cases the algorithms are also more efficient than more conventional ones.†

The Problem of Multiplying Long Integers

Consider the problem of multiplying two n-bit integers X and Y. Recall that the algorithm for multiplication of n-bit (or n-digit) integers usually taught in elementary school involves computing n partial products of size n and thus is an $O(n^2)$ algorithm, if we count single bit or digit multiplications and addi-tions as one step. One divide-and-conquer approach to integer multiplication would break each of X and Y into two integers of $n/2$ bits each as shown in

---

† In the towers of Hanoi case, the divide-and-conquer algorithm is really the same as the one given initially.

Fig. 10.2. (For simplicity we assume n is a power of 2 here.)

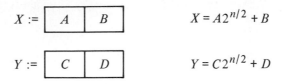

$$X = A2^{n/2} + B$$

$$Y = C2^{n/2} + D$$

Fig. 10.2. Breaking n-bit integers into $\dfrac{n}{2}$-bit pieces.

The product of X and Y can now be written

$$XY = AC2^n + (AD+BC)\,2^{n/2} + BD \qquad (10.1)$$

If we evaluate XY in this straightforward way, we have to perform four multiplications of $(n/2)$-bit integers (AC, AD, BC, and BD), three additions of integers with at most $2n$ bits (corresponding to the three $+$ signs in (10.1)), and two shifts (multiplication by 2^n and $2^{n/2}$). As these additions and shifts take $O(n)$ steps, we can write the following recurrence for $T(n)$, the total number of bit operations needed to multiply n-bit integers according to (10.1).

$$T(1) = 1$$

$$T(n) = 4T(n/2) + cn \qquad (10.2)$$

Using reasoning like that in Example 9.4, we can take the constant c in (10.2) to be 1, so the driving function $d(n)$ is just n, and then deduce that the homogeneous and particular solutions are both $O(n^2)$.

In the case that formula (10.1) is used to multiply integers, the asymptotic efficiency is thus no greater than for the elementary school method. But recall that for equations like (10.2) we get an asymptotic improvement if we decrease the number of subproblems. It may be a surprise that we can do so, but consider the following formula for multiplying X by Y.

$$XY = AC2^n + [(A-B)(D-C) + AC + BD]2^{n/2} + BD \qquad (10.3)$$

Although (10.3) looks more complicated than (10.1) it requires only three multiplications of $(n/2)$-bit integers, AC, BD, and $(A-B)(D-C)$, six additions or subtractions, and two shifts. Since all but the multiplications take $O(n)$ steps, the time $T(n)$ to multiply n-bit integers by (10.3) is given by

$$T(1) = 1$$

$$T(n) = 3T(n/2) + cn$$

whose solution is $T(n) = O(n^{\log_2 3}) = O(n^{1.59})$.

The complete algorithm, including the details implied by the fact that (10.3) requires multiplication of negative, as well as positive, $(n/2)$-bit integers, is given in Fig. 10.3. Note that lines $(8)-(11)$ are performed by copying bits, and the multiplication by 2^n and $2^{n/2}$ in line (16) by shifting. Also, the multiplication by s in line (16) simply introduces the proper sign into the result.

```
       function mult ( X, Y, n: integer ): integer;
           { X and Y are signed integers ≤ 2ⁿ.
             n is a power of 2. The function returns XY }
           var
               s: integer;   { holds the sign of XY }
               m1, m2, m3: integer;  { hold the three products }
               A, B, C, D: integer;  { hold left and right halves of X and Y }
           begin
(1)            s := sign(X) * sign(Y);
(2)            X := abs(X);
(3)            Y := abs(Y); { make X and Y positive }
(4)            if n = 1 then
(5)                if (X = 1) and (Y = 1) then
(6)                    return (s)
                   else
(7)                    return (0)
               else begin
(8)                A := left n/2 bits of X;
(9)                B := right n/2 bits of X;
(10)               C := left n/2 bits of Y;
(11)               D := right n/2 bits of Y;
(13)               m1 := mult(A, C, n/2);
(14)               m2 := mult(A−B, D−C, n/2);
(15)               m3 := mult(B, D, n/2);
(16)               return (s * (m1*2ⁿ + (m1 + m2 + m3) * 2^(n/2) + m3))
               end
       end; { mult }
```

Fig. 10.3. Divide-and-conquer integer multiplication algorithm.

Observe that the divide-and-conquer algorithm of Fig. 10.3 is asymptotically faster than the method taught in elementary school, taking $O(n^{1.59})$ steps against $O(n^2)$. We may thus raise the question: if this algorithm is so superior why don't we teach it in elementary school? There are two answers. First, while easy to implement on a computer, the description of the algorithm is sufficiently complex that if we attempted to teach it in elementary school students would not learn to multiply. Furthermore, we have ignored constants of proportionality. While procedure *mult* of Fig. 10.3 is asymptotically superior to the usual method, the constants are such that for small problems (actually up to about 500 bits) the elementary school method is superior, and we rarely ask elementary school children to multiply such numbers.

Constructing Tennis Tournaments

The technique of divide-and-conquer has widespread applicability, not only in algorithm design but in designing circuits, constructing mathematical proofs and in other walks of life. We give one example as an illustration. Consider the design of a round robin tennis tournament schedule, for $n = 2^k$ players. Each player must play every other player, and each player must play one match per day for $n-1$ days, the minimum number of days needed to complete the tournament.

The tournament schedule is thus an n row by $n-1$ column table whose entry in row i and column j is the player i must contend with on the jth day.

The divide-and-conquer approach constructs a schedule for one-half of the players. This schedule is designed by a recursive application of the algorithm by finding a schedule for one half of these players and so on. When we get down to two players, we have the base case and we simply pair them up.

Suppose there are eight players. The schedule for players 1 through 4 fills the upper left corner (4 rows by 3 columns) of the schedule being constructed. The lower left corner (4 rows by 3 columns) of the schedule must pit the high numbered players (5 through 8) against one another. This subschedule is obtained by adding 4 to each entry in the upper left.

We have now simplified the problem. All that remains is to have lower-numbered players play high-numbered players. This is easily accomplished by having players 1 through 4 play 5 through 8 respectively on day 4 and cyclically permuting 5 through 8 on subsequent days. The process is illustrated in Fig. 10.4. The reader should now be able to generalize the ideas of this figure to provide a schedule for 2^k players for any k.

Balancing Subproblems

In designing algorithms one is always faced with various trade-offs. One rule that has emerged is that it is generally advantageous to balance competing costs wherever possible. For example in Chapter 5 we saw that the 2-3 tree balanced the costs of searching with those of inserting, while more straightforward methods take $O(n)$ steps either for each lookup or for each insertion, even though the other operation can be done in a constant number of steps.

Similarly, for divide-and-conquer algorithms, we are generally better off if the subproblems are of approximately equal size. For example, insertion sort can be viewed as partitioning a problem into two subproblems, one of size 1 and one of size $n-1$, with a maximum cost of n steps to merge. This gives a recurrence

$$T(n) = T(1) + T(n-1) + n$$

which has an $O(n^2)$ solution. Mergesort, on the other hand, partitions the problems into two subproblems each of size $n/2$ and has $O(n\log n)$ performance. As a general principle, we often find that partitioning a problem into

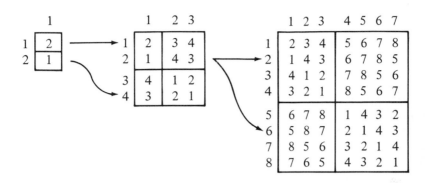

Fig. 10.4. A round-robin tournament for eight players.

equal or nearly equal subproblems is a crucial factor in obtaining good performance.

10.2 Dynamic Programming

Often there is no way to divide a problem into a small number of subproblems whose solution can be combined to solve the original problem. In such cases we may attempt to divide the problem into as many subproblems as necessary, divide each subproblem into smaller subproblems and so on. If this is all we do, we shall likely wind up with an exponential-time algorithm.

Frequently, however, there are only a polynomial number of subproblems, and thus we must be solving some subproblem many times. If instead we keep track of the solution to each subproblem solved, and simply look up the answer when needed, we would obtain a polynomial-time algorithm.

It is sometimes simpler from an implementation point of view to create a table of the solutions to all the subproblems we might ever have to solve. We fill in the table without regard to whether or not a particular subproblem is actually needed in the overall solution. The filling-in of a table of subproblems to get a solution to a given problem has been termed *dynamic programming*, a name that comes from control theory.

The form of a dynamic programming algorithm may vary, but there is the common theme of a table to fill and an order in which the entries are to be filled. We shall illustrate the techniques by two examples, calculating odds on a match like the World Series, and the "triangulation problem."

World Series Odds

Suppose two teams, A and B, are playing a match to see who is the first to win n games for some particular n. The World Series is such a match, with $n = 4$. We may suppose that A and B are equally competent, so each has a 50% chance of winning any particular game. Let $P(i, j)$ be the probability that if A needs i games to win, and B needs j games, that A will eventually win the match. For example, in the World Series, if the Dodgers have won two games and the Yankees one, then $i = 2$, $j = 3$, and $P(2, 3)$, we shall discover, is 11/16.

To compute $P(i, j)$, we can use a recurrence equation in two variables. First, if $i = 0$ and $j > 0$, then team A has won the match already, so $P(0, j) = 1$. Similarly, $P(i, 0) = 0$ for $i > 0$. If i and j are both greater than 0, at least one more game must be played, and the two teams each win half the time. Thus, $P(i, j)$ must be the average of $P(i-1, j)$ and $P(i, j-1)$, the first of these being the probability A will win the match if it wins the next game and the second being the probability A wins the match even though it loses the next game. To summarize:

$$P(i, j) = 1 \quad \text{if } i = 0 \text{ and } j > 0$$

$$= 0 \quad \text{if } i > 0 \text{ and } j = 0$$

$$= (P(i-1, j) + P(i, j-1))/2 \quad \text{if } i > 0 \text{ and } j > 0 \qquad (10.4)$$

If we use (10.4) recursively as a function, we can show that $P(i, j)$ takes no more than time $O(2^{i+j})$. Let $T(n)$ be the maximum time taken by a call to $P(i, j)$, where $i+j = n$. Then from (10.4),

$$T(1) = c$$

$$T(n) = 2T(n-1)+d$$

for some constants c and d. The reader may check by the means discussed in the previous chapter that $T(n) \le 2^{n-1}c + (2^{n-1}-1)d$, which is $O(2^n)$ or $O(2^{i+j})$.

We have thus proven an exponential upper bound on the time taken by the recursive computation of $P(i, j)$. However, to convince ourselves that the recursive formula for $P(i, j)$ is a bad way to compute it, we need to get a big-omega lower bound. We leave it as an exercise to show that when we call $P(i, j)$, the total number of calls to P that gets made is $\binom{i+j}{i}$, the number of ways to choose i things out of $i+j$. If $i = j$, that number is $\Omega(2^n/\sqrt{n})$, where $n = i+j$. Thus, $T(n)$ is $\Omega(2^n/\sqrt{n})$, and in fact, we can show it is $O(2^n/\sqrt{n})$ also. While $2^n/\sqrt{n}$ grows asymptotically more slowly than 2^n, the difference is not great, and $T(n)$ grows far too fast for the recursive calculation of $P(i, j)$ to be practical.

The problem with the recursive calculation is that we wind up computing the same $P(i, j)$ repeatedly. For example, if we want to compute $P(2, 3)$, we compute, by (10.4), $P(1, 3)$ and $P(2, 2)$. $P(1, 3)$ and $P(2, 2)$ both require the computation of $P(1, 2)$, so we compute that value twice.

   A better way to compute $P(i, j)$ is to fill in the table suggested by Fig. 10.5. The bottom row is all 0's and the rightmost column all 1's by the first two lines of (10.4). By the last line of (10.4), each other entry is the average of the entry below it and the entry to the right. Thus, an appropriate way to fill in the table is to proceed in diagonals beginning at the lower right corner, and proceeding up and to the left along diagonals representing entries with a constant value of $i+j$, as suggested in Fig. 10.6. This program is given in Fig. 10.7, assuming it works on a two-dimensional array P of suitable size.

1/2	21/32	13/16	15/16	1	4	
11/32	1/2	11/16	7/8	1	3	↑
3/16	5/16	1/2	3/4	1	2	j
1/16	1/8	1/4	1/2	1	1	
0	0	0	0		0	
4	3	2	1	0		

$\leftarrow i$

Fig. 10.5. Table of odds.

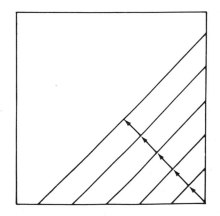

Fig. 10.6. Pattern of computation.

   The analysis of function *odds* is easy. The loop of lines (4)–(5) takes $O(s)$ time, and that dominates the $O(1)$ time for lines (2)–(3). Thus, the outer loop takes time $O(\sum_{s=1}^{n} s)$ or $O(n^2)$, where $i+j = n$. Thus dynamic pro-

```
      function odds ( i, j: integer ) : real;
         var
            s, k: integer;
         begin
(1)         for s := 1 to i + j do begin
               { compute diagonal of entries whose indices sum to s }
(2)            P[0, s] := 1.0;
(3)            P[s, 0] := 0.0;
(4)            for k := 1 to s−1 do
(5)               P[k, s−k] := (P[k−1, s−k] + P[k, s−k−1])/2.0
            end;
(6)         return (P[i, j])
         end; { odds }
```

Fig. 10.7. Odds calculation.

gramming takes $O(n^2)$ time, compared with $O(2^n/\sqrt{n})$ for the straightforward approach. Since $2^n/\sqrt{n}$ grows wildly faster than n^2, we would prefer dynamic programming to the recursive approach under essentially any circumstances.

The Triangulation Problem

As another example of dynamic programming, consider the problem of *tri-angulating* a polygon. We are given the vertices of a polygon and a distance measure between each pair of vertices. The distance may be the ordinary (Euclidean) distance in the plane, or it may be an arbitrary cost function given by a table. The problem is to select a set of *chords* (lines between nonadjacent vertices) such that no two chords cross each other, and the entire polygon is divided into triangles. The total length (distance between endpoints) of the chords selected must be a minimum. We call such a set of chords a *minimal triangulation*.

Example 10.1. Figure 10.8 shows a seven-sided polygon and the (x, y) coordinates of its vertices. The distance function is the ordinary Euclidean distance. A triangulation, which happens not to be minimal, is shown by dashed lines. Its cost is the sum of the lengths of the chords (v_0, v_2), (v_0, v_3), (v_0, v_5), and (v_3, v_5), or $\sqrt{8^2+16^2} + \sqrt{15^2+16^2} + \sqrt{22^2+2^2} + \sqrt{7^2+14^2} = 77.56$. □

As well as being interesting in its own right, the triangulation problem has a number of useful applications. For example, Fuchs, Kedem, and Uselton [1977] used a generalization of the triangulation problem for the following purpose. Consider the problem of shading a two-dimensional picture of an object whose surface is defined by a collection of points in 3-space. The light source comes from a given direction, and the brightness of a point on the surface depends on the angles between the direction of light, the direction of the viewer's eye, and a perpendicular to the surface at that point. To estimate the direction of the surface at a point, we can compute a minimum triangulation

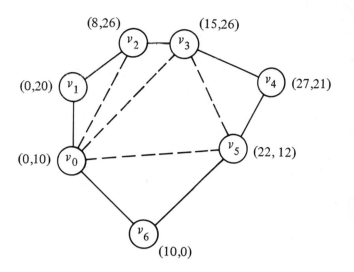

Fig. 10.8. A heptagon and a triangulation.

of the points defining the surface.

Each triangle defines a plane in a 3-space, and since a minimum triangulation was found, the triangles are expected to be very small. It is easy to find the direction of a perpendicular to a plane, so we can compute the light intensity for the points of each triangle, on the assumption that the surface can be treated as a triangular plane in a given region. If the triangles are not sufficiently small to make the light intensity look smooth, then local averaging can improve the picture.

Before proceeding with the dynamic programming solution to the triangulation problem, let us state two observations about triangulations that will help us design the algorithm. Throughout we assume we have a polygon with n vertices $v_0, v_1, \ldots, v_{n-1}$, in clockwise order.

Fact 1. In any triangulation of a polygon with more than three vertices, every pair of adjacent vertices is touched by at least one chord. To see this, suppose neither v_i nor v_{i+1}† were touched by a chord. Then the region that edge (v_i, v_{i+1}) bounds would have to include edges (v_{i-1}, v_i), (v_{i+1}, v_{i+2}) and at least one additional edge. This region then would not be a triangle.

Fact 2. If (v_i, v_j) is a chord in a triangulation, then there must be some v_k

---

† In what follows, we take all subscripts to be computed modulo n. Thus, in Fig. 10.8, v_i and v_{i+1} could be v_6 and v_0, respectively, since $n = 7$.

such that (v_i, v_k) and (v_k, v_j) are each either edges of the polygon or chords. Otherwise, (v_i, v_j) would bound a region that was not a triangle.

To begin searching for a minimum triangulation, we pick two adjacent vertices, say v_0 and v_1. By the two facts we know that in any triangulation, and therefore in the minimum triangulation, there must be a vertex v_k such that (v_1, v_k) and (v_k, v_0) are chords or edges in the triangulation. We must therefore consider how good a triangulation we can find after selecting each possible value for k. If the polygon has n vertices, there are a total of $(n-2)$ choices to make.

Each choice of k leads to at most two *subproblems*, which we define to be polygons formed by one chord and the edges in the original polygon from one end of the chord to the other. For example, Fig. 10.9 shows the two subproblems that result if we select the vertex v_3.

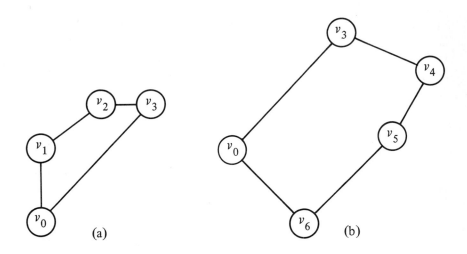

Fig. 10.9. The two subproblems after selecting v_3.

Next, we must find minimum triangulations for the polygons of Fig. 10.9(a) and (b). Our first instinct is that we must again consider all chords emanating from two adjacent vertices. For example, in solving Fig. 10.9(b), we might consider choosing chord (v_3, v_5), which leaves subproblem (v_0, v_3, v_5, v_6), a polygon two of whose sides, (v_0, v_3) and (v_3, v_5), are chords of the original polygon. This approach leads to an exponential-time algorithm.

However, by considering the triangle that involves the chord (v_0, v_k) we never have to consider polygons more than one of whose sides are chords of the original polygon. Fact 2 tells us that, in the minimal triangulation, the chord in the subproblem, such as (v_0, v_3) in Fig. 10.9(b), must make a triangle with one of the other vertices. For example, if we select v_4, we get the triangle (v_0, v_3, v_4) and the subproblem (v_0, v_4, v_5, v_6) which has only one chord of the original polygon. If we try v_5, we get the subproblems (v_3, v_4, v_5) and (v_0, v_5, v_6), with chords (v_3, v_5) and (v_0, v_5) only.

In general, define *the subproblem of size s beginning at vertex v_i*, denoted S_{is}, to be the minimal triangulation problem for the polygon formed by the s vertices beginning at v_i and proceeding clockwise, that is, $v_i, v_{i+1}, \ldots, v_{i+s-1}$. The chord in S_{is} is (v_i, v_{i+s-1}). For example, Fig. 10.9(a) is S_{04} and Fig. 10.9(b) is S_{35}. To solve S_{is} we must consider the following three options.

1. We may pick vertex v_{i+s-2} to make a triangle with the chords (v_i, v_{i+s-1}) and (v_i, v_{i+s-2}) and third side (v_{i+s-2}, v_{i+s-1}), and then solve the subproblem $S_{i,s-1}$.

2. We may pick vertex v_{i+1} to make a triangle with the chords (v_i, v_{i+s-1}) and (v_{i+1}, v_{i+s-1}) and third side (v_i, v_{i+1}), and then solve the subproblem $S_{i+1,s-1}$.

3. For some k between 2 and $s-3$ we may pick vertex v_{i+k} and form a triangle with sides (v_i, v_{i+k}), (v_{i+k}, v_{i+s-1}), and (v_i, v_{i+s-1}) and then solve subproblems $S_{i,k+1}$ and $S_{i+k,s-k}$.

If we remember that "solving" any subproblem of size three or less requires no action, we can summarize (1)–(3) by saying that we pick some k between 1 and $s-2$ and solve subproblems $S_{i,k+1}$ and $S_{i+k,s-k}$. Figure 10.10 illustrates this division into subproblems.

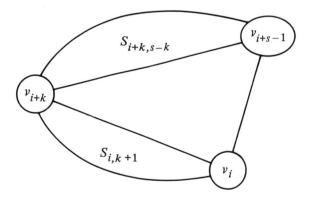

Fig. 10.10. Division of S_{is} into subproblems.

If we use the obvious recursive algorithm implied by the above rules to solve subproblems of size four or more, then it is possible to show that each call on a subproblem of size s gives rise to a total of 3^{s-4} recursive calls, if we "solve" subproblems of size three or less directly and count only calls on subproblems size four or more. Thus the number of subproblems to be solved is exponential in s. Since our initial problem is of size n, where n is the number of vertices in the given polygon, the total number of steps performed by this recursive procedure is exponential in n.

Yet something is clearly wrong in this analysis, because we know that besides the original problem, there are only $n(n-4)$ different subproblems that ever need to be solved. They are represented by S_{is}, where $0 \le i < n$ and $4 \le s < n$. Evidently not all the subproblems solved by the recursive procedure are different. For example, if in Fig. 10.8 we choose chord (v_0, v_3), and then in the subproblem of Fig. 10.9(b) we pick v_4, we have to solve subproblem S_{44}. But we would also have to solve this problem if we first picked chord (v_0, v_4), or if we picked chord (v_1, v_4) and then, when solving subproblem S_{45}, picked vertex v_0 to complete a triangle with v_1 and v_4.

This suggests an efficient way to solve the triangulation problem. We make a table giving the cost C_{is} of triangulating S_{is} for all i and s. Since the solution to any given problem depends only on the solution to problems of smaller size, the logical order in which to fill in the table is in size order. That is, for sizes $s = 4, 5, \ldots, n-1$ we fill in the minimum cost for problems S_{is}, for all vertices i. It is convenient to include problems of size $0 \le s < 4$ as well, but remember that S_{is} has cost 0 if $s < 4$.

By rules (1)–(3) above for finding subproblems, the formula for computing C_{is} for $s \ge 4$ is:

$$C_{is} = \min_{1 \le k \le s-2} \left[C_{i,k+1} + C_{i+k,s-k} + D(v_i, v_{i+k}) + D(v_{i+k}, v_{i+s-1}) \right] \quad (10.5)$$

where $D(v_p, v_q)$ is the length of the chord between vertices v_p and v_q, if v_p and v_q are not adjacent points on the polygon; $D(v_p, v_q)$ is 0 if v^p and v_q are adjacent.

Example 10.2. Figure 10.11 holds the table of costs for $S_{i,s}$ for $0 \le i \le 6$ and $4 \le s \le 6$, based on the polygon and distances of Fig. 10.8. The costs for the rows with $s < 3$ are all zero. We have filled in the entry C_{07}, in column 0 and the row for $s = 7$. This entry, like all in that row, represents the triangulation of the entire polygon. To see that, just notice that we can, if we wish, consider the edge (v_0, v_6) to be a chord of a larger polygon and the polygon of Fig. 10.8 to be a subproblem of this polygon, which has a series of additional vertices extending clockwise from v_6 to v_0. Note that the entire row for $s = 7$ has the same value as C_{07}, to within the accuracy of the computation.

Let us, as an example, show how the entry 38.09 in the column for $i = 6$ and row for $s = 5$ is filled in. According to (10.5) the value of this entry, C_{65}, is the minimum of three sums, corresponding to $k = 1, 2$, or 3. These sums are:

$$C_{62} + C_{04} + D(v_6, v_0) + D(v_0, v_3)$$

$$C_{63} + C_{13} + D(v_6, v_1) + D(v_1, v_3)$$

$$C_{64} + C_{22} + D(v_6, v_2) + D(v_2, v_3)$$

7	$C_{07}=$ 75.43						
6	$C_{06}=$ 53.54	$C_{16}=$ 55.22	$C_{26}=$ 57.58	$C_{36}=$ 64.69	$C_{46}=$ 59.78	$C_{56}=$ 59.78	$C_{66}=$ 63.62
5	$C_{05}=$ 37.54	$C_{15}=$ 31.81	$C_{25}=$ 35.49	$C_{35}=$ 37.74	$C_{45}=$ 45.50	$C_{55}=$ 39.98	$C_{65}=$ 38.09
4	$C_{04}=$ 16.16	$C_{14}=$ 16.16	$C_{24}=$ 15.65	$C_{34}=$ 15.65	$C_{44}=$ 22.69	$C_{54}=$ 22.69	$C_{64}=$ 17.89
s	$i=0$	1	2	3	4	5	6

Fig. 10.11. Table of C_{is}'s.

The distances we need are calculated from the coordinates of the vertices as:

$$D(v_2, v_3) = D(v_6, v_0) = 0$$

(since these are polygon edges, not chords, and are present "for free")

$$D(v_6, v_2) = 26.08$$

$$D(v_1, v_3) = 16.16$$

$$D(v_6, v_1) = 22.36$$

$$D(v_0, v_3) = 21.93$$

The three sums above are 38.09, 38.52, and 43.97, respectively. We may conclude that the minimum cost of the subproblem S_{65} is 38.09. Moreover, since the first sum was smallest, we know that to achieve this minimum we must utilize the subproblems S_{62} and S_{04}, that is, select chord (v_0, v_3) and then solve S_{64} as best we can; chord (v_1, v_3) is the preferred choice for that subproblem. □

There is a useful trick for filling out the table of Fig. 10.11 according to the formula (10.5). Each term of the min operation in (10.5) requires a pair of entries. The first pair, for $k = 1$, can be found in the table (a) at the "bottom" (the row for $s = 2$)† of the column of the element being computed, and (b) just below and to the right‡ of the element being computed. The second pair is (a) next to the bottom of the column, and (b) two positions down and to the right. Fig. 10.12 shows the two lines of entries we follow to get all the pairs of entries we need to consider simultaneously. The pattern — up the

---

† Remember that the table of Fig. 10.11 has rows of 0's below those shown.
‡ By "to the right" we mean in the sense of a table that wraps around. Thus, if we are at the rightmost column, the column "to the right" is the leftmost column.

column and down the diagonal — is a common one in filling tables during dynamic programming.

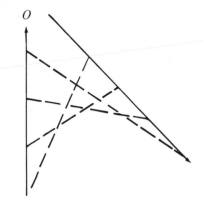

Fig. 10.12. Pattern of table scan to compute one element.

Finding Solutions from the Table

While Fig. 10.11 gives us the cost of the minimum triangulation, it does not immediately give us the triangulation itself. What we need, for each entry, is the value of k that produced the minimum in (10.5). Then we can deduce that the solution consists of chords (v_i, v_{i+k}), and (v_{i+k}, v_{i+s-1}) (unless one of them is not a chord, because $k = 1$ or $k = s-2$), plus whatever cords are implied by the solutions to $S_{i,k+1}$ and $S_{i+k,s-k}$. It is useful, when we compute an element of the table, to include with it the value of k that gave the best solution.

Example 10.3. In Fig. 10.11, the entry C_{07}, which represents the solution to the entire problem of Fig. 10.8, comes from the terms for $k = 5$ in (10.5). That is, the problem S_{07} is split into S_{06} and S_{52}; the former is the problem with six vertices $v_0, v_1, \ldots, v_5$, and the latter is a trivial "problem" of cost 0. Thus we introduce the chord (v_0, v_5) of cost 22.09 and must solve S_{06}.

The minimum cost for C_{06} comes from the terms for $k = 2$ in (10.5), whereby the problem S_{06} is split into S_{03} and S_{24}. The former is the triangle with vertices v_0, v_1, and v_2, while the latter is the quadrilateral defined by v_2, v_3, v_4, and v_5. S_{03} need not be solved, but S_{24} must be, and we must include the costs of chords (v_0, v_2) and (v_2, v_5) which are 17.89 and 19.80, respectively. We find the minimum value for C_{24} is assumed when $k = 1$ in (10.5), giving us the subproblems C_{22} and C_{33}, both of which have size less than or equal to three and therefore cost 0. The chord (v_3, v_5) is introduced, with a cost of 15.65. □

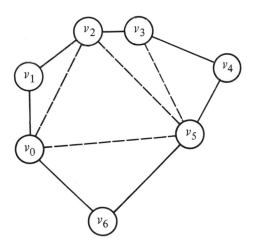

Fig. 10.13. A minimal cost triangulation.

10.3 Greedy Algorithms

Consider the problem of making change. Assume coins of values 25¢ (quarter), 10¢ (dime), 5¢ (nickel) and 1¢ (penny), and suppose we want to return 63¢ in change. Almost without thinking we convert this amount to two quarters, one dime and three pennies. Not only were we able to determine quickly a list of coins with the correct value, but we produced the shortest list of coins with that value.

The algorithm the reader probably used was to select the largest coin whose value was not greater than 63¢ (a quarter), add it to the list and subtract its value from 63 getting 38¢. We then selected the largest coin whose value was not greater than 38¢ (another quarter) and added it to the list, and so on.

This method of making change is a *greedy algorithm*. At any individual stage a greedy algorithm selects that option which is "locally optimal" in some particular sense. Note that the greedy algorithm for making change produces an overall optimal solution only because of special properties of the coins. If the coins had values 1¢, 5¢, and 11¢ and we were to make change of 15¢, the greedy algorithm would first select an 11¢ coin and then four 1¢ coins, for a total of five coins. However, three 5¢ coins would suffice.

We have already seen several greedy algorithms, such as Dijkstra's shortest path algorithm and Kruskal's minimum cost spanning tree algorithm. Dijkstra's shortest path algorithm is "greedy" in the sense that it always chooses the closest vertex to the source among those whose shortest path is not yet known. Kruskal's algorithm is also "greedy"; it picks from the remaining edges the shortest among those that do not create a cycle.

We should emphasize that not every greedy approach succeeds in producing the best result overall. Just as in life, a greedy strategy may produce a good result for a while, yet the overall result may be poor. As an example, we might consider what happens when we allow negative-weight edges in Dijkstra's and Kruskal's algorithms. It turns out that Kruskal's spanning tree algorithm is not affected; it still produces the minimum cost tree. But Dijkstra's algorithm fails to produce shortest paths in some cases.

Example 10.4. We see in Fig. 10.14 a graph with a negative cost edge between b and c. If we apply Dijkstra's algorithm with source s, we correctly discover first that the minimum path to a has length 1. Now, considering only edges from s or a to b or c, we expect that b has the shortest path from s, namely $s \rightarrow a \rightarrow b$, of length 3. We then discover that c has a shortest path from s of length 1.

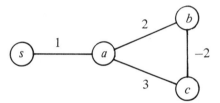

Fig. 10.14. Graph with negative weight edge.

However, our "greedy" selection of b before c was wrong from a global point of view. It turns out that the path $s \rightarrow a \rightarrow c \rightarrow b$ has length only 2, so our minimum distance of 3 for b was wrong.† □

Greedy Algorithms as Heuristics

For some problems no known greedy algorithm produces an optimal solution, yet there are greedy algorithms that can be relied upon to produce "good" solutions with high probability. Frequently, a suboptimal solution with a cost a few percent above optimal is quite satisfactory. In these cases, a greedy algorithm often provides the fastest way to get a "good" solution. In fact, if the problem is such that the only way to get an optimal solution is to use an

---

† In fact, we should be careful what we mean by "shortest path" when there are negative edges. If we allow negative cost cycles, then we could traverse such a cycle repeatedly to get arbitrarily large negative distances, so presumably we want to restrict ourselves to acyclic paths.

exhaustive search technique, then a greedy algorithm or other heuristic for getting a good, but not necessarily optimal, solution may be our only real choice.

Example 10.5. Let us introduce a famous problem where the only known algorithms that produce optimal solutions are of the "try-all-possibilities" variety and can have running times that are exponential in the size of the input. The problem, called the *traveling salesman problem*, or *TSP*, is to find, in an undirected graph with weights on the edges, a *tour* (a simple cycle that includes all the vertices) the sum of whose edge-weights is a minimum. A tour is often called a Hamilton (or Hamiltonian) cycle.

Figure 10.15(a) shows one instance of the traveling salesman problem, a graph with six vertices (often called "cities"). The coordinates of each vertex are given, and we take the weight of each edge to be its length. Note that, as is conventional with the TSP, we assume all edges exist, that is, the graph is complete. In more general instances, where the weight of edges is not based on Euclidean distance, we might find a weight of infinity on an edge that really was not present.

Figure 10.15(b)−(e) shows four tours of the six "cities" of Fig. 10.15(a). The reader might ponder which of these four, if any, is optimal. The lengths of these four tours are 50.00, 49.73, 48.39, and 49.78, respectively; (d) is the shortest of all possible tours.

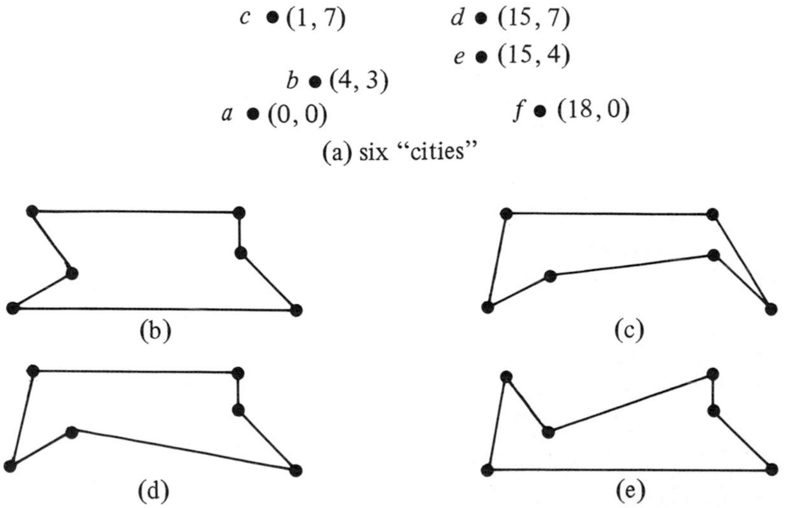

$$c \bullet (1,7) \qquad d \bullet (15,7)$$
$$e \bullet (15,4)$$
$$b \bullet (4,3)$$
$$a \bullet (0,0) \qquad\qquad f \bullet (18,0)$$

(a) six "cities"

(b)

(c)

(d)

(e)

Fig. 10.15. An instance of the traveling salesman problem.

The TSP has a number of practical applications. As its name implies, it can be used to route a person who must visit a number of points and return to his starting point. For example, the TSP has been used to route collectors of coins from pay phones. The vertices are the phones and the "home base." The cost of each edge is the travel time between the two points in question.

Another "application" of the TSP is in solving the *knight's tour problem*: find a sequence of moves whereby a knight can visit each square of the chessboard exactly once and return to its starting point. Let the vertices be the chessboard squares and let the edges between two squares that are a knight's move apart have weight 0; all other edges have weight infinity. An optimal tour has weight 0 and must be a knight's tour. Surprisingly, good heuristics for the TSP have no trouble finding knight's tours, although finding one "by hand" is a challenge.

The greedy algorithm for the TSP we have in mind is a variant of Kruskal's algorithm. Like that algorithm, we shall consider edges shortest first. In Kruskal's algorithm we accept an edge in its turn if it does not form a cycle with the edges already accepted, and we reject the edge otherwise. For the TSP, the acceptance criterion is that an edge under consideration, together with already selected edges,

1. does not cause a vertex to have degree three or more, and
2. does not form a cycle, unless the number of selected edges equals the number of vertices in the problem.

Collections of edges selected under these criteria will form a collection of unconnected paths, until the last step, when the single remaining path is closed to form a tour.

In Fig. 10.15(a), we would first pick edge (d, e), since it is the shortest, having length 3. Then we consider edges (b, c), (a, b), and (e, f), all of which have length 5. It doesn't matter in which order we consider them; all meet the conditions for selection, and we must select them if we are to follow the greedy approach. Next shortest edge is (a, c), with length 7.08. However, this edge would form a cycle with (a, b) and (b, c), so we reject it. Edge (d, f) is next rejected on similar grounds. Edge (b, e) is next to be considered, but it must be rejected because it would raise the degrees of b and e to three, and could then never form a tour with what we had selected. Similarly we reject (b, d). Next considered is (c, d), which is accepted. We now have one path, $a \rightarrow b \rightarrow c \rightarrow d \rightarrow e \rightarrow f$, and eventually accept (a, f) to complete the tour. The resulting tour is Fig. 10.15(b), which is fourth best of all the tours, but less than 4% more costly than the optimal. □

10.4 Backtracking

Sometimes we are faced with the task of finding an optimal solution to a problem, yet there appears to be no applicable theory to help us find the optimum, except by resorting to exhaustive search. We shall devote this section to a systematic, exhaustive searching technique called backtracking and a technique called alpha-beta pruning, which frequently reduces the search substantially.

Consider a game such as chess, checkers, or tic-tac-toe, where there are
two players. The players alternate moves, and the state of the game can be
represented by a board position. Let us assume that there are a finite number
of board positions and that the game has some sort of stopping rule to ensure
termination. With each such game, we can associate a tree called the *game
tree*. Each node of the tree represents a board position. With the root we
associate the starting position. If board position x is associated with node n,
then the children of n correspond to the set of allowable moves from board
position x, and with each child is associated the resulting board position. For
example, Fig. 10.16 shows part of the tree for tic-tac-toe.

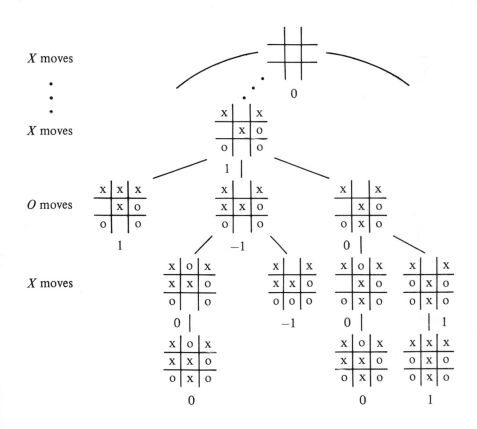

Fig. 10.16. Part of the tic-tac-toe game tree.

The leaves of the tree correspond to board positions where there is no
move, either because one of the players has won or because all squares are
filled and a draw resulted. We associate a value with each node of the tree.

First we assign values to the leaves. Say the game is tic-tac-toe. Then a leaf is assigned -1, 0 or 1 depending on whether the board position corresponds to a loss, draw or win for player 1 (playing X).

The values are propagated up the tree according to the following rule. If a node corresponds to a board position where it is player 1's move, then the value is the maximum of the values of the children of that node. That is, we assume player 1 will make the move most favorable to himself i.e., that which produces the highest-valued outcome. If the node corresponds to player 2's move, then the value is the minimum of the values of the children. That is, we assume player 2 will make his most favorable move, producing a loss for player 1 if possible, and a draw as next preference.

Example 10.6. The values of the boards have been marked in Fig. 10.16. The leaves that are wins for O get value -1, while those that are draws get 0, and wins for X get $+1$. Then we proceed up the tree. On level 8, where only one empty square remains, and it is X's move, the values for the unresolved boards is the "maximum" of the one child at level 9.

On level 7, where it is O's move and there are two choices, we take as a value for an interior node the minimum of the values of its children. The leftmost board shown on level 7 is a leaf and has value 1, because it is a win for X. The second board on level 7 has value $\min(0, -1) = -1$, while the third board has value $\min(0, 1) = 0$. The one board shown at level 6, it being X's move on that level, has value $\max(1, -1, 0) = 1$, meaning that there is some choice X can make that will win; in this case the win is immediate. $\square$

Note that if the root has value 1, then player 1 has a winning strategy. Whenever it is his turn he is guaranteed that he can select a move that leads to a board position of value 1. Whenever it is player 2's move he has no real choice but to select a moving leading to a board position of value 1, a loss for him. The fact that a game is assumed to terminate guarantees an eventual win for the first player. If the root has value 0, as it does in tic-tac-toe, then neither player has a winning strategy but can only guarantee himself a draw by playing as well as possible. If the root has value -1, then player 2 has a winning strategy.

Payoff Functions

The idea of a game tree, where nodes have values -1, 0, and 1, can be generalized to trees where leaves are given any number (called the *payoff*) as a value, and the same rules for evaluating interior nodes applies: take the maximum of the children on those levels where player 1 is to move, and the minimum of the children on levels where player 2 moves.

As an example where general payoffs are useful, consider a complex game, like chess, where the game tree, though finite, is so huge that evaluating it from the bottom up is not feasible. Chess programs work, in essence, by building for each board position from which it must move, the game tree with that board as root, extending downward for several levels; the exact number of levels depends on the speed with which the computer can work. As most of the leaves of the tree will be ambiguous, neither wins, losses, nor draws, each program uses a function of board positions that attempts to estimate the probability of the computer winning in that position. For example, the difference in material figures heavily into such an estimation, as do such factors as the defensive strength around the kings. By using this payoff function, the computer can estimate the probability of a win after making each of its possible next moves, on the assumption of subsequent best play by each side, and chose the move with the highest payoff.†

Implementing Backtrack Search

Suppose we are given the rules for a game,‡ that is, its legal moves and rules for termination. We wish to construct its game tree and evaluate the root. We could construct the tree in the obvious way, and then visit the nodes in postorder. The postorder traversal assures that we visit an interior node n after all its children, whereupon we can evaluate n by taking the min or max of the values of its children, as appropriate.

The space to store the tree can be prohibitively large, but by being careful we need never store more than one path, from the root to some node, at any one time. In Fig. 10.17 we see the sketch of a recursive program that represents the path in the tree by the sequence of active procedure calls at any time. That program assumes the following:

1. Payoffs are real numbers in a limited range, for example -1 to $+1$.
2. The constant ∞ is larger than any positive payoff and its negation is smaller than any negative payoff.
3. The type modetype is declared

     type
              modetype = (MIN, MAX)

4. There is a type boardtype declared in some manner suitable for the representation of board positions.
5. There is a function PAYOFF that computes the payoff for any board that is a *leaf* (i.e., won, lost, or drawn position).

---

† Incidentally, some of the other things good chessplaying programs do are:
1. Use heuristics to eliminate from consideration certain moves that are unlikely to be good. This helps expand the tree to more levels in a fixed time.
2. Expand "capture chains", which are sequences of capturing moves beyond the last level to which the tree is normally expanded. This helps estimate the relative material strength of positions more accurately.
3. Prune the tree search by alpha-beta pruning, as discussed later in this section.

‡ We should not imply that only "games" can be solved in this manner. As we shall see in subsequent examples, the "game" could really represent the solution to a practical problem.

```
        function search ( B: boardtype; mode: modetype ) : real;
            { evaluates the payoff for board B, assuming it is
              player 1's move if mode = MAX and player 2's move
              if mode = MIN. Returns the payoff }
        var
                C: boardtype; { a child of board B }
                value: real; { temporary minimum or maximum value }
        begin
(1)         if B is a leaf then
(2)             return (payoff(B))
            else begin
                { initialize minimum or maximum value of children }
(3)             if mode = MAX then
(4)                 value := −∞
                else
(5)                 value := ∞;
(6)             for each child C of board B do
(7)                 if mode = MAX then
(8)                     value := max(value, search(C, MIN))
                    else
(9)                     value := min(value, search(C, MAX));
(10)            return (value)
            end
        end; { search }
```

Fig. 10.17. Recursive backtrack search program.

Another implementation we might consider is to use a nonrecursive program that keeps a stack of boards corresponding to the sequence of active calls to *search*. The techniques discussed in Section 2.6 can be used to construct such a program.

Alpha-Beta Pruning

There is a simple observation that allows us to eliminate from consideration much of a typical game tree. In terms of Fig. 10.17, the for-loop of line (6) can skip over certain children, often many of the children. Suppose we have a node n, as in Fig. 10.18, and we have already determined that c_1, the first of n's children, has a value of 20. As n is a max node, we know its value is at least 20. Now suppose that continuing with our search we find that d, a child of c_2 has value 15. As c_2, another child of n, is a min node, we know the value of c_2 cannot exceed 15. Thus, whatever value c_2 has, it cannot affect the value of n or any parent of n.

It is thus possible in the situation of Fig. 10.18, to skip consideration of the children of c_2 that we have not already examined. The general rules for

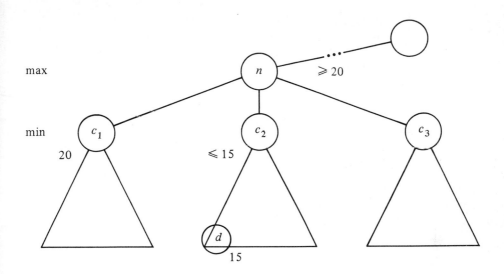

Fig. 10.18. Pruning the children of a node.

skipping or "pruning" nodes involves the notion of final and tentative values for nodes. The *final* value is what we have simply been calling the "value." A *tentative* value is an upper bound on the value of a min node, or a lower bound on the value of a max node. The rules for computing final and tentative values are the following.

1. If all the children of a node n have been considered or pruned, make the tentative value of n final.

2. If a max node n has tentative value v_1 and a child with final value v_2, then set the tentative value of n to $\max(v_1, v_2)$. If n is a min node, set its tentative value to $\min(v_1, v_2)$.

3. If p is a min node with parent q (a max node), and p and q have tentative values v_1 and v_2, respectively, with $v_1 \le v_2$, then we may prune all the unconsidered children of p. We may also prune the unconsidered children of p if p is a max node (and therefore q is a min node) and $v_2 \le v_1$.

Example 10.7. Consider the tree in Fig. 10.19. Assuming values for the leaves as shown, we wish to calculate the value for the root. We begin a post-order traversal. After reaching node D, by rule (2) we assign a tentative value of 2, which is the final value of D, to node C. We then search E and return to C and then to B. By rule (1), the final value of C is fixed at 2 and the value of B is tentatively set to 2. The search continues down to G and then back to F. The value F is tentatively set to 6. By rule (3), with p and q equal to F and B, respectively, we may prune H. That is, there is no need to search node H, since the tentative value of F can never decrease and it is already greater than the value of B, which can never increase.

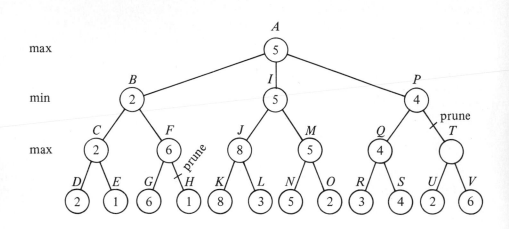

Fig. 10.19. A game tree.

Continuing our example, A is assigned a tentative value of 2 and the search proceeds to K. J is assigned a tentative value of 8. L does not determine the value of max node J. I is assigned a tentative value of 8. The search goes down to N, and M is assigned a tentative value of 5. Node O must be searched, since 5, the tentative value of M, is less than the tentative value of I. The tentative values of I and A are revised, and the search goes down to R. Eventually R and S are searched, and P is assigned a tentative value of 4. We need not search T or below, since that can only lower P's value and it is already too low to affect the value of A. □

Branch-and-Bound Search

Games are not the only sorts of "problems" that can be solved exhaustively by searching a complete tree of possibilities. A wide variety of problems where we are asked to find a minimum or maximum configuration of some sort are amenable to solution by backtracking search over a tree of all possibilities. The nodes of the tree can be thought of as sets of configurations, and the children of a node n each represent a subset of the configurations that n represents. Finally, the leaves each represent single configurations, or solutions to the problem, and we may evaluate each such configuration to see if it is the best solution found so far.

If we are reasonably clever in how we design the search, the children of a node will each represent far fewer configurations than the node itself, so we need not go to too great a depth before reaching leaves. Lest this notion of searching appears too vague, let us take a concrete example.

Example 10.8. Recall from the previous section our discussion of the traveling salesman problem. There we gave a "greedy algorithm" for finding a good but not necessarily optimum tour. Now let us consider how we might find the optimum tour by systematically considering all tours. One way is to consider all permutations of the nodes, and evaluate the tour that visits the nodes in that order, remembering the best found so far. The time for such an approach is $O(n!)$ on an n node graph, since we must consider $(n-1)!$ different permutations,† and each permutation takes $O(n)$ time to evaluate.

We shall consider a somewhat different approach that is no better than the above in the worst case, but on the average, when coupled with a technique called "branch-and-bound" that we shall discuss shortly, produces the answer far more rapidly than the "try all permutations" method. Start constructing a tree, with a root that represents all tours. Tours are what we called "configurations" in the prefatory material. Each node has two children, and the tours that a node represents are divided by these children into two groups — those that have a particular edge and those that do not. For example, Fig. 10.20 shows portions of the tree for the TSP instance from Fig. 10.15(a).

In Fig. 10.20 we have chosen to consider the edges in lexicographic order $(a, b), (a, c), \ldots, (a, f), (b, c), \ldots, (b, f), (c, d)$, and so on. We could, of course pick any other order. Observe that not every node in the tree has two children. We can eliminate some children because the edges selected do not form part of a tour. Thus, there is no node for "tours containing $(a, b), (a, c)$, and (a, d)," because a would have degree 3 and the result would not be a tour. Similarly, as we go down the tree we shall eliminate some nodes because some city would have degree less than 2. For example, we shall find no node for tours without $(a, b), (a, c), (a, d)$, or (a, e). □

Bounding Heuristics Needed for Branch-and-Bound

Using ideas similar to those in alpha-beta pruning, we can eliminate far more nodes of the search tree than would be suggested by Example 10.8. Suppose, to be specific, that our problem is to minimize some function, e.g., the cost of a tour in the TSP. Suppose also that we have a method for getting a lower bound on the cost of any solution among those in the set of solutions represented by some node n. If the best solution found so far costs less than the lower bound for node n, we need not explore any of the nodes below n.

Example 10.9. We shall discuss one way to get a lower bound on certain sets of solutions for the TSP, those sets represented by nodes in a tree of solutions as suggested in Fig. 10.20. First of all, suppose we wish a lower bound on all solutions to a given instance of the TSP. Observe that the cost of any tour can be expressed as one half the sum over all nodes n of the cost of the two tour edges adjacent to n. This remark leads to the following rule. The sum

---

† Note that we need not consider all $n!$ permutations, since the starting point of a tour is immaterial. We may therefore consider only those permutations that begin with 1.

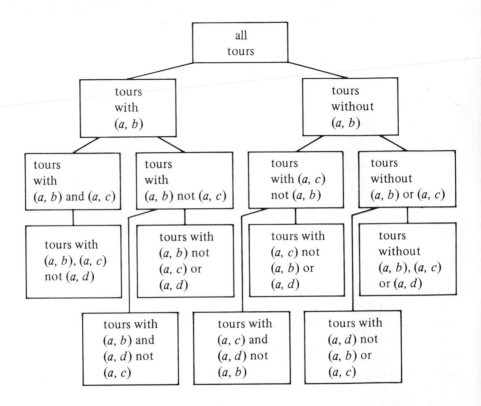

Fig. 10.20. Beginning of a solution tree for a TSP instance.

of the two tour edges adjacent to node n is no less than the sum of the two edges of least cost adjacent to n. Thus, no tour can cost less than one half the sum over all nodes n of the two lowest cost edges incident upon n.

For example, consider the TSP instance in Fig. 10.21. Unlike the instance in Fig. 10.15, the distance measure for edges is not Euclidean; that is, it bears no relation to the distance in the plane between the cities it connects. Such a cost measure might be traveling time or fare, for example. In this instance, the least cost edges adjacent to node a are (a, d), and (a, b), with a total cost of 5. For node b, we have (a, b) and (b, e), with a total cost of 6. Similarly, the two lowest cost edges adjacent to c, d, and e, total 8, 7, and 9, respectively. Our lower bound on the cost of a tour is thus $(5+6+8+7+9)/2 = 17.5$.

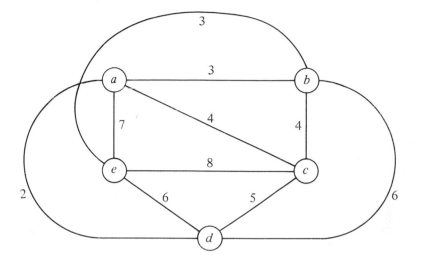

Fig. 10.21. An instance of TSP.

Now suppose we want a lower bound on the cost of a subset of tours defined by some node in the search tree. If the search tree is constructed as in Example 10.8, each node represents tours defined by a set of edges that must be in the tour and a set of edges that may not be in the tour. These *constraints* alter our choices for the two lowest-cost edges at each node. Obviously an edge constrained to be in any tour must be included among the two edges selected, no matter whether they are or are not lowest or second lowest in cost.† Similarly, an edge constrained to be out cannot be selected, even if its cost is lowest.

Thus, if we are constrained to include edge (a, e), and exclude (b, c), the two edges for node a are (a, d) and (a, e), with a total cost of 9. For b we select (u, b) and (b, e), as before, with a total cost of 6. For c, we cannot select (b, c), and so select (a, c) and (c, d), with a total cost of 10. For d we select (a, d) and (c, d), as before, while for e we must select (a, e), and choose to select (b, e). The lower bound for these constraints is thus $(9+6+9+7+10)/2 = 20.5$. □

Now let us construct the search tree along the lines suggested in Example 10.8. We consider the edges in lexicographic order, as in that example. Each

---

† The rules for constructing the search tree will be seen to eliminate any set of constraints that cannot yield any tour, e.g., because three edges adjacent to one node are required to be in the tour.

time we *branch*, by considering the two children of a node, we try to infer additional decisions regarding which edges must be included or excluded from tours represented by those nodes. The rules we use for these inference are:

1. If excluding an edge (x, y) would make it impossible for x or y to have as many as two adjacent edges in the tour, then (x, y) must be included.

2. If including (x, y) would cause x or y to have more than two edges adjacent in the tour, or would complete a non-tour cycle with edges already included, then (x, y) must be excluded.

When we branch, after making what inferences we can, we compute lower bounds for both children. If the lower bound for a child is as high or higher than the lowest cost tour found so far, we can "prune" that child and need not construct or consider its descendants. Interestingly, there are situations where the lower bound for a node n is lower than the best solution so far, yet both children of n can be pruned because their lower bounds exceed the cost of the best solution so far.

If neither child can be pruned, we shall, as a heuristic, consider the child with the smaller lower bound first, in the hope of more rapidly reaching a solution that is cheaper than the one so far found best.† After considering one child, we must consider again whether its sibling can be pruned, since a new best solution may have been found. For the instance of Fig. 10.21, we get the search tree of Fig. 10.22. To interpret nodes of that tree, it helps to understand that the capital letters are names of the search tree nodes. The numbers are the lower bounds, and we list the constraints applying to that node but none of its ancestors by writing xy if edge (x, y) must be included and $\overline{xy}$ if (x, y) must be excluded. Also note that the constraints introduced at a node apply to all its descendants. Thus to get all the constraints applying at a node we must follow the path from that node to the root.

Lastly, let us remark that as for backtrack search in general, we construct the tree one node at a time, retaining only one path, as in the recursive algorithm of Fig. 10.17, or its nonrecursive counterpart. The nonrecursive version is probably to be preferred, so that we can maintain the list of constraints conveniently on a stack.

Example 10.10. Figure 10.22 shows the search tree for the TSP instance of Fig. 10.21. To see how it is constructed, we begin at the root A of Fig. 10.22. The first edge in lexicographic order is (a, b), so we consider the two children B and C, corresponding to the constraints ab and $\overline{ab}$, respectively. There is, as yet, no "best solution so far," so we shall consider both B and C eventually.‡ Forcing (a, b) to be included does not raise the lower bound, but

---

† An alternative is to use a heuristic to obtain a good solution using the constraints required for each child. For example, the reader should be able to modify the greedy TSP algorithm to respect constraints.

‡ We could start with some heuristically found solution, say the greedy one, although that would not affect this example. The greedy solution for Fig. 10.21 has cost 21.

excluding it raises the bound to 18.5, since the two cheapest legal edges for nodes a and b now total 6 and 7, respectively, compared with 5 and 6 with no constraints. Following our heuristic, we shall consider the descendants of node B first.

The next edge in lexicographic order is (a, c). We thus introduce children D and E corresponding to tours where (a, c) is included and excluded, respectively. In node D, we can infer that neither (a, d) nor (a, e) can be in a tour, else a would have too many edges incident. Following our heuristic we consider E before D, and branch on edge (a, d). The children F and G are introduced with lower bounds 18 and 23, respectively. For each of these children we know about three of the edges incident upon a, and so can infer something about the remaining edge (a, e).

Consider the children of F first. The first remaining edge in lexicographic order is (b, c). If we include (b, c), then, as we have included (a, b), we cannot include (b, d) or (b, e). As we have eliminated (a, e) and (b, e), we must have (c, e) and (d, e). We cannot have (c, d) or c and d would have three incident edges. We are left with one tour (a, b, c, e, d, a), whose cost is 23. Similarly, node I, where (b, c) is excluded, can be proved to represent only the tour (a, b, e, c, d, a), of cost 21. That tour has the lowest cost found so far.

We now backtrack to E and consider its second child, G. But G has a lower bound of 23, which exceeds the best cost so far, 21. Thus we prune G. We now backtrack to B and consider its other child, D. The lower bound on D is 20.5, but since costs are integers, we know no tour represented by D can have cost less than 21. Since we already have a tour that cheap, we need not explore the descendants of D, and therefore prune D. Now we backtrack to A and consider its second child, C.

At the level of node C, we have only considered edge (a, b). Nodes J and K are introduced as children of C. J corresponds to those tours that have (a, c) but not (a, b), and its lower bound in 18.5. K corresponds to tours having neither (a, b) nor (a, c), and we may infer that those tours have (a, d) and (a, e). The lower bound for K is 21, and we may immediately prune K, since we already know a tour that is low in cost.

We next consider the children of J, which are L and M, and we prune M because its lower bound exceeds the best tour cost so far. The children of L are N and P, corresponding to tours that have (b, c), and that exclude (b, c). By considering the degree of nodes b and c, and remembering that the selected edges cannot form a cycle of fewer than all five cities, we can infer that nodes N and P each represent single tours. One of these, (a, c, b, e, d, a), has the lowest cost of any tour, 19. We have explored or pruned the entire tree and therefore end. □

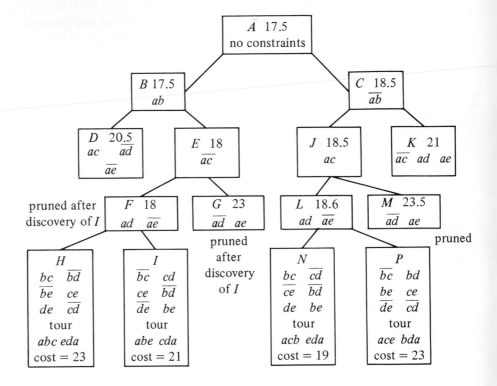

Fig. 10.22. Search tree for TSP solution.

10.5 Local Search Algorithms

Sometimes the following strategy will produce an optimal solution to a problem.

1. Start with a random solution.
2. Apply to the current solution a transformation from some given set of transformations to improve the solution. The improvement becomes the new "current" solution.
3. Repeat until no transformation in the set improves the current solution.

    The resulting solution may or may not be optimal. In principle, if the "given set of transformations" includes all the transformations that take one solution and replace it by any other, then we shall never stop until we reach an optimal solution. But then the time to apply (2) above is the same as the time needed to examine all solutions, and the whole approach is rather pointless.

The method makes sense when we can restrict our set of transformations to a small set, so we can consider all transformations in a short time; perhaps $O(n^2)$ or $O(n^3)$ transformations should be allowed when the problem is of "size" n. If the transformation set is small, it is natural to view the solutions that can be transformed to one another in one step as "close." The transformations are called "local transformations," and the method is called *local search*.

Example 10.11. One problem we can solve exactly by local search is the minimal spanning tree problem. The local transformations are those in which we take some edge not in the current spanning tree, add it to the tree, which must produce a unique cycle, and then eliminate exactly one edge of the cycle (presumably that of highest cost) to form a new tree.

For example, consider the graph of Fig. 10.21. We might start with the tree shown in Fig. 10.23(a). One transformation we could perform is to add edge (d, e) and remove another edge in the cycle formed, which is (e, a, c, d, e). If we remove edge (a, e), we decrease the cost of the tree from 20 to 19. That transformation can be made, leaving the tree of Fig. 10.23(b), to which we again try to apply an improving transformation. One such is to insert edge (a, d) and delete edge (c, d) from the cycle formed. The result is shown in Fig. 10.23(c). Then we might introduce (a, b) and delete (b, c) as in Fig. 10.23(d), and subsequently introduce (b, e) in favor of (d, e). The resulting tree of Fig. 10.23(e) is minimal. We can check that every edge not in that tree has the highest cost of any edge in the cycle it forms. Thus no transformation is applicable to Fig. 10.23(e). □

The time taken by the algorithm of Example 10.11 on a graph of n nodes and e edges depends on the number of times we need to improve the solution. Just testing that no transformation is applicable could take $O(ne)$ time, since e edges must be tried, and each could form a cycle of length nearly n. Thus the algorithm is not as good as Prim's or Kruskal's algorithms, but serves as an example where an optimal solution can be obtained by local search.

Local Search Approximation Algorithms

Local search algorithms have had their greatest effectiveness as heuristics for the solution to problems whose exact solutions require exponential time. A common method of search is to start with a number of random solutions, and apply the local transformations to each, until reaching a *locally optimal* solution, one that no transformation can improve. We shall frequently reach different locally optimal solutions, from most or all of the random starting solutions, as suggested in Fig. 10.24. If we are lucky, one of them will be

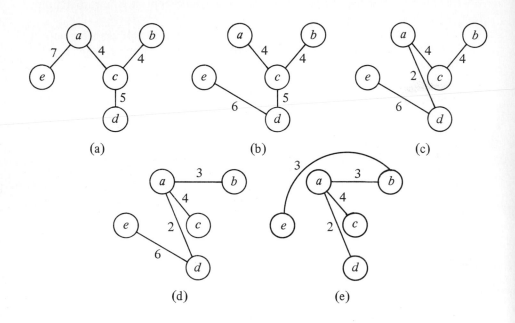

Fig. 10.23. Local search for a minimal spanning tree.

globally optimal, that is, as good as any solution.

In practice, we may not find a globally optimal solution as suggested in Fig. 10.24, since the number of locally optimal solutions may be enormous. However, we may at least choose that locally optimal solution that has the least cost among all those found. As the number of kinds of local transformations that have been used to solve various problems is great, we shall close the section with two examples — the TSP, and a simple problem of "package placement."

The Traveling Salesman Problem

The TSP is one for which local search techniques have been remarkably successful. The simplest transformation that has been used is called "2-opting." It consists of taking any two edges, such as (A, B) and (C, D) in Fig. 10.25, removing them, and reconnecting their endpoints to form a new tour. In Fig. 10.25, the new tour runs from B, clockwise to C, then along the edge (C, A), counterclockwise from A to D, and finally along the edge (D, B). If the sum of the lengths of (A, C) and (B, D) is less than the sum of the lengths of (A, B) and (C, D), then we have an improved tour.† Note that we cannot

---

† Do not be fooled by the picture of Fig. 10.25. True, if lengths of edges are distances in the plane, then the dashed edges in Fig. 10.25 must be longer than those they replace. In

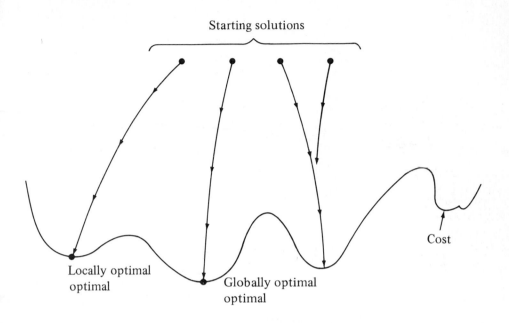

Starting solutions

Cost

Locally optimal
optimal

Globally optimal
optimal

Fig. 10.24. Local search in the space of solutions.

connect A to D and B to C, as the result would not be a tour, but two disjoint cycles.

To find a locally optimal tour, we start with a random tour, and consider all pairs of nonadjacent edges, such as (A, B) and (C, D) in Fig. 10.25. If the tour can be improved by replacing these edges with (A, C) and (B, D), do so, and continue considering pairs of edges that have not been considered before. Note that the introduced edges (A, C) and (B, D) must each be paired with all the other edges of the tour, as additional improvements could result.

Example 10.12. Reconsider Fig. 10.21, and suppose we start with the tour of Fig. 10.26(a). We might replace (a, e) and (c, d), with a total cost of 12, by (a, d) and (c, e), with a total cost of 10, as shown in Fig. 10.26(b). Then we might replace (a, b) and (c, e) by (a, c) and (b, e), giving the optimal tour shown in Fig. 10.26(c). One can check that no pair of edges can be removed from Fig. 10.26(c) and be profitably replaced by crossing edges with the same endpoints. As one case in point, (b, c) and (d, e) together have the relatively high cost of 10. But (c, e) and (b, d) are worse, costing 14 together. □

---

fact, 2-opts in the plane improve the cost exactly when the removed edges, (A, B) and (C, D) cross each other while the replacing ones do not. However, there is no reason to assume the distances in Fig. 10.25 are distances in the plane, or if they are, it could have been (A, B) and (C, D) that crossed, not (A, C) and (B, D).

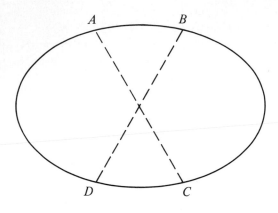

Fig. 10.25. 2-opting.

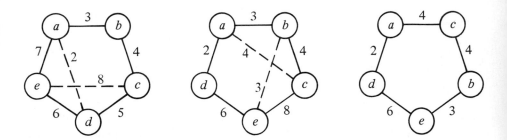

Fig. 10.26. Optimizing a TSP instance by 2-opting.

We can generalize 2-opting to k-opting for any constant k, where we remove up to k edges and reconnect the remaining pieces in any order so that result is a tour. Note that we do not require the removed edges to be nonadjacent in general, although for the 2-opting case there was no point in considering the removal of two adjacent edges. Also note that for $k>2$, there is more than one way to connect the pieces. For example, Fig. 10.27 shows the general process of 3-opting using any of the following eight sets of edges.

1. $(A, F), (D, E), (B, C)$ (as the tour was)
2. $(A, F), (C, E), (D, B)$ (a 2-opt)
3. $(A, E), (F, D), (B, C)$ (another 2-opt)
4. $(A, E), (F, C), (B, D)$ (a true 3-opt)
5. $(A, D), (C, E), (B, F)$ (another true 3-opt)
6. $(A, D), (C, F), (B, E)$ (another true 3-opt)
7. $(A, C), (D, E), (B, F)$ (a 2-opt)
8. $(A, C), (D, F), (B, E)$ (a 3-opt)

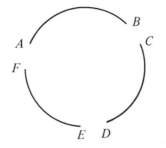

Fig. 10.27. Pieces of a tour after removing three edges.

It is easy to check that, for fixed k, the number of different k-opting transformations we need to consider if there are n vertices is $O(n^k)$. For example, the exact number is $n(n-3)/2$ for $k = 2$. The time taken to obtain a locally optimal tour may be considerably higher than this, however, since we could make many local transformations before reaching a locally optimum tour, and each improving transformation introduces new edges that may participate in later transformations that improve the tour still further. Lin and Kernighan [1973] have found that variable-depth-opting is in practice a very powerful method and has a good chance of getting the optimum tour on 40-100 city problems.

Package Placement

The *one-dimensional package placement* problem can be stated as follows. We have an undirected graph, whose vertices we call "packages." The edges are labeled by "weights," and the weight $w(a, b)$ of edge (a, b) is the number of "wires" between packages a and b. The problem is to order the vertices $p_1, p_2, \ldots, p_n$, such that the sum of $|i-j| \, w(p_i, p_j)$ over all pairs i and j is minimized. That is, we want to minimize the sum of the lengths of the wires needed to connect all the packages with the required number of wires.

The package placement problem has had a number of applications. For example, the "packages" could be logic cards on a rack, and the weight of an interconnection between cards is the number of wires connecting them. A similar problem comes up when we try to design integrated circuits from arrays of standard modules and interconnections between them. A generalization of the one-dimensional package placement problem allows placement of "packages," which have height and width, in a two-dimensional region, while minimizing the sum of the lengths of the wires between packages. This problem also has application to the design of integrated circuits, among other areas.

There are a number of local transformations we could use to find local optima for instances of the one-dimensional package placement problem. Here are several.

1. Interchange adjacent packages p_i and p_{i+1} if the resulting order is less costly. Let $L(j)$ be the sum of the weights of the edges extending to the left of p_j, i.e., $\sum_{k=1}^{j-1} w(p_k, p_j)$. Similarly, let $R(j)$ be $\sum_{k=j+1}^{n} w(p_k, p_j)$. Improvement results if $L(i) - R(i) + R(i+1) - L(i+1) + 2w(p_i, p_{i+1})$ is negative. The reader should verify this formula by computing the costs before and after the interchange and taking the difference.

2. Take a package p_i and insert it between p_j and p_{j+1} for some i and j.

3. Interchange any two packages p_i and p_j.

Example 10.13. Suppose we take the graph of Fig. 10.21 to represent a package placement instance. We shall restrict ourselves to the simple transformation set (1). An initial placement, a, b, c, d, e, is shown in Fig. 10.28(a); it has a cost of 97. Note that the cost function weights edges by their distance, so (a, e) contributes $4 \times 7 = 28$ to the cost of 97. Let us consider interchanging d with e. We have $L(d) = 13$, $R(d) = 6$, $L(e) = 24$, and $R(e) = 0$. Thus $L(d) - R(d) + R(e) - L(e) + 2w(d, e) = -5$, and we can interchange d and e to improve the placement to (a, b, c, e, d), with a cost of 92 as shown in Fig. 10.28(b).

In Fig. 10.28(b), we can interchange c with e profitably, producing Fig. 10.28(c), whose placement (a, b, e, c, d) has a cost of 91. Fig. 10.28(c) is locally optimal for set of transformations (1). It is not globally optimal; (a, c, e, d, b) has a cost of 84. □

As with the TSP, we cannot estimate closely the time it takes to reach a local optimum. We can only observe that for set of transformations (1), there are only $n-1$ transformations to consider. Further, if we compute $L(i)$ and $R(i)$ once, we only have to change them when p_i is interchanged with p_{i-1} or p_{i+1}. Moreover, the recalculation is easy. If p_i and p_{i+1} are interchanged, for example, then the new $L(i)$ and $R(i)$ are, respectively, $L(i+1) - w(p_i, p_{i+1})$ and $R(i+1) + w(p_i, p_{i+1})$. Thus $O(n)$ time suffices to test for an improving transformation and to recalculate the $L(i)$'s and $R(i)$'s. We also need only $O(n)$ time to initialize the $L(i)$'s and $R(i)$'s, if we use the recurrence

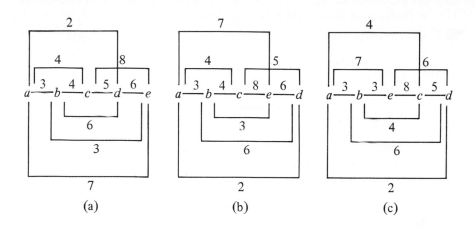

Fig. 10.28. Local optimizations.

$$L(1) = 0$$

$$L(i) = L(i-1) + w(p_{i-1}, p_i)$$

and a similar recurrence for R.

In comparison, the sets of transformations (2) and (3) each have $O(n^2)$ members. It will therefore take $O(n^2)$ time just to confirm that we have a locally optimal solution. However, as for set (1), we cannot closely bound the total time taken when a sequence of improvements are made, since each improvement can create additional opportunities for improvement. □

Exercises

10.1 How many moves do the algorithms for moving n disks in the Towers of Hanoi problem take?

*10.2 Prove that the recursive (divide and conquer) algorithm for the Tower of Hanoi and the simple nonrecursive algorithm described at the beginning of Section 10.1 perform the same steps.

10.3 Show the actions of the divide-and-conquer integer multiplication algorithm of Fig. 10.3 when multiplying 1011 by 1101.

*10.4 Generalize the tennis tournament construction of Section 10.1 to tournaments where the number of players is not a power of two. *Hint.* If the number of players n is odd, then one player must receive a *bye* (not play) on any given day, and it will take n days, rather than $n-1$ to complete the tournament. However, if there are two groups with an odd number of players, then the players receiving the bye from each group may actually play each other.

10.5 We can recursively define the number of combinations of m things out of n, denoted $\binom{n}{m}$, for $n \geq 1$ and $0 \leq m \leq n$, by

$$\binom{n}{m} = 1 \quad \text{if } m = 0 \text{ or } m = n$$

$$\binom{n}{m} = \binom{n-1}{m} + \binom{n-1}{m-1} \quad \text{if } 0 < m < n$$

a) Give a recursive function to compute $\binom{n}{m}$.

b) What is its worst-case running time as a function of n?

c) Give a dynamic programming algorithm to compute $\binom{n}{m}$. *Hint.* The algorithm constructs a table generally known as Pascal's triangle.

d) What is the running time of your answer to (c) as a function of n.

10.6 Another way to compute the number of combinations of m things out of n is to calculate $(n) (n-1) (n-2) \cdots (n-m+1)/(1) (2) \cdots (m)$.

a) What is the worst case running time of this algorithm as a function of n?

*b) Is it possible to compute the "World Series Odds" function $P(i, j)$ from Section 10.2 in a similar manner? How fast can you perform this computation?

10.7 a) Rewrite the odds calculation of Fig. 10.7 to take into account the fact that the first team has a probability p of winning any given game.

b) If the Dodgers have won one game and the Yankees two, but the Dodgers have a .6 probability of winning any given game, who is more likely to win the World Series?

10.8 The odds calculation of Fig. 10.7 requires $O(n^2)$ space. Rewrite the algorithm to use only $O(n)$ space.

*10.9** Prove that Equation (10.4) results in exactly $\binom{i+j}{i}$ calls to P.

10.10 Find a minimal triangulation for a regular octagon, assuming distances are Euclidean.

10.11 The *paragraphing problem*, in a very simple form, can be stated as follows: We are given a sequence of words $w_1, w_2, \ldots, w_k$ of lengths $l_1, l_2, \ldots, l_k$, which we wish to break into lines of length L. Words are separated by blanks whose ideal width is b, but blanks can stretch or shrink if necessary (but without overlapping words), so that a line $w_i\ w_{i+1}\ \cdots\ w_j$ has length exactly L. However, the *penalty* for stretching or shrinking is the magnitude of the total amount by which blanks are stretched or shrunk. That is, the cost of setting line $w_i \, w_{i+1} \, \cdots \, w_j$ is $(j-i) \, |b'-b|$, where b', the actual width of the blanks, is $(L-l_i-l_{i+1}- \cdots -l_j)/(j-i)$. However, if $j=k$ (we have the last line), the cost is zero unless $b' < b$, since we do not have to stretch the last line. Give a dynamic programming algorithm to find a least-cost separation of $w_1, w_2, \ldots, w_k$ into lines of length L. *Hint.* For $i=k, k-1, \ldots, 1$, compute the least cost of setting $w_i, w_{i+1}, \ldots, w_k$.

10.12 Suppose we are given n elements $x_1, x_2, \ldots, x_n$ related by a linear order $x_1 < x_2 < \cdots < x_n$, and that we wish to arrange these elements m in a binary search tree. Suppose that p_i is the probability that a request to find an element will be for x_i. Then for any given binary search tree, the average cost of a lookup is $\sum_{i=1}^{n} p_i(d_i + 1)$, where d_i is the depth of the node holding x_i. Given the p_i's, and assuming the x_i's never change, we can find a binary search tree that minimizes the lookup cost. Find a dynamic programming algorithm to do so. What is the running time of your algorithm? *Hint.* Compute for all i and j the optimal lookup cost among all trees containing only $x_i, x_{i+1}, \ldots, x_{i+j-1}$, that is, the j elements beginning with x_i.

**10.13 For what values of coins does the greedy change-making algorithm of Section 10.3 produce an optimal solution?

10.14 a) Write the recursive triangulation algorithm discussed in Section 10.2.

  b) Show that the recursive algorithm results in exactly 3^{s-4} calls on nontrivial problems when started on a problem of size $s \geq 4$.

10.15 Describe a greedy algorithm for

  a) The one-dimensional package placement problem.

  b) The paragraphing problem (Exercise 10.11).

Give an example where your algorithm does not produce an optimal answer, or show that no such example exists.

10.16 Give a nonrecursive version of the tree search algorithm of Fig. 10.17.

10.17 Consider a game tree in which there are six marbles, and players 1 and 2 take turns picking from one to three marbles. The player who takes the last marble *loses* the game.

  a) Draw the complete game tree for this game.

  b) If the game tree were searched using the alpha-beta pruning technique, and nodes representing configurations with the smallest number of marbles are searched first, which nodes are pruned?

    c) Who wins the game if both play their best?

*10.18 Develop a branch and bound algorithm for the TSP based on the idea that we shall begin a tour at vertex 1, and at each level, branch based on what node comes next in the tour (rather than on whether a particular edge is chosen as in Fig. 10.22). What is an appropriate lower bound estimator for configurations, which are lists of vertices $1, v_1, v_2, \ldots$ that begin a tour? How does your algorithm behave on Fig. 10.21, assuming a is vertex 1?

*10.19 A possible local search algorithm for the paragraphing problem is to allow local transformations that move the first word of one line to the previous line or the last word of a line to the line following. Is this algorithm locally optimal, in the sense that every locally optimal solution is a globally optimal solution?

10.20 If our local transformations consist of 2-opts only, are there any locally optimal tours in Fig. 10.21 that are not globally optimal?

Bibliographic Notes

There are many important examples of divide-and-conquer algorithms including the $O(n \log n)$ Fast Fourier Transform of Cooley and Tukey [1965], the $O(n \log n \log \log n)$ integer multiplication algorithm of Schonhage and Strassen [1971], and the $O(n^{2.81})$ matrix multiplication algorithm of Strassen [1969]. The $O(n^{1.59})$ integer multiplication algorithm is from Karatsuba and Ofman [1962]. Moenck and Borodin [1972] develop several efficient divide-and-conquer algorithms for modular arithmetic and polynomial interpolation and evaluation.

    Dynamic programming was popularized by Bellman [1957]. The application of dynamic programming to triangulation is due to Fuchs, Kedem, and Uselton [1977]. Exercise 10.11 is from Knuth [1981]. Knuth [1971] contains a solution to the optimal binary search tree problem in Exercise 10.12.

    Lin and Kernighan [1973] describe an effective heuristic for the traveling salesman problem.

    See Aho, Hopcroft, and Ullman [1974] and Garey and Johnson [1979] for a discussion of NP-complete and other computationally difficult problems.

CHAPTER 11

Data Structures
and
Algorithms
for
External Storage

We begin this chapter by considering the differences in access characteristics between main memory and external storage devices such as disks. We then present several algorithms for sorting files of externally stored data. We conclude the chapter with a discussion of data structures and algorithms, such as indexed files and B-trees, that are well suited for storing and retrieving information on secondary storage devices.

11.1 A Model of External Computation

In the algorithms discussed so far, we have assumed that the amount of input data is sufficiently small to fit in main memory at the same time. But what if we want to sort all the employees of the government by length of service or store all the information in the nation's tax returns? In such problems the amount of data to be processed exceeds the capacity of the main memory. Most large computer systems have on-line external storage devices, such as disks or mass storage devices, on which vast quantities of data can be stored. These on-line storage devices, however, have access characteristics that are quite different from those of main memory. A number of data structures and algorithms have been developed to utilize these devices more effectively. This chapter discusses data structures and algorithms for sorting and retrieving information stored in secondary memory.

Pascal, and some other languages, provide the file data type, which is intended to represent data stored in secondary memory. Even if the language being used does not have a file data type, the operating system undoubtedly supports the notion of files in secondary memory. Whether we are talking about Pascal files or files maintained by the operating system directly, we are faced with limitations on how files may be accessed. The operating system divides secondary memory into equal-sized *blocks*. The size of blocks varies among operating systems, but 512 to 4096 bytes is typical.

We may regard a file as stored in a linked list of blocks, although more typically the operating system uses a tree-like arrangement, where the blocks holding the file are leaves of the tree, and interior nodes each hold pointers to many blocks of the file. If, for example, 4 bytes suffice to hold the address of a block, and blocks are 4096 bytes long, then a root block can hold pointers to up to 1024 blocks. Thus, files of up to 1024 blocks, i.e., about four million bytes, could be represented by a root block and blocks holding the file. Files of up to 2^{20} blocks, or 2^{32} bytes could be represented by a root block pointing to 1024 blocks at an intermediate level, each of which points to 1024 leaf blocks holding a part of the file, and so on.

The basic operation on files is to bring a single block to a *buffer* in main memory; a buffer is simply a reserved area of main memory whose size is the same as the size of a block. A typical operating system facilitates reading the blocks in the order in which they appear in the list of blocks that holds the file. That is, we initially read the first block into the buffer for that file, then replace it by the second block, which is written into the same buffer, and so on.

We can now see the rationale behind the rules for reading Pascal files. Each file is stored in a sequence of blocks, with a whole number of records in each block. (Space may be wasted, as we avoid having one record split across block boundaries.) The read-cursor always points to one of the records in the block that is currently in the buffer. When that cursor must move to a record not in the buffer, it is time to read the next block of the file.

Similarly, we can view the Pascal file-writing process as one of creating a file in a buffer. As records are "written" into the file, they are placed in the buffer for that file, in the position immediately following any previously placed records. When the buffer cannot hold another complete record, the buffer is copied into an available block of secondary storage and that block is appended to the end of the list of blocks for that file. We can now regard the buffer as empty and write more records into it.

The Cost Measure for Secondary Storage Operations

It is the nature of secondary storage devices such as disks that the time to find a block and read it into main memory is large compared with the time to process the data in that block in simple ways. For example, suppose we have a block of 1000 integers on a disk rotating at 1000 rpm. The time to position the head over the track holding this block (*seek time*) plus the time spent waiting for the block to come around to the head (*latency time*) might average 100 milliseconds. The process of writing a block into a particular place on secondary storage takes a similar amount of time. However, the machine could typically do 100,000 instructions in those 100 milliseconds. This is more than enough time to do simple processing to the thousand integers once they are in main memory, such as summing them or finding their maximum. It might even be sufficient time to quicksort the integers.

When evaluating the running time of algorithms that operate on data stored as files, we are therefore forced to consider as of primary importance the number of times we read a block into main memory or write a block onto secondary storage. We call such an operation a *block access*. We assume the size of blocks is fixed by the operating system, so we cannot appear to make an algorithm run faster by increasing the block size, thereby decreasing the number of block accesses. As a consequence, the figure of merit for algorithms dealing with external storage will be the number of block accesses. We begin our study of algorithms for external storage by looking at external sorting.

11.2 External Sorting

Sorting data organized as files, or more generally, sorting data stored in secondary memory, is called "external" sorting. Our study of external sorting begins with the assumption that the data are stored on a Pascal file. We show how a "merge sorting" algorithm can sort a file of n records with only $O(\log n)$ passes through the file; that figure is substantially better than the $O(n)$ passes needed by the algorithms studied in Chapter 8. Then we consider how utilization of certain powers of the operating system to control the reading and writing of blocks at appropriate times can speed up sorting by reducing the time that the computer is idle, waiting for a block to be read into or written out of main memory.

Merge Sorting

The essential idea behind merge sort is that we organize a file into progressively larger *runs*, that is, sequences of records $r_1, \ldots, r_k$, where the key of r_i is no greater than the key of r_{i+1} for $1 \le i < k$. We say a file $r_1, \ldots, r_m$ of records is *organized into runs of length k* if for all $i \ge 0$ such that $ki \le m$, $r_{k(i-1)+1}, r_{k(i-1)+2}, \ldots, r_{ki}$ is a run of length k, and furthermore if m is not divisible by k, and $m = pk+q$, where $q<k$, then the sequence of records $r_{m-q+1}, r_{m-q+2}, \ldots, r_m$, called the *tail*, is a run of length q. For example, the sequence of integers shown in Fig. 11.1 is organized into runs of length 3 as shown. Note that the tail is of length less than 3, but consists of records in sorted order, namely 5, 12.

| 7 | 15 | 29 | 8 | 11 | 13 | 16 | 22 | 31 | 5 | 12 |

Fig. 11.1. File with runs of length three.

The basic step of a merge sort on files is to begin with two files, say f_1 and f_2, organized into runs of length k. Assume that

1. the numbers of runs, including tails, on f_1 and f_2 differ by at most one,
2. at most one of f_1 and f_2 has a tail, and

3. the one with a tail has at least as many runs as the other.

Then it is a simple process to read one run from each of f_1 and f_2, merge the runs and append the resulting run of length $2k$ onto one of two files g_1 and g_2, which are being organized into runs of length $2k$. By alternating between g_1 and g_2, we can arrange that these files are not only organized into runs of length $2k$, but satisfy (1), (2), and (3) above. To see that (2) and (3) are satisfied it helps to observe that the tail among the runs of f_1 and f_2 gets merged into (or perhaps is) the last run created.

We begin by dividing all n of our records into two files f_1 and f_2, as evenly as possible. Any file can be regarded as organized into runs of length 1. Then we can merge the runs of length 1 and distribute them into files g_1 and g_2, organized into runs of length 2. We make f_1 and f_2 empty, and merge g_1 and g_2 into f_1 and f_2, which will then be organized into runs of length 4. Then we merge f_1 and f_2 to create g_1 and g_2 organized into runs of length 8, and so on.

After i passes of this nature, we have two files consisting of runs of length 2^i. If $2^i \geq n$, then one of the two files will be empty and the other will contain a single run of length n, i.e., it will be sorted. As $2^i \geq n$ when $i \geq \log n$, we see that $\lceil \log n \rceil$ passes suffice. Each pass requires the reading of two files and the writing of two files, all of length about $n/2$. The total number of blocks read or written on a pass is thus about $2n/b$, where b is the number of records that fit on one block. The number of block reads and writes for the entire sorting process is thus $O((n \log n)/b)$, or put another way, the amount of reading and writing is about the same as that required by making $O(\log n)$ passes through the data stored on a single file. This figure is a large improvement over the $O(n)$ passes required by many of the sorting algorithms discussed in Chapter 8.

Figure 11.2 shows the merge process in Pascal. We read two files organized into runs of length k and write two files organized into runs of length $2k$. We leave to the reader the specification of an algorithm, following the ideas above, that uses the procedure *merge* of Fig. 11.2 $\log n$ times to sort a file of n records.

```
procedure merge ( k: integer; { the input run length }
        f1, f2, g1, g2: file of recordtype );
    var
        outswitch: boolean; { tells if writing g1 (true) or g2 (false) }
        winner: integer; { selects file with smaller key in current record }
        used: array [1..2] of integer; { used[j] tells how many
            records have been read so far from the current run of file fⱼ }
        fin: array [1..2] of boolean; { fin[j] is true if we have
            finished the run from fⱼ – either we have read k records,
            or reached the end of the file of fⱼ }
        current: array [1..2] of recordtype; { the current records
            from the two files }
```

procedure *getrecord* (*i*: integer); { advance file f_i, but
        not beyond the end of the file or the end of the run.
        Set *fin*[*i*] if end of file or run found }
  begin
      used[*i*] := *used*[*i*] + 1;
      if (*used*[*i*] = *k*) **or**
          (*i* = 1) **and** *eof*(*f*1) **or**
          (*i* = 2) **and** *eof*(*f*2) **then** *fin*[*i*] := true
      else if *i* = 1 **then** *read*(*f*1, *current*[1])
      else *read*(*f*2, *current*[2])
  end; { *getrecord* }

begin { *merge* }
  outswitch := true; { first merged run goes to *g*1 }
  rewrite(*g*1); *rewrite*(*g*2);
  reset(*f*1); *reset*(*f*2);
  while not *eof*(*f*1) **or not** *eof*(*f*2) **do begin** { merge two files }
    { initialize }
    used[1] := 0; *used*[2] := 0;
    fin[1] := false; *fin*[2] := false;
    getrecord(1); *getrecord*(2);
    while not *fin*[1] **or not** *fin*[2] **do begin** { merge two runs }
      { select winner }
      if *fin*[1] **then** *winner* := 2
          { *f*2 wins by "default" – run from *f*1 exhausted }
      else if *fin*[2] **then** *winner* := 1
          { *f*1 wins by default }
      else { neither run exhausted }
          if *current*[1].*key* < *current*[2].*key* **then** *winner* := 1
          else *winner* := 2;
      { write winning record }
      if *outswitch* **then** *write*(*g*1, *current*[*winner*])
      else *write*(*g*2, *current*[*winner*]);
      { advance winning file }
      getrecord(*winner*)
    end;
    { we have finished merging two runs - switch output
      file and repeat }
    outswitch := **not** *outswitch*;
  end;
end; { *merge* }

Fig. 11.2. The procedure *merge*.

Notice that the procedure *merge*, of Fig. 11.2, is not required ever to have a complete run in memory; it reads and writes a record at a time. It is our desire not to store whole runs in main memory that forces us to have two input files. Otherwise, we could read two runs at a time from one file.

Example 11.1. Suppose we have the list of 23 numbers shown divided into two files in Fig. 11.3(a). We begin by merging runs of length 1 to create the two files of Fig. 11.3(b). For example, the first runs of length 1 are 28 and 31; these are merged by picking 28 and then 31. The next two runs of length one, 3 and 5, are merged to form a run of length two, and that run is placed on the second file of Fig. 11.3(b). The runs are separated in Fig. 11.3(b) by vertical lines that are not part of the file. Notice that the second file in Fig. 11.3(b) has a tail of length one, the record 22, while the first file has no tail.

We go from Fig. 11.3(b) to (c) by merging runs of length two. For example, the two runs 28, 31 and 3, 5 are merged to form 3, 5, 28, 31 in Fig. 11.3(c). By the time we get to runs of length 16 in Fig. 11.3(e), one file has one complete run and the other has only a tail, of length 7. At the last stage, where the files are ostensibly organized as runs of length 32, we in fact have one file with a tail only, of length 23, and the second file is empty. The single run of length 23 is, of course, the desired sorted order. □

Speeding Up Merge Sort

We have, for the sake of a simple example, shown the merge sort process as starting from runs of length one. We shall save considerable time if we begin with a pass that, for some appropriate k, reads groups of k records into main memory, sorts them, say by quicksort, and writes them out as a run of length k.

For example, if we have a million records, it would take 20 passes through the data to sort starting with runs of length one. If, however, we can hold 10,000 records at once in main memory, we can, in one pass, read 100 groups of 10,000 records, sort each group, and be left with 100 runs of length 10,000 distributed evenly between two files. Seven more merging passes results in a file organized as one run of length $10,000 \times 2^7 = 1,280,000$, which is more than a million and means that the data are sorted.

Minimizing Elapsed Time

In modern time-shared computer systems, one is generally not charged for the time one's program spends waiting for blocks of a file to be read in, as must happen during the merge sort process. However, the fact is that the elapsed time of such a sorting process is greater, often substantially so, than the time spent computing with data found in the main memory. If we are sorting really large files, where the whole operation takes hours, the elapsed time becomes important, even if we do not "pay" for the time, and we should consider how the merging process might be performed in minimum elapsed time.

As was mentioned, it is typical that the time to read data from a disk or tape is greater than the time spent doing simple computations with that data,

```
28   3   93   10   54   65   30   90   10   69    8   22
31   5   96   40   85    9   39   13    8   77   10
```

(a) initial files

```
28   31 | 93   96 | 54   85 | 30   39 |  8   10 |  8   10
 3    5 | 10   40 |  9   65 | 13   90 | 69   77 | 22
```

(b) organized into runs of length 2

```
 3    5   28   31 |  9   54   65   85 |  8   10   69   77
10   40   93   06 | 13   30   39   90 |  8   10   22
```

(c) organized into runs of length 4

```
 3    5   10   28   31   40   93   96 |  8    8   10   10   22   69   77
 9   13   30   39   54   65   85   90 |
```

(d) organized into runs of length 8

```
 3   5    9   10   13   28   30   31   39   40   54   65   85   90   93   96
 8   8   10   10   22   69   77
```

(e) organized into runs of length 16

3 5 8 8 9 10 10 10 13 22 28 30 31 39 40 54 65 69 77 85 90 93 96

(f) organized into runs of length 32

Fig. 11.3. Merge-sorting a list.

such as merging. We should therefore expect that if there is only one channel over which data can be transferred into or out of main memory, then this channel will form a bottleneck; the data channel will be busy all the time, and the total elapsed time equals the time spent moving data into and out of main memory. That is, all the computation will be done almost as soon as the data becomes available, while additional data are being read or written.

Even in this simple environment, we must exercise a bit of care to make sure that we are done in the minimum amount of time. To see what can go wrong, suppose we read the two input files f_1 and f_2 one block at a time, alternately. The files are organized into runs of some length much larger than the size of a block, so to merge two runs we must read many blocks from each file. However, suppose it happens that all the records in the run from file f_1 happen to precede all the records from file f_2. Then as we read blocks alternately, all the blocks from f_2 have to remain in memory. There may not be space to hold all these blocks in main memory, and even if there is, we must, after reading all the blocks of the run, wait while we copy and write the whole run from f_2.

To avoid these problems, we consider the keys of the last records in the last blocks read from f_1 and f_2, say keys k_1 and k_2, respectively. If either run is exhausted, we naturally read next from the other file. However, if a run is not exhausted, we next read a block from f_1 if $k_1 < k_2$, and we read from f_2 otherwise. That is, we determine which of the two runs will first have all its records currently in main memory selected, and we replenish the supply of records for that run first. If selection of records proceeds faster than reading, we know that when we have read the last block of the two runs, there cannot be more than two "blockfuls" of records left to merge; perhaps the records will be distributed over three blocks, but never more.

Multiway Merge

If reading and writing between main and secondary memory is the bottleneck, perhaps we could save time if we had more than one data channel. Suppose, for example, that we have $2m$ disk units, each with its own channel. We could place m files, $f_1, f_2, \ldots, f_m$ on m of the disk units, say organized as runs of length k. We can read m runs, one from each file, and merge them into one run of length mk. This run is placed on one of m output files $g_1, g_2, \ldots, g_m$, each getting a run in turn.

The merging process in main memory can be carried out in $O(\log m)$ steps per record if we organize the m *candidate records*, that is, the currently smallest unselected records from each file, into a partially ordered tree or other data structure that supports the priority queue operations INSERT and DELETEMIN in logarithmic time. To select from the priority queue the record with the smallest key, we perform DELETEMIN, and then INSERT into the priority queue the next record from the file of the winner, as a replacement for the selected record.

If we have n records, and the length of runs is multiplied by m with each pass, then after i passes runs will be of length m^i. If $m^i \geq n$, that is, after $i = \log_m n$ passes, the entire list will be sorted. As $\log_m n = \log_2 n / \log_2 m$, we save by a factor of $\log_2 m$ in the number of times we read each record. Moreover, if m is the number of disk units used for input files, and m are used for output, we can process data m times as fast as if we had only one disk unit for input and one for output, or $2m$ times as fast as if both input and output files were stored on one disk unit. Unfortunately, increasing m indefinitely does not speed processing by a factor of $\log m$. The reason is that for large enough

m, the time to do the merging in main memory, which is actually increasing as $\log m$, will exceed the time to read or write the data. At that point, further increases in m actually increase the elapsed time, as computation in main memory becomes the bottleneck.

Polyphase Sorting

We can perform an m-way merge sort with only $m+1$ files, as an alternative to the $2m$-file strategy described above. A sequence of passes is made, merging runs from m of the files into longer runs on the remaining file. The insights needed are the following:

1. In one pass, when runs from each of m files are merged into runs of the $m+1^{st}$ file, we need not use all the runs on each of the m input files. Rather, each file, when it becomes the output file, is filled with runs of a certain length. It uses some of these runs to help fill each of the other m files when it is their turn to be the output file.

2. Each pass produces files of a different length. Since each of the files loaded with runs on the previous m passes contributes to the runs of the current pass, the length on one pass is the sum of the lengths of the runs produced on the previous m passes. (If fewer than m passes have taken place, regard passes prior to the first as having produced runs of length 1.)

Such a merge-sorting process is called a *polyphase sort*. The exact calculation of the numbers of passes needed, as a function of m and n (the number of records), and calculation of the initial distribution of the runs into m files are left as exercises. However, we shall give one example here to suggest the general case.

Example 11.2. If $m = 2$, we start with two files f_1 and f_2, organized into runs of length 1. Records from f_1 and f_2 are merged to make runs of length 2 on the third file, f_3. Just enough runs are merged to empty f_1. We then merge the remaining runs of length 1 from f_2 with an equal number of runs of length 2 from f_3. These yield runs of length 3, and are placed on f_1. Then we merge runs of length 2 on f_3 with runs of length 3 on f_1. These runs, of length 5, are placed on f_2, which was emptied at the previous pass.

The run length sequence: 1, 1, 2, 3, 5, 8, 13, 21, . . . , is the *Fibonacci* sequence. This sequence is generated by the recurrence relation $F_0 = F_1 = 1$, and $F_i = F_{i-1} + F_{i-2}$, for $i \geq 2$. Note that the ratio of consecutive Fibonacci numbers F_{i+1}/F_i approaches the "golden ratio" $(\sqrt{5}+1)/2 = 1.618\ldots$ as i gets large.

It turns out that in order to keep this process going, until the list is sorted, the initial numbers of records on f_1 and f_2 must be two consecutive Fibonacci numbers. For example, Fig. 11.4 shows what happens when we start with $n = 34$ records (34 is the Fibonacci number F_8) distributed 13 on f_1 and 21 on f_2. (13 and 21 are F_6 and F_7, so the ratio F_7/F_6 is very close to 1.618. It is 1.615, in fact). The status of a file is represented in Fig. 11.4 as $a(b)$, meaning it has a runs of length b. □

after pass	f_1	f_2	f_3
initial	13(1)	21(1)	empty
1	empty	8(1)	13(2)
2	8(3)	empty	5(2)
3	3(3)	5(5)	empty
4	empty	2(5)	3(8)
5	2(13)	empty	1(8)
6	1(13)	1(21)	empty
7	empty	empty	1(34)

Fig. 11.4. Example of polyphase sorting.

When Input/Output Speed is not a Bottleneck

When reading files is the bottleneck, the next block read must be carefully chosen. As we have mentioned, the situation to avoid is one where we have to store many blocks of one run because that run happened to have records with high keys, which get selected after most or all of the records of the other run. The trick to avoiding this problem is to determine quickly which run will first exhaust those of its records currently in main memory, by comparing the last such records in each file.

When the time taken to read data into main memory is comparable to, or less than, the time taken to process the data, it becomes even more critical that we select the input file from which to read a block carefully, since there is no hope of building a reserve of records in main memory in case the merge process suddenly starts taking more records from one run than the other. The "trick" mentioned above helps us in a variety of situations, as we shall see.

We consider the case where merging, rather than reading or writing, is a bottleneck for two reasons.

1. As we have seen, if we have many disk or tape units available, we may speed input/output sufficiently that the time to do the merge exceeds input time or output time.

2. Higher speed channels may yet become commercially available.

We therefore shall consider a simple model of the problem that might be encountered when merging becomes the bottleneck in a merge sort performed on data stored in secondary memory. Specifically, we assume that

a) We are merging runs that are much larger than blocks.

b) There are two input files and two output files. The input files are stored on one disk (or other device connected to main memory by a single channel) and the output files are on another similar unit with one channel.

c)   The times to
      i)   read a block
     ii)   write a block, and
   iii)   select enough of the lowest-keyed records from among those of
          two runs presently in main memory to fill a block,

are all the same.

Under these assumptions, we can consider a class of merging strategies where several input buffers (space to hold a block) are allocated in main memory. At all times some of these buffers will hold the unselected records from the two input runs, and one of them will be in the process of being read into from one of the input files. Two other buffers will hold output, that is, the selected records in their properly merged order. At all times, one of these buffers is in the process of being written into one of the output files and the other is being filled from records selected from the input buffers.

A *stage* consists of doing the following (possibly at the same time):

1.   reading an input block into an input buffer,

2.   filling one of the output buffers with selected records, that is, records with the smallest keys among those currently held in the input buffer, and

3.   writing the other output buffer into one of the two output files being formed.

By our assumptions, (1), (2), and (3) all take the same time. For maximum efficiency, we must do them in parallel. We can do so, unless (2), the selection of records with the smallest keys, includes some of the records currently being read in.† We must thus devise a strategy for selecting buffers to be read so that at the beginning of each stage the b unselected records with smallest keys are already in input buffers, where b is the number of records that fills a block or buffer.

The conditions under which merging can be done in parallel with reading are simple. Let k_1 and k_2 be the highest keys among unselected records in main memory from the first and second runs, respectively. Then there must be in main memory at least b unselected records whose keys do not exceed $\min(k_1, k_2)$. We shall first show how to do the merge with six buffers, three used for each file, then show that four buffers suffice if we share them between the two files.

---

† It is tempting to assume that if (1) and (2) take the same time, then selection could never catch up with reading; if the whole block were not yet read, we would select from the first records of the block, those that had the lower keys, anyway. However, the nature of reading from disks is that a long period elapses before the block is found and anything at all is read. Thus our only safe assumption is that nothing of the block being read in a stage is available for selection during that stage.

A Six-Input Buffer Scheme

Our first scheme is represented by the picture in Fig. 11.5. The two output buffers are not shown. There are three buffers for each file; each buffer has capacity b records. The shaded area represents available records, and keys are in increasing order clockwise around the circles. At all times, the total number of unselected records is $4b$ (unless fewer than that number of records remain from the runs being merged). Initially, we read the first two blocks from each run into buffers.† As there are always $4b$ records available, and at most $3b$ can be from one file, we know there are b records at least from each file. If k_1 and k_2 are the largest available keys from the two runs, there must be b records with keys equal to or less than k_1 and b with keys equal to or less than k_2. Thus, there are b records with keys equal to or less than $\min(k_1, k_2)$.

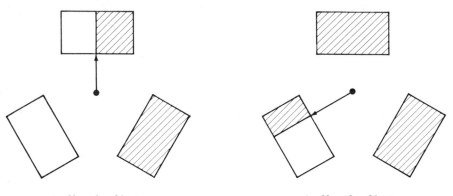

buffers for file 1   buffers for file 2

Fig. 11.5. A six-input buffer merge.

The question of which file to read from next is trivial. Usually, since two buffers will be partially full, as in Fig. 11.5, there will be only one empty buffer, and this must be filled. If it happens, as in the beginning, that each run has two completely filled and one empty buffer, pick either empty buffer to fill. Note that our proof that we could not exhaust a run [b records with keys equal to or less than $\min(k_1, k_2)$ exist] depended only on the fact that $4b$ records were present.

As an aside, the arrows in Fig. 11.5 represent pointers to the first (lowest-keyed) available records from the two runs. In Pascal, we could

---

† If these are not the first runs from each file, then this initialization can be done after the previous runs were read and the last $4b$ records from these runs are being merged.

represent such a pointer by two integers. The first, in the range 1..3, represents the buffer pointed to, and the second, in the range 1..b, represents the record within the buffer. Alternatively, we could let the buffers be the first, middle, and last thirds of one array and use one integer in the range 1..3b. In other languages, where pointers can point to elements of arrays, a pointer of type ↑ recordtype would be preferred.

A Four-Buffer Scheme

Figure 11.6 suggests a four-buffer scheme. At the beginning of each stage, 2b records are available. Two of the input buffers are assigned to one of the files; B_1 and B_2 in Fig. 11.6 are assigned to file one. One of these buffers will be partially full (empty in the extreme case) and the other full. A third buffer is assigned to the other file, as B_3 is assigned to file two in Fig. 11.6. It is partially full (completely full in the extreme case). The fourth buffer is uncommitted, and will be filled from one of the files during the stage.

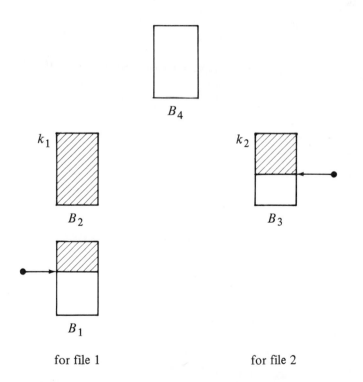

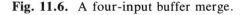

Fig. 11.6. A four-input buffer merge.

We shall maintain, of course, the property that allows us to merge in parallel with reading; at least b of the records in Fig. 11.6 must have keys equal to or less than $\min(k_1, k_2)$, where k_1 and k_2 are the keys of the last available records from the two files, as indicated in Fig. 11.6. We call a configuration that obeys the property *safe*. Initially, we read one block from each run (this is the extreme case, where B_1 is empty and B_3 is full in Fig. 11.6), so that the initial configuration is safe. We must, on the assumption that Fig. 11.6 is safe, show that the configuration will be safe after the next stage is complete.

If $k_1 < k_2$, we choose to fill B_4 with the next block from file one, and otherwise, we fill it from file two. Suppose first that $k_1 < k_2$. Since B_1 and B_3 in Fig. 11.6 have together exactly b records, we must, during the next stage, exhaust B_1, else we would exhaust B_3 and contradict the safety of Fig. 11.6. Thus after a stage the configuration looks like Fig. 11.7(a).

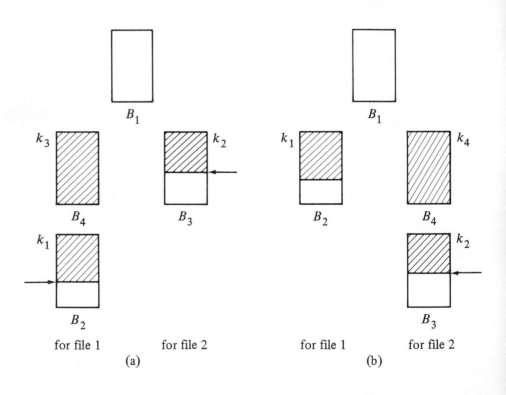

Fig. 11.7. Configuration after one stage.

To see that Fig. 11.7(a) is safe, consider two cases. First, if k_3, the last key in the newly-read block B_4, is less than k_2, then as B_4 is full, there are surely b records equal to or less than $\min(k_3, k_2)$, and the configuration is safe. If $k_2 \leq k_3$, then since $k_1 < k_2$ was assumed (else we would have filled B_4 from file two), the b records in B_2 and B_3 have keys equal to or less than $\min(k_2, k_3) = k_2$.

Now let us consider the case where $k_1 \geq k_2$ in Fig. 11.6. Then we choose to read the next block from file two. Figure 11.7(b) shows the resulting situation. As in the case $k_1 < k_2$, we can argue that B_1 must exhaust, which is why we show file one as having only buffer B_2 in Fig. 11.7(b). The proof that Fig. 11.7(b) is safe is just like the proof for Fig. 11.7(a).

Note that, as in the six buffer scheme, we do not read a file past the end of a run. However, if there is no need to read in a block from one of the present runs, we can read a block from the next run on that file. In this way we have the opportunity to read one block from each of the next runs, and we are ready to begin merging runs as soon as we have selected the last records of the previous run.

11.3 Storing Information in Files

In this section we consider data structures and algorithms for storing and retrieving information in externally stored files. We shall view a file as a sequence of records, where each record consists of the same sequence of fields. Fields can be either *fixed length,* having a predetermined number of bytes, or *variable length,* having an arbitrary size. Files with fixed length records are commonly used in database management systems to store highly structured data. Files with variable length records are typically used to store textual information; they are not available in Pascal. In this section we shall assume fixed-length fields; the fixed-length techniques can be modified simply to work for variable-length records.

The operations on files we shall consider are the following.

1. INSERT a particular record into a particular file.

2. DELETE from a particular file all records having a designated value in each of a designated set of fields.

3. MODIFY all records in a particular file by setting to designated values certain fields in those records that have a designated value in each of another set of fields.

4. RETRIEVE all records having designated values in each of a designated set of fields.

Example 11.3. For example, suppose we have a file whose records consist of three fields: *name, address,* and *phone.* We might ask to retrieve all records with *phone* = 555-1234, to insert the record (Fred Jones, 12 Apple St., 555-1234), or to delete all records with *name* = "Fred Jones" and *address* = "12 Apple St." As another example, we might wish to modify all records with *name* = "Fred Jones" by setting the *phone* field to 555-1234. □

To a great extent we can view operations on files as if the files were sets of records and the operations were those discussed in Chapters 4 and 5. There are two important differences, however. First, when we talk of files on external storage devices, we must use the cost measure discussed in Section 11.1 when evaluating strategies for organizing the files. That is, we assume that files are stored in some number of physical blocks, and the cost of an operation is the number of blocks that we must read into main memory or write from main memory onto external storage.

The second difference is that records, being concrete data types of most programming languages, can be expected to have pointers to them, while the abstract elements of a set would not normally have "pointers" to them. In particular, database systems frequently make use of pointers to records when organizing data. The consequence of such pointers is that records frequently must be regarded as *pinned*; they cannot be moved around in storage because of the possibility that a pointer from some unknown place would fail to point to the record after it was moved.

A simple way to represent pointers to records is the following. Each block has a *physical address,* which is the location on the external storage device of the beginning of the block. It is the job of the file system to keep track of physical addresses. One way to represent record addresses is to use the physical address of the block holding the record together with an *offset*, giving the number of bytes in the block preceding the beginning of the record. These physical address—offset pairs can then be stored in fields of type "pointer to record."

A Simple Organization

The simplest, and also least efficient, way to implement the above file operations is to use the file reading and writing primitives such as found in Pascal. In this "organization" (which is really a "lack of organization"), records can be stored in any order. Retrieving a record with specified values in certain fields is achieved by scanning the file and looking at each record to see if it has the specified values. An insertion into a file can be performed by appending the record to the end of the file.

For modification of records, scan the file and check each record to see if it matches the designated fields, and if so, make the required changes to the record. A deletion operation works almost the same way, but when we find a record whose fields match the values required for the deletion to take place, we must find a way to delete the record. One possibility is to shift all subsequent records one position forward in their blocks, and move the first record of each subsequent block into the last position of the previous block of the file. However, this approach will not work if records are pinned, because a pointer to the ith record in the file would then point to the $i+1^{st}$ record.

If records are pinned, we must use a somewhat different approach. We mark deleted records in some way, but we do not move records to fill their space, nor do we ever insert a new record into their space. Thus, the record becomes deleted logically from the file, but its space is still used for the file.

This is necessary so that if we ever follow a pointer to a deleted record, we shall discover that the record pointed to was deleted and take some appropriate action, such as making the pointer NIL so it will not be followed again. Two ways to mark records as deleted are:

1. Replace the record by some value that could never be the value of a "real" record, and when following a pointer, assume the record is deleted if it has that value.

2. Let each record have a *deletion bit,* a single bit that is 1 in records that have been deleted and 0 otherwise.

Speeding Up File Operations

The obvious disadvantage of a sequential file is that file operations are slow. Each operation requires us to read the entire file, and some blocks may have to be rewritten as well. Fortunately, there are file organizations that allow us to access a record by reading into main memory only a small fraction of the entire file.

To make such organizations possible, we assume each record of a file has a *key*, a set of fields that uniquely identifies each record. For example, the *name* field of the *name-address-phone* file might be considered a key by itself. That is, we might assume that two records with the same *name* field value cannot exist simultaneously in the file. Retrieval of a record, given values for its key fields, is a common operation, and one that is made especially easy by many common file organizations.

Another element we need for fast file operations is the ability to access blocks directly, rather than running sequentially through the blocks holding a file. Many of the data structures we use for fast file operations will use pointers to the blocks themselves, which are the physical addresses of the blocks, as described above. Unfortunately, we cannot write in Pascal, or in many other languages, programs that deal with data on the level of physical blocks and their addresses; such operations are normally done by file system commands. However, we shall briefly give an informal description of how organizations that make use of direct block access work.

Hashed Files

Hashing is a common technique used to provide fast access to information stored on secondary files. The basic idea is similar to open hashing discussed in Section 4.7. We divide the records of a file among *buckets*, each consisting of a linked list of one or more blocks of external storage. The organization is similar to that portrayed in Fig. 4.10. There is a bucket table containing B pointers, one for each bucket. Each pointer in the bucket table is the physical address of the first block of the linked-list of blocks for that bucket.

The buckets are numbered $0, 1, \ldots, B-1$. A hash function h maps each key value into one of the integers 0 through $B-1$. If x is a key, $h(x)$ is the number of the bucket that contains the record with key x, if such a record is present at all. The blocks making up each bucket are chained together in a linked list. Thus, the header of the ith block of a bucket contains a pointer to the physical address of the $i+1^{st}$ block. The last block of a bucket contains a NIL pointer in its header.

This arrangement is illustrated in Fig. 11.8. The major difference between Figs. 11.8 and 4.10 is that here, elements stored in one block of a bucket do not need to be chained by pointers; only the blocks need to be chained.

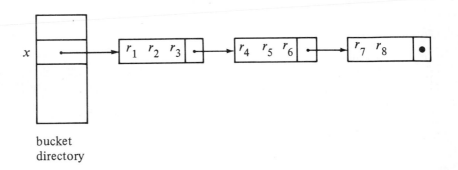

bucket
directory

Fig. 11.8. Hashing with buckets consisting of chained blocks.

If the size of the bucket table is small, it can be kept in main memory. Otherwise, it can be stored sequentially on as many blocks as necessary. To look for the record with key x, we compute $h(x)$, and find the block of the bucket table containing the pointer to the first block of bucket $h(x)$. We then read the blocks of bucket $h(x)$ successively, until we find a block that contains the record with key x. If we exhaust all blocks in the linked list for bucket $h(x)$, we conclude that x is not the key of any record.

This structure is quite efficient if the operation is one that specifies values for the fields in the key, such as retrieving the record with a specified key value or inserting a record (which, naturally, specifies the key value for that record). The average number of block accesses required for an operation that specifies the key of a record is roughly the average number of blocks in a bucket, which is n/bk if n is the number of records, a block holds b records, and k is the number of buckets. Thus, on the average, operations based on keys are k times faster with this organization than with the unorganized file. Unfortunately, operations not based on keys are not speeded up, as we must examine essentially all the buckets during these other operations. The only general way to speed up operations not based on keys seems to be the use of secondary indices, which are discussed at the end of this section.

To insert a record with key value x, we first check to see if there is already a record with key x. If there is, we report error, since we assume that the key uniquely identifies each record. If there is no record with key x, we insert the new record in the first block in the chain for bucket $h(x)$ into which the record can fit. If the record cannot fit into any existing block in the chain for bucket $h(x)$, we call upon the file system to find a new block into which the record is placed. This new block is then added to the end of the chain for bucket $h(x)$.

To delete a record with key x, we first locate the record, and then set its deletion bit. Another possible deletion strategy (which cannot be used if the records are pinned) is to replace the deleted record with the last record in the chain for $h(x)$. If the removal of the last record makes the last block in the chain for $h(x)$ empty, we can then return the empty block to the file system for later re-use.

A well-designed hashed-access file organization requires only a few block accesses for each file operation. If our hash function is good, and the number of buckets is roughly equal to the number of records in the file divided by the number of records that can fit on one block, then the average bucket consists of one block. Excluding the number of block accesses to search the bucket table, a typical retrieval based on keys will then take one block access, and a typical insertion, deletion, or modification will take two block accesses. If the average number of records per bucket greatly exceeds the number that will fit on one block, we can periodically reorganize the hash table by doubling the number of buckets and splitting each bucket into two. The ideas were covered at the end of Section 4.8.

Indexed Files

Another common way to organize a file of records is to maintain the file sorted by key values. We can then search the file as we would a dictionary or telephone directory, scanning only the first name or word on each page. To facilitate the search we can create a second file, called a *sparse index*, which consists of pairs (x, b), where x is a key value and b is the physical address of the block in which the first record has key value x. This sparse index is maintained sorted by key values.

Example 11.4. In Fig. 11.9 we see a file and its sparse index file. We assume that three records of the main file, or three pairs of the index file, fit on one block. Only the key values, assumed to be single integers, are shown for records in the main file. □

To retrieve a record with a given key x, we first search the index file for a pair (x, b). What we actually look for is the largest z such that $z \leq x$ and there is a pair (z, b) in the index file. Then key x appears in block b if it is present in the main file at all.

There are several strategies that can be used to search the index file. The simplest is *linear search*. We read the index file from the beginning until we encounter a pair (x, b), or until we encounter the first pair (y, b) where $y > x$. In the latter case, the preceding pair (z, b') must have $z < x$, and if the record with key x is anywhere, it is in block b'.

Linear search is only suitable for small index files. A faster method is

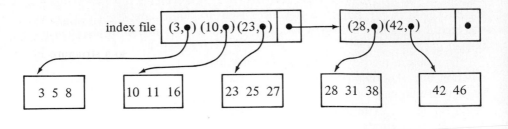

Fig. 11.9. A main file and its sparse index.

binary search. Assume the index file is stored on blocks $b_1, b_2, \ldots, b_n$. To search for key value x, we take the middle block $b_{\lceil n/2 \rceil}$ and compare x with the key value y in the first pair in that block. If $x < y$, we repeat the search on blocks $b_1, b_2, \ldots, b_{\lceil n/2 \rceil - 1}$. If $x \geq y$, but x is less than the key of block $b_{\lceil n/2 \rceil + 1}$ (or if $n = 1$, so there is no such block), we use linear search to see if x matches the first component of an index pair on block $b_{\lceil n/2 \rceil}$. Otherwise, we repeat the search on blocks $b_{\lceil n/2 \rceil + 1}, b_{\lceil n/2 \rceil + 2}, \ldots, b_n$. With binary search we need examine only $\lceil \log_2(n+1) \rceil$ blocks of the index file.

To initialize an indexed file, we sort the records by their key values, and then distribute the records to blocks, in that order. We may choose to pack as many as will fit into each block. Alternatively, we may prefer to leave space for a few extra records that may be inserted later. The advantage is that then insertions are less likely to overflow the block into which the insertion takes place, with the resultant requirement that adjacent blocks be accessed. After partitioning the records into blocks in one of these ways, we create the index file by scanning each block in turn and finding the first key on each block. Like the main file, some room for growth may be left on the blocks holding the index file.

Suppose we have a sorted file of records that are stored on blocks $B_1, B_2, \ldots, B_m$. To insert a new record into this sorted file, we use the index file to determine which block B_i should contain the new record. If the new record will fit in B_i, we place it there, in the correct sorted order. We then adjust the index file, if the new record becomes the first record on B_i.

If the new record cannot fit in B_i, a variety of strategies are possible. Perhaps the simplest is to go to block B_{i+1}, which can be found through the index file, to see if the last record of B_i can be moved to the beginning of B_{i+1}. If so, this last record is moved to B_{i+1}, and the new record can then be inserted in the proper position in B_i. The index file entry for B_{i+1}, and possibly for B_i, must be adjusted appropriately.

If B_{i+1} is also full, or if B_i is the last block ($i = m$), then a new block is obtained from the file system. The new record is inserted in this new block, and the new block is to follow block B_i in the order. We now use this same procedure to insert a record for the new block in the index file.

Unsorted Files with a Dense Index

Another way to organize a file of records is to maintain the file in random order and have another file, called a *dense index*, to help locate records. The dense index consists of pairs (x, p), where p is a pointer to the record with key x in the main file. These pairs are sorted by key value, so a structure like the sparse index mentioned above, or the B-tree mentioned in the next section, could be used to help find keys in the dense index.

With this organization we use the dense index to find the location in the main file of a record with a given key. To insert a new record, we keep track of the last block of the main file and insert the new record there, getting a new block from the file system if the last block is full. We also insert a pointer to that record in the dense index file. To delete a record, we simply set the deletion bit in the record and delete the corresponding entry in the dense index (perhaps by setting a deletion bit there also).

Secondary Indices

While the hashed and indexed structures speed up operations based on keys substantially, none of them help when the operation involves a search for records given values for fields other than the key fields. If we wish to find the records with designated values in some set of fields $F_1, \ldots, F_k$ we need a *secondary index* on those fields. A secondary index is a file consisting of pairs (v, p), where v is a list of values, one for each of the fields $F_1, \ldots, F_k$, and p is a pointer to a record. There may be more than one pair with a given v, and each associated pointer is intended to indicate a record of the main file that has v as the list of values for the fields $F_1, \ldots, F_k$.

To retrieve records given values for the fields $F_1, \ldots, F_k$, we look in the secondary index for a record or records with that list of values. The secondary index itself can be organized in any of the ways discussed for the organization of files by key value. That is, we pretend that v is a key for (v, p).

For example, a hashed organization does not really depend on keys being unique, although if there were very many records with the same "key" value, then records might distribute themselves into buckets quite nonuniformly, with the effect that hashing would not speed up access very much. In the extreme, say there were only two values for the fields of a secondary index. Then all but two buckets would be nonempty, and the hash table would only speed up operations by a factor of two at most, no matter how many buckets there were. Similarly, a sparse index does not require that keys be unique, but if they are not, then there may be two or more blocks of the main file that have the same lowest "key" value, and all such blocks will be searched when we are looking for records with that value.

With either the hashed or sparse index organization for our secondary index file, we may wish to save space by bunching all records with the same value. That is, the pairs $(v, p_1), (v, p_2), \ldots, (v, p_m)$ can be replaced by v followed by the list $p_1, p_2, \ldots, p_m$.

One might naturally wonder whether the best response time to random operations would not be obtained if we created a secondary index for each field, or even for all subsets of the fields. Unfortunately, there is a penalty to be paid for each secondary index we choose to create. First, there is the space needed to store the secondary index, and that may or may not be a problem, depending on whether space is at a premium.

In addition, each secondary index we create slows down all insertions and all deletions. The reason is that when we insert a record, we must also insert an entry for that record into each secondary index, so the secondary indices will continue to represent the file accurately. Updating a secondary index takes at least two block accesses, since we must read and write one block. However, it may take considerably more than two block accesses, since we have to find that block, and any organization we use for the secondary index file will require a few extra accesses, on the average, to find any block. Similar remarks hold for each deletion. The conclusion we draw is that selection of secondary indices requires judgment, as we must determine which sets of fields will be specified in operations sufficiently frequently that the time to be saved by having that secondary index available more than balances the cost of updating that index on each insertion and deletion.

11.4 External Search Trees

The tree data structures presented in Chapter 5 to represent sets can also be used to represent external files. The B-tree, a generalization of the 2-3 tree discussed in Chapter 5, is particularly well suited for external storage, and it has become a standard organization for indices in database systems. This section presents the basic techniques for retrieving, inserting, and deleting information in B-trees.

Multiway Search Trees

An m-ary search tree is a generalization of a binary search tree in which each node has at most m children. Generalizing the binary search tree property, we require that if n_1 and n_2 are two children of some node, and n_1 is to the left of n_2, then the elements descending from n_1 are all less than the elements descending from n_2. The operations of MEMBER, INSERT, and DELETE for an m-ary search tree are implemented by a natural generalization of those operations for binary search trees, as discussed in Section 5.1.

However, we are interested here in the storage of records in files, where the files are stored in blocks of external storage. The correct adaptation of the multiway tree idea is to think of the nodes as physical blocks. An interior node holds pointers to its m children and also holds the $m-1$ key values that separate the descendants of the children. Leaf nodes are also blocks; these blocks hold the records of the main file.

If we used a binary search tree of n nodes to represent an externally stored file, then it would require $\log_2 n$ block accesses to retrieve a record from the file, on the average. If instead, we use an m-ary search tree to represent the file, it would take an average of only $\log_m n$ block accesses to retrieve a record. For $n = 10^6$, the binary search tree would require about 20 block accesses, while a 128-way search tree would take only 3 block accesses.

We cannot make m arbitrarily large, because the larger m is, the larger the block size must be. Moreover, it takes longer to read and process a larger block, so there is an optimum value for m to minimize the amount of time needed to search an external m-ary search tree. In practice a value close to the minimum is obtained for a broad range of m's. (See Exercise 11.18).

B-trees

A B-tree is a special kind of balanced m-ary tree that allows us to retrieve, insert, and delete records from an external file with a guaranteed worst-case performance. It is a generalization of the 2-3 tree discussed in Section 5.4. Formally, a *B-tree of order m* is an m-ary search tree with the following properties:

1. The root is either a leaf or has at least two children.
2. Each node, except for the root and the leaves, has between $\lceil m/2 \rceil$ and m children.
3. Each path from the root to a leaf has the same length.

Note that every 2-3 tree is a B-tree of order 3. Figure 11.10 shows a B-tree of order 5, in which we assume that at most three records fit in a leaf block.

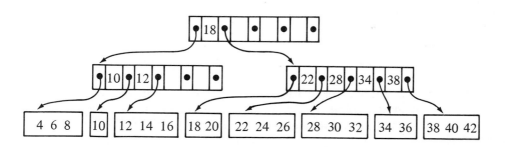

Fig. 11.10. B-tree of order 5.

We can view a B-tree as a hierarchical index in which each node occupies a block in external storage. The root of the B-tree is the first level index. Each non-leaf node in the B-tree is of the form

$$(p_0, k_1, p_1, k_2, p_2, \ldots, k_n, p_n)$$

where p_i is a pointer to the ith child of the node, $0 \le i \le n$, and k_i is a key, $1 \le i \le n$. The   keys   within   a   node   are   in   sorted   order   so

$k_1 < k_2 < \cdots < k_n$. All keys in the subtree pointed to by p_0 are less than k_1. For $1 \le i < n$, all keys in the subtree pointed to by p_i have values greater than or equal to k_i and less than k_{i+1}. All keys in the subtree pointed to by p_n are greater than k_n.

There are several ways to organize the leaves. Here we shall assume that the main file is stored only in the leaves. Each leaf is assumed to occupy one block.

Retrieval

To retrieve a record r with key value x, we trace the path from the root to the leaf that contains r, if it exists in the file. We trace this path by successively fetching interior nodes $(p_0, k_1, p_1, \ldots, k_n, p_n)$ from external storage into main memory and finding the position of x relative to the keys $k_1, k_2, \ldots, k_n$. If $k_i \le x < k_{i+1}$, we next fetch the node pointed to by p_i and repeat the process. If $x < k_1$, we use p_0 to fetch the next node; if $x \ge k_n$, we use p_n. When this process brings us to a leaf, we search for the record with key value x. If the number of entries in a node is small, we can use linear search within the node; otherwise, it would pay to use binary search.

Insertion

Insertion into a B-tree is similar to insertion into a 2-3 tree. To insert a record r with key value x into a B-tree, we first apply the lookup procedure to locate the leaf L at which r should belong. If there is room for r in L, we insert r into L in the proper sorted order. In this case no modifications need to be made to the ancestors of L.

If there is no room for r in L, we ask the file system for a new block L' and move the last half of the records from L to L', inserting r into its proper place in L or L'.† Let node P be the parent of node L. P is known, since the lookup procedure traced a path from the root to L through P. We now apply the insertion procedure recursively to place in P a key k' and a pointer l' to L'; k' and l' are inserted immediately after the key and pointer for L. The value of k' is the smallest key value in L'.

If P already has m pointers, insertion of k' and l' into P will cause P to be split and require an insertion of a key and pointer into the parent of P. The effects of this insertion can ripple up through the ancestors of node L back to the root, along the path that was traced by the original lookup procedure. It may even be necessary to split the root, in which case we create a new root with the two halves of the old root as its two children. This is the only situation in which a node may have fewer than $m/2$ children.

---

† This strategy is the simplest of a number of responses that can be made to the situation where a block has to be split. Some other choices, providing higher average occupancy of blocks at the cost of extra work with each insertion, are mentioned in the exercises.

Deletion

To delete a record r with key value x, we first find the leaf L containing r. We then remove r from L, if it exists. If r is the first record in L, we then go to P, the parent of L, to set the key value in P's entry for L to be the new first key value of L. However, if L is the first child of P, the first key of L is not recorded in P, but rather will appear in one of the ancestors of P, specifically, the lowest ancestor A such that L is not the leftmost descendant of A. Therefore, we must propagate the change in the lowest key value of L backwards along the path from the root to L.

If L becomes empty after deletion, we give L back to the file system.† We now adjust the keys and pointers in P to reflect the removal of L. If the number of children of P is now less than $m/2$, we examine the node P' immediately to the left (or the right) of P at the same level in the tree. If P' has at least $\lceil m/2 \rceil + 1$ children, we distribute the keys and pointers in P and P' evenly between P and P', keeping the sorted order of course, so that both nodes will have at least $\lceil m/2 \rceil$ children. We then modify the key values for P and P' in the parent of P, and, if necessary, recursively ripple the effects of this change to as many ancestors of P as are affected.

If P' has exactly $\lceil m/2 \rceil$ children, we combine P and P' into a single node with $2\lceil m/2 \rceil - 1$ children (this is m children at most). We must then remove the key and pointer to P' from the parent for P'. This deletion can be done with a recursive application of the deletion procedure.

If the effects of the deletion ripple all the way back to the root, we may have to combine the only two children of the root. In this case the resulting combined node becomes the new root, and the old root can be returned to the file system. The height of the B-tree has now been reduced by one.

Example 11.5. Consider the B-tree of order 5 in Fig. 11.10. Inserting the record with key value 23 into this tree produces the B-tree in Fig. 11.11. To insert 23, we must split the block containing 22, 23, 24, and 26, since we assume that at most three records fit in one block. The two smaller stay in that block, and the larger two are placed in a new block. A pointer-key pair for the new node must be inserted into the parent, which then splits because it cannot hold six pointers. The root receives the pointer-key pair for the new node, but the root does not split because it has excess capacity.

Removing record 10 from the B-tree of Fig. 11.11 results in the B-tree of Fig. 11.12. Here, the block containing 10 is discarded. Its parent now has only two children, and the right sibling of the parent has the minimum number, three. Thus we combine the parent with its sibling, making one node with five children. □

---

† We can use a variety of strategies to prevent leaf blocks from ever becoming completely empty. In particular, we describe below a scheme for preventing interior nodes from getting less than half full, and this technique can be applied to the leaves as well, with a value of m equal to the largest number of records that will fit in one block.

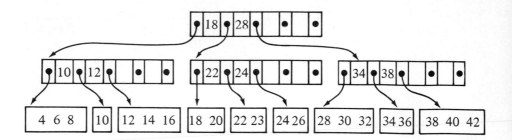

Fig. 11.11. B-tree after insertion.

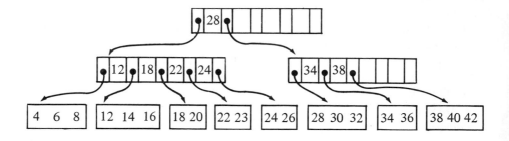

Fig. 11.12. B-tree after deletion.

Time Analysis of B-tree Operations

Suppose we have a file with n records organized into a B-tree of order m. If each leaf contains b records on the average, then the tree has about $\lceil n/b \rceil$ leaves. The longest possible paths in such a tree will occur if each interior node has the fewest children possible, that is, $m/2$ children. In this case there will be about $2\lceil n/b \rceil/m$ parents of leaves, $4\lceil n/b \rceil/m^2$ parents of parents of leaves, and so on.

If there are j nodes along the path from the root to a leaf, then $2^{j-1}\lceil n/b \rceil/m^{j-1} \geq 1$, or else there would be fewer than one node at the root's level. Therefore, $\lceil n/b \rceil \geq (m/2)^{j-1}$, and $j \leq 1 + \log_{m/2}\lceil n/b \rceil$. For example, if $n = 10^6$, $b = 10$, and $m = 100$, then $j \leq 3.5$. Note that b is not the maximum number of records we can put in a block, but an average or expected number. However, by redistributing records among neighboring blocks whenever one gets less than half full, we can ensure that b is at least half the maximum value. Also note that we have assumed in the above that each interior node has the minimum possible number of children. In practice, the average interior node will have more than the minimum, and the above analysis is therefore conservative.

For an insertion or deletion, j block accesses are needed to locate the appropriate leaf. The exact number of additional block accesses that are needed to accomplish the insertion or deletion, and to ripple its effects through the tree, is difficult to compute. Most of the time only one block, the leaf storing the record of interest, needs to be rewritten. Thus, $2 + \log_{m/2}\lceil n/b \rceil$ can be taken as the approximate number of block accesses for an insertion or deletion.

Comparison of Methods

We have discussed hashing, sparse indices, and B-trees as possible methods for organizing external files. It is interesting to compare, for each method, the number of block accesses involved in a file operation.

Hashing is frequently the fastest of the three methods, requiring two block accesses on average for each operation (excluding the block accesses required to search the bucket table), if the number of buckets is sufficiently large that the typical bucket uses only one block. With hashing, however, we cannot easily access the records in sorted order.

A sparse index on a file of n records allows the file operations to be done in about $2 + \log(n/bb')$ block accesses using binary search; here b is the number of records that fit on a block, and b' is the number of key-pointer pairs that fit on a block for the index file. B-trees allow file operations in about $2 + \log_{m/2}\lceil n/b \rceil$ block accesses, where m, the maximum degree of the interior nodes, is approximately b'. Both sparse indices and B-trees allow records to be accessed in sorted order.

All of these methods are remarkably good compared to the obvious sequential scan of a file. The timing differences among them, however, are small and difficult to determine analytically, especially considering that the relevant parameters such as the expected file length and the occupancy rates of blocks are hard to predict in advance.

It appears that the B-tree is becoming increasingly popular as a means of accessing files in database systems. Part of the reason lies in its ability to handle queries asking for records with keys in a certain range (which benefit from the fact that the records appear in sorted order in the main file). The sparse index also handles such queries efficiently, but is almost sure to be less efficient than the B-tree. Intuitively, the reason B-trees are superior to sparse indices is that we can view a B-tree as a sparse index on a sparse index on a sparse index, and so on. (Rarely, however, do we need more than three levels of indices.)

B-trees also perform relatively well when used as secondary indices, where "keys" do not really define a unique record. Even if the records with a given value for the designated fields of a secondary index extend over many blocks, we can read them all with a number of block accesses that is just equal

to the number of blocks holding these records plus the number of their ancestors in the B-tree. In comparison, if these records plus another group of similar size happen to hash to the same bucket, then retrieval of either group from
a hash table would require a number of block accesses about double the
number of blocks on which either group would fit. There are possibly other
reasons for favoring the B-tree, such as their performance when several
processes are accessing the structure simultaneously, that are beyond the scope
of this book.

Exercises

11.1 Write a program *concatenate* that takes a sequence of file names as
arguments and writes the contents of the files in turn onto the standard output, thereby concatenating the files.

11.2 Write a program *include* that copies its input to its output except when
it encounters a line of the form #include *filename*, in which case it is
to replace this line with the contents of the named file. Note that
included files may also contain #include statements.

11.3 How does your program for Exercise 11.2 behave when a file includes
itself?

11.4 Write a program *compare* that will compare two files record-by-record
to determine whether the two files are identical.

∗**11.5** Rewrite the file comparison program of Exercise 11.4 using the LCS
algorithm of Section 5.6 to find the longest common subsequence of
records in both files.

11.6 Write a program *find* that takes two arguments consisting of a pattern
string and a file name, and prints all lines of the file containing the
pattern string as a substring. For example, if the pattern string is
"ufa" and the file is a word list, then *find* prints all words containing
the trigram "ufa."

11.7 Write a program that reads a file and writes on its standard output
the records of the file in sorted order.

11.8 What are the primitives Pascal provides for dealing with external
files? How would you improve them?

∗**11.9** Suppose we use a three-file polyphase sort, where at the ith phase we
create a file with r_i runs of length l_i. At the nth phase we want one
run on one of the files and none on the other two. Explain why each
of the following must be true

a) $l_i = l_{i-1} + l_{i-2}$ for $i \geq 1$, where l_0 and l_{-1} are taken to be the
lengths of runs on the two initially occupied files.

b) $r_i = r_{i-2} - r_{i-1}$ (or equivalently, $r_{i-2} = r_{i-1} + r_i$ for $i \geq 1$), where r_0 and r_{-1} are the number of runs on the two initial files.

c) $r_n = r_{n-1} = 1$, and therefore, $r_n, r_{n-1}, \ldots, r_1$, forms a Fibonacci sequence.

*11.10 What additional condition must be added to those of Exercise 11.9 to make a polyphase sort possible

a) with initial runs of length one (i.e., $l_0 = l_{-1} = 1$)

b) running for k phases, but with initial runs other than one allowed.

Hint. Consider a few examples, like $l_n = 50$, $l_{n-1} = 31$, or $l_n = 50$, $l_{n-1} = 32$.

**11.11 Generalize Exercises 11.9 and 11.10 to polyphase sorts with more than three files.

**11.12 Show that:

a) Any external sorting algorithm that uses only one tape as external storage must take $\Omega(n^2)$ time to sort n records.

b) $O(n \log n)$ time suffices if there are two tapes to use as external storage.

11.13 Suppose we have an external file of directed arcs $x \rightarrow y$ that form a directed acyclic graph. Assume that there is not enough space in internal memory to hold the entire set of vertices or edges at one time.

a) Write an external topological sort program that prints out a linear ordering of the vertices such that if $x \rightarrow y$ is a directed arc, then vertex x appears before vertex y in the linear ordering.

b) What is the time and space complexity of your program as a function of the number of block accesses?

c) What does your program do if the directed graph is cyclic?

**d) What is the minimum number of block accesses needed to topologically sort an externally stored dag?

11.14 Suppose we have a file of one million records, where each record takes 100 bytes. Blocks are 1000 bytes long, and a pointer to a block takes 4 bytes. Devise a hashed organization for this file. How many blocks are needed for the bucket table and the buckets?

11.15 Devise a B-tree organization for the file of Exercise 11.14.

11.16 Write programs to implement the operations RETRIEVE, INSERT, DELETE, and MODIFY on

a) hashed files,

b) indexed files,

c) B-tree files.

11.17 Write a program to find the kth largest element in

    a) a sparse-indexed file

    b) a B-tree file

*11.18 Assume that it takes $a + bm$ milliseconds to read a block containing a node of an m-ary search tree. Assume that it takes $c + d\log_2 m$ milliseconds to process each node in internal memory. If there are n nodes in the tree, we need to read about $\log_m n$ nodes to locate a given record. Therefore, the total time taken to find a given record in the tree is

$$(\log_m n)(a + bm + c + d\log_2 m) = (\log_2 n)((a + c + bm)/\log_2 m) + d)$$

milliseconds. Make reasonable estimates for the values of $a, b, c,$ and d and plot this quantity as a function of m. For what value of m is the minimum attained?

11.19 A B-tree is a B-tree in which each interior node is at least 2/3 full (rather than just 1/2 full). Devise an insertion scheme for B*-trees that delays splitting interior nodes until two sibling nodes are full. The two full nodes can then be divided into three, each 2/3 full. What are the advantages and disadvantages of B*-trees compared with B-trees?

*11.20 When the key of a record is a string of characters, we can save space by storing only a prefix of the key as the key separator in each interior node of the B-tree. For example, "cat" and "dog" could be separated by the prefix "d" or "do" of "dog." Devise a B-tree insertion algorithm that uses prefix key separators that at all times are as short as possible.

*11.21 Suppose that the operations on a certain file are insertions and deletions fraction p of the time, and the remaining $1-p$ of the time are retrievals where exactly one field is specified. There are k fields in records, and a retrieval specifies the ith field with probability q_i. Assume that a retrieval takes a milliseconds if there is no secondary index for the specified field, and b milliseconds if the field has a secondary index. Also assume that an insertion or deletion takes $c + sd$ milliseconds, where s is the number of secondary indices. Determine, as a function of $a, b, c, d, p,$ and the q_i's, which secondary indices should be created for the file in order that the average time per operation be minimized.

**11.22 Suppose that keys are of a type that can be linearly ordered, such as real numbers, and that we know the probability distribution with which keys of given values will appear in the file. We could use this knowledge to outperform binary search when looking for a key in a sparse index. One scheme, called *interpolation search*, uses this statistical information to predict where in the range of index blocks $B_i, \ldots, B_j$ to which the search has been limited, a key x is most likely to lie. Give

a) an algorithm to take advantage of statistical knowledge in this way, and

b) a proof that $O(\log\log n)$ block accesses suffice, on the average, to find a key.

11.23 Suppose we have an external file of records, each consisting of an edge of a graph G and a cost associated with that edge.

a) Write a program to construct a minimum-cost spanning tree for G, assuming that there is enough memory to store all the vertices of G in core but not all the edges.

b) What is the time complexity of your program as a function of the number of vertices and edges?

Hint. One approach to this problem is to maintain a forest of currently connected components in core. Each edge is read and processed as follows: If the next edge has ends in two different components, add the edge and merge the components. If the edge creates a cycle in an existing component, add the edge and remove the highest cost edge from that cycle (which may be the current edge). This approach is similar to Kruskal's algorithm but does not require the edges to be sorted, an important consideration in this problem.

11.24 Suppose we have a file containing a sequence of positive and negative numbers $a_1, a_2, \ldots, a_n$. Write an $O(n)$ program to find a contiguous subsequence $a_i, a_{i+1}, \ldots, a_j$ that has the largest sum $a_i + a_{i+1} + \cdots + a_j$ of any such subsequence.

Bibliographic Notes

For additional material on external sorting see Knuth [1973]. Further material on external data structures and their use in database systems can be found there and in Ullman [1982] and Wiederhold [1982]. Polyphase sorting in discussed by Shell [1971]. The six-buffer merging scheme in Section 11.2 is from Friend [1956] and the four-buffer scheme from Knuth [1973].

Secondary index selection, of which Exercise 11.21 is a simplification, is discussed by Lum and Ling [1970] and Schkolnick [1975]. B-trees originated with Bayer and McCreight [1972]. Comer [1979] surveys the many variations, and Gudes and Tsur [1980] discusses their performance in practice.

Information about Exercise 11.12, one- and two-tape sorting, can be found in Floyd and Smith [1973]. Exercise 11.22 on interpolation search is discussed in detail by Yao and Yao [1976] and Perl, Itai, and Avni [1978].

An elegant implementation of the approach suggested in Exercise 11.23 to the external minimum-cost spanning tree problem was devised by V. A. Vyssotsky around 1960 (unpublished). Exercise 11.24 is due to M. I. Shamos.

CHAPTER 12

Memory Management

This chapter discusses the basic strategies whereby memory space can be reused, or shared by different objects that grow and shrink in arbitrary ways. For example, we shall discuss methods that maintain linked lists of available space, and "garbage collection" techniques, where we figure out what is available only when it seems we have run out of available space.

12.1 Issues in Memory Management

There are numerous situations in computer system operation when a limited memory resource is *managed*, that is, shared among several "competitors." A programmer who does not engage in the implementation of systems programs (compilers, operating systems, and so on) may be unaware of such activities, since they frequently are carried out "behind the scenes." As a case in point, Pascal programmers are aware that the procedure *new(p)* will make pointer *p* point to a new object of the correct type. But where does space for that object come from? The procedure *new* has access to a large region of memory, called the "heap," that the program variables do not use. From that region, an unused block of consecutive bytes sufficient to hold an object of the type that *p* points to is selected, and *p* is made to hold the address of the first byte of that block. But how does the procedure *new* know which bytes of the memory are "unused"? Section 12.4 suggests the answer.

Even more mysterious is what happens if the value of *p* is changed, either by an assignment or by another call to *new(p)*. The block of memory *p* pointed to may now be *inaccessible*, in the sense that there is no way to get to it through the program's data structures, and we could reuse its space. On the other hand, before *p* was changed, the value of *p* may have been copied into some other variable. In that case, the memory block is still part of the program's data structures. How do we know whether a block in the memory region used by procedure *new* is no longer needed by the program?

Pascal's sort of memory management is only one of several different types. For example, in some situations, like Pascal, objects of different sizes share the same memory space. In others, all objects sharing the space are of the same size. This distinction regarding object sizes is one way we can classify the kinds of memory management problems we face. Some more examples follow.

1. In the programming language Lisp, memory space is divided into *cells*, which are essentially records consisting of two fields; each field can hold either an *atom* (object of elementary type, e.g., an integer) or a pointer to a cell. Atoms and pointers are the same size, so all cells require the same number of bytes. All known data structures can be made out of these cells. For example, linked lists of atoms can use the first fields of cells to hold atoms and second fields to hold pointers to the next cells on the list. Binary trees can be represented by using the first field of each cell to point to the left child and the second field to point to the right child. As a Lisp program runs, the memory space used to hold a cell may find itself part of many different structures at different times, either because one cell is moved among structures, or because the cell becomes detached from all structures and its space is reused.

2. A file system generally divides secondary storage devices, like disks, into fixed length blocks. For example, UNIX typically uses blocks of 512 bytes. Files are stored in a sequence of (not necessarily consecutive) blocks. As files are created and destroyed, blocks of secondary storage are made available to be reused.

3. A typical multiprogramming operating system allows several programs to share main memory at one time. Each program has a required amount of memory, which is known to the operating system, that requirement being part of the request for service issued when it is desired to run the program. While in examples (1) and (2), the objects sharing memory (cells and blocks, respectively) were each of the same size, different programs require different amounts of memory. Thus, when a program using say 100K bytes terminates, it may be replaced by two programs using 50K each, or one using 20K and another 70K (with 10K left unused). Alternatively, the 100K bytes freed by the program terminating may be combined with an adjacent 50K that are unused and a program needing up to 150K may be run. Another possibility is that no new program can fit in the space made available, and that 100K bytes is left free temporarily.

4. There are a large number of programming languages, like Snobol, APL, or SETL, that allocate space to objects of arbitrary size. These objects, which are values assigned to variables, are allocated a block of space from a larger block of memory, which is often called the *heap*. When a variable changes value, the new value is allocated space in the heap, and a pointer for the variable is set to point to the new value. Possibly the old value of the variable is now unused, and its space can be reused. However, languages like Snobol or SETL implement assignments like $A=B$ by making the pointer for A point to the same object that B's pointer points to. If either A or B is reassigned, the previous object is not freed, and its space cannot be reclaimed.

Example 12.1. In Fig. 12.1(a) we see the heap that might be used by a Snobol program with three variables, A, B, and C. The value of any variable in Snobol is a character string, and in this case, the value of both A and B is

'OH HAPPY DAY' and the value of C is 'PASS THE SALT'.

We have chosen to represent character strings by pointers to blocks of memory in the heap. These blocks have their first 2 bytes (the number 2 is a typical value that could be changed) devoted to an integer giving the length of the string. For example, 'OH HAPPY DAY' has length 12, counting the blanks between words, so the value of A (and of B) occupies 14 bytes.

If the value of B were changed to 'OH HAPPY DAZE', we would find an empty block in the heap 15 bytes long to store the new value of B, including the 2 bytes for the length. The pointer for B is made to point to the new value, as shown in Fig. 12.1(b). The block holding integer 12 and 'OH HAPPY DAY' is still useful, since A points to it. If the value of A now changes, that block would become useless and could be reused. How one tells conveniently that there are no pointers to such a block is a major subject of this chapter. □

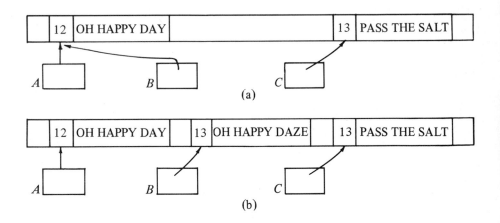

Fig. 12.1. String variables in a heap.

In the four examples above, we can see differences along at least two orthogonal "dimensions." The first issue is whether objects sharing storage are or are not of equal length. In the first two examples, Lisp programs and file storage, the objects, which are Lisp cells in one case and blocks holding parts of files in the other, are of the same size. This fact allows certain simplifications of the memory management problem. For example, in Lisp implementation, a region of memory is divided into spaces, each of which can hold exactly one cell. The management problem is to find empty spaces to hold newly-created cells, and it is never necessary to store a cell in such a position that it overlaps two spaces. Similarly, in the second example, a disk is divided into equal sized blocks, and each block is assigned to hold part of one file; we never use a block to store parts of two or more files, even if a file ends in the middle of a block.

In contrast, the third and fourth examples, covering memory allocation by a multiprogramming system and heap management for those languages that deal with variables whose values are "big" objects, speak of allocating space in blocks of different sizes. This requirement presents certain problems that do not appear in the fixed-length case. For example, we fear *fragmentation*, a situation in which there is much unused space, but it is in such small pieces that space for one large object cannot be found. We shall say more about heap management in Sections 12.4 and 12.5.

The second major issue is whether *garbage collection*, a charming term for the recovery of unused space, is done explicitly or implicitly, that is, by program command or only in response to a request for space that cannot be satisfied otherwise. In the case of file management, when a file is deleted, the blocks used to hold it are known to the file system. For example, the file system could record the address of one or more "master blocks" for each file in existence; the master blocks list the addresses of all the blocks used for the file. Thus, when a file is deleted, the file system can explicitly make available for reuse all the blocks used for that file.

In contrast, Lisp cells, when they become detached from the data structures of the program, continue to occupy their memory space. Because of the possibility of multiple pointers to a cell, we cannot tell when a cell is completely detached; therefore we cannot explicitly collect cells as we do blocks of a deleted file. Eventually, all memory spaces will become allocated to useful or useless cells, and the next request for space for another cell implicitly will trigger a "garbage collection." At that time, the Lisp interpreter marks all the useful cells, by an algorithm such as the one we shall discuss in Section 12.3, and then links all the blocks holding useless cells into an available space list, so they can be reused.

Figure 12.2 illustrates the four kinds of memory management and gives an example of each. We have already discussed the fixed block size examples in Fig. 12.2. The management of main memory by a multiprogramming system is an example of explicit reclamation of variable length blocks. That is, when a program terminates, the operating system, knowing what area of memory was given to the program, and knowing no other program could be using that space, makes the space available immediately to another program.

The management of a heap in Snobol or many other languages is an example of variable length blocks with garbage collection. As for Lisp, a typical Snobol interpreter does not try to reclaim blocks of memory until it runs out of space. At that time the interpreter performs a garbage collection as the Lisp interpreter does, but with the additional possibility that strings will be moved around the heap to reduce fragmentation, and that adjacent free blocks will be combined to make larger blocks. Notice that the latter two steps are pointless in the Lisp environment.

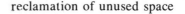

reclamation of unused space

		explicit	garbage collection
block size	fixed	file system	Lisp
	variable	multiprogramming system	Snobol

Fig. 12.2. Examples of the four memory management strategies.

12.2 Managing Equal-Sized Blocks

Let us imagine we have a program that manipulates cells each consisting of a pair of fields; each field can be a pointer to a cell or can hold an "atom." The situation, of course, is like that of a program written in Lisp, but the program may be written in almost any programming language, even Pascal, if we define cells to be of a variant record type. Empty cells available for incorporation into a data structure are kept on an available space list, and each program variable is represented by a pointer to a cell. The cell pointed to may be one cell of a large data structure.

Example 12.2. In Fig. 12.3 we see a possible structure. A, B, and C are variables; lower case letters represent atoms. Notice some interesting phenomena. The cell holding atom a is pointed to by the variable A and by another cell. The cell holding atom c is pointed to by two different cells. The cells holding g and h are unusual in that although each points to the other, they cannot be reached from any of the variables A, B, or C, nor are they on the available space list. □

Let us assume that as the program runs, new cells may on occasion be seized from the available space list. For example, we might wish to have the null pointer in the cell with atom c in Fig. 12.3 replaced by a pointer to a new cell that holds atom i and a null pointer. This cell will be removed from the top of the available space list. It is also possible that from time to time, pointers will change in such a way that cells become detached from the program variables, as the cells holding g and h in Fig. 12.3 have been. For example, the cell holding c may, at one time, have pointed to the cell holding g. As another example, the value of variable B may at some time change, which would, if nothing else has changed, detach the cell now pointed to by B in Fig. 12.3 and also detach the cell holding d and e (but not the cell holding c, since it would still be reached from A). We call cells not reachable from any variable and not on the available space list *inaccessible*.

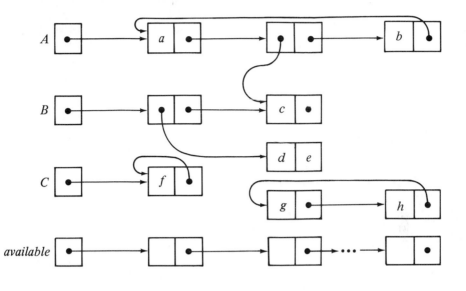

Fig. 12.3. A network of cells.

When cells are detached, and therefore are no longer needed by the program, it would be nice if they found their way back onto the available space list, so they could be reused. If we don't reclaim such cells, we shall eventually reach the unacceptable situation where the program is not using all the cells, yet it wants a cell from available space, and the available space list is empty. It is then that a time-consuming garbage collection must be performed. This garbage collection step is "implicit," in the sense that it was not explicitly called for by the request for space.

Reference Counts

One seemingly attractive approach to detecting inaccessible cells is to include in each cell a *reference count*, that is, an integer-valued field whose value equals the number of pointers to the cell. It is easy to maintain reference counts. When making some pointer point to a cell, add one to the reference count for that cell. When a non-null pointer is reassigned, first decrease by one the reference count for the cell pointed to. If a reference count reaches zero, the cell is inaccessible, and it can be returned to the available list.

Unfortunately, reference counts don't always work. The cells with g and h in Fig. 12.3 are inaccessible cells linked in a cycle. Their reference counts are each 1, so we would not return them to the available list. One can attempt to detect cycles of inaccessible cells in a variety of ways, but it is probably not worth doing so. Reference counts are useful for structures that do not have pointer cycles. One example of a structure with no possibility of cycles is a collection of variables pointing to blocks holding data, as in Fig. 12.1. There, we can do explicit garbage collection simply by collecting a block when its reference count reaches zero. However, when data structures allow pointer cycles, the reference count strategy is usually inferior, both in terms of the space needed in cells and the time taken dealing with the issue of inaccessible cells, to another approach which we shall discuss in the next section.

12.3 Garbage Collection Algorithms for Equal-Sized Blocks

Let us now give an algorithm for finding which of a collection of cells of the types suggested in Fig. 12.3 are accessible from the program variables. We shall define the setting for the problem precisely by defining a cell type in Pascal that is a variant record type; the four variants, which we call PP, PA, AP, and AA, are determined by which of the two data fields are pointers and which are atoms. For example, PA means the left field is a pointer and the right field an atom. An additional boolean field in cells, called *mark*, indicates whether the cell has been found accessible. That is, by setting *mark* to true when garbage collecting, we "mark" the cell, indicating it is accessible. The important type definitions are shown in Fig. 12.4.

```
type
     atomtype = { some appropriate type;
                  preferably of the same size as pointers }
     patterns = (PP, PA, AP, AA);
     celltype = record
          mark: boolean;
          case pattern: patterns of
               PP: (left: ↑ celltype; right: ↑ celltype);
               PA: (left: ↑ celltype; right: atomtype);
               AP: (left: atomtype; right: ↑ celltype);
               AA: (left: atomtype; right: atomtype);
     end;
```

Fig. 12.4. Definition of the type for cells.

We assume there is an array of cells, taking up most of the memory, and some collection of variables, that are pointers to cells. For simplicity, we assume there is only one variable, called *source*, pointing to a cell, but the extension to many variables is not hard.† That is, we declare

---

† Each programming language must provide for itself a method of representing the current set of variables, and any of the methods discussed in Chapters 4 and 5 is appropriate. For example, most implementations use a hash table to hold the variables.

```
        var
            source: ↑ celltype;
            memory: array [1..memorysize] of celltype;
```

To mark the cells accessible from *source*, we first "unmark" all cells, accessible or not, by running down the array *memory* and setting the *mark* field to false. Then we perform a depth-first search of the graph emanating from *source*, marking all cells visited. The cells visited are exactly those that are accessible. We then traverse the array *memory* and add to the available space list all unmarked cells. Figure 12.5 shows a procedure *dfs* to perform the depth-first search; *dfs* is called by the procedure *collect* that unmarks all cells, and then marks accessible cells by calling *dfs*. We do not show the code linking the available space list because of the peculiarities of Pascal. For example, while we could link available cells using either all left or all right cells, since pointers and atoms are assumed the same size, we are not permitted to replace atoms by pointers in cells of variant type *AA*.

```
(1)     procedure dfs ( currentcell: ↑ celltype );
                    { If current cell was marked, do nothing. Otherwise, mark
                      it and call dfs on any cells pointed to by the current cell }
            begin
(2)             with currentcell↑ do
(3)                 if mark = false then begin
(4)                     mark := true;
(5)                     if (pattern = PP) or (pattern = PA) then
(6)                         if left <> nil then
(7)                             dfs(left);
(8)                     if (pattern = PP) or (pattern = AP) then
(9)                         if right <> nil then
(10)                            dfs(right);
                    end
            end;   { dfs }


(11)    procedure collect;
            var
                i: integer;
            begin
(12)            for i := 1 to memorysize do  { "unmark" all cells }
(13)                memory[i].mark := false;
(14)            dfs(source);  { mark accessible cells }
(15)            { the code for the collection goes here }
            end;  { collect }
```

Fig. 12.5. Algorithm for marking accessible cells.

Collection in Place

The algorithm of Fig. 12.5 has a subtle flaw. In a computing environment where memory is limited, we may not have the space available to store the stack required for recursive calls to *dfs*. As discussed in Section 2.6, each time *dfs* calls itself, Pascal (or any other language permitting recursion) creates on "activation record" for that particular call to *dfs*. In general, an activation record contains space for parameters and variables local to the procedure, of which each call needs its own copy. Also necessary in each activation record is a "return address," the place to which control must go when this recursive call to the procedure terminates.

In the particular case of *dfs*, we only need space for the parameter and the return address (i.e., was it called by line (14) in *collect* or by line (7) or line (10) of another invocation of *dfs*). However, this space requirement is enough that, should all of memory be linked in a single chain extending from *source* (and therefore the number of active calls to *dfs* would at some time equal the length of this chain), considerably more space would be required for the stack of activation records than was allotted for memory. Should such space not be available, it would be impossible to carry out the marking.

Fortunately, there is a clever algorithm, known as the *Deutsch-Schorr-Waite Algorithm*, for doing the marking "in place." The reader should convince himself that the sequence of cells upon which a call to *dfs* has been made but not yet terminated indeed forms a path from *source* to the cell upon which the current call to *dfs* was made. Thus we can use a nonrecursive version of *dfs*, and instead of a stack of activation records to record the path of cells from *source* to the cell currently being examined, we can use the pointer fields along that path to hold the path itself. That is, each cell on the path, except the last, holds in either the *left* or *right* field, a pointer to its *predecessor*, the cell immediately closer to *source*. We shall describe the Deutsch-Schorr-Waite algorithm with an extra, one-bit field, called *back*. Field *back* is of an enumerated type (L, R), and tells whether field *left* or field *right* points to the predecessor. Later, we shall discuss how the information in *back* can be held in the *pattern* field, so no extra space in cells is required.

The new procedure for nonrecursive depth-first search, which we call *nrdfs*, uses a pointer *current* to point to the current cell, and a pointer *previous* to the predecessor of the current cell. The variable *source* points to a cell *source1*, which has a pointer in its right field only.† Before marking, *source1* is initialized to have *back* = R, and its right field pointing to itself. The cell ordinarily pointed to by *source1*, is pointed to by *current*, and *source1* is pointed to by *previous*. We halt the marking operation when *current* = *previous*, which can only occur when they both point to *source1*, the entire structure having been searched.

Example 12.3. Figure 12.6(a) shows a possible structure emanating from

---

† This awkwardness is made necessary by peculiarities of Pascal.

source. If we depth-first search this structure, we visit (1), (2), (3) and (4), in that order. Figure 12.6(b) shows the pointer modifications made when cell (4) is current. The value of the *back* field is shown, although fields *mark* and *pattern* are not. The current path is from (4) to (3) to (2) to (1) back to *source*1; it is represented by dashed pointers. For example, cell (1) has *back* = *L*, since the *left* field of (1), which in Fig. 12.6(a) held a pointer to (2), is being used in Fig. 12.6(b) to hold part of the path. Now the *left* field in (1) points backwards, rather than forwards along the path; however, we shall restore that pointer when the depth-first search finally retreats from cell (2) back to (1). Similarly, in cell (2), *back* = *R*, and the right field of (2) points backwards along the path to (1), rather than ahead to (3), as it did in Fig. 12.6(a). □

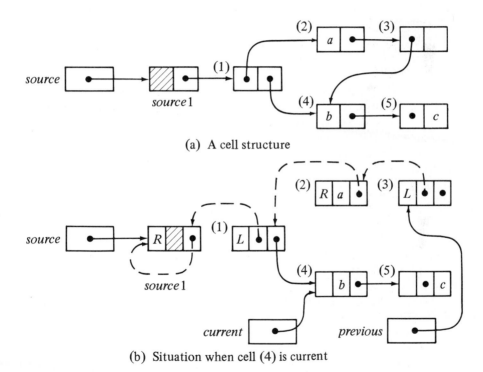

(a) A cell structure

(b) Situation when cell (4) is current

Fig. 12.6. Using pointers to represent the path back to *source*.

There are three basic steps to performing the depth-first search. They are:

1. *Advance.* If we determine that the current cell has one or more nonnull pointers, advance to the first of these, that is, follow the pointer in *left*, or

if there is none, the pointer in *right*. By "advance," we mean to make the cell pointed to become the current cell and the current cell become previous. To help find the way back, the pointer just followed is made to point to the old previous cell. These changes are shown in Fig. 12.7(a), on the assumption that the left pointer is followed. In that figure, the old pointers are solid lines and the new ones dashed.

2. *Switch*. If we determine that the cells emanating from the current cell have been searched (for example, the current cell may have only atoms, may be marked, or we may just have "retreated" to the current cell from the cell pointed to by the *right* field of the current cell), we consult the field *back* of the previous cell. If that value is L, and the *right* field of the previous cell holds a nonnull pointer to some cell C, we make C become the current cell, while the identity of the previous cell is not changed. However, the value of *back* in the previous cell is set to R, and the left pointer in that cell receives its correct value; that is, it is made to point to the old current cell. To maintain the path back to *source* from the previous cell, the pointer to C in field *right* is made to point where left used to point. Figure 12.7(b) shows these changes.

3. *Retreat*. If we determine, as in (2), that the cells emanating from the current cell have been searched, but field *back* of the previous cell is R, or it is L but field right holds an atom or null pointer, then we have now searched all cells emanating from the previous cell. We retreat, by making the previous cell be the current cell and the next cell along the path from the previous cell to *source* be the new previous cell. These changes are shown in Fig. 12.7(c), on the assumption that $back = R$ in the previous cell.

One fortuitous coincidence is that each of the steps of Fig. 12.7 can be viewed as the simultaneous rotation of three pointers. For example, in Fig. 12.7(a), we simultaneously replace (*previous*, *current*, *current↑.left*) by (*current*, *current↑.left*, *previous*), respectively. The simultaneity must be emphasized; the location of *current↑.left* does not change when we assign a new value to *current*. To perform these pointer modifications, a procedure *rotate*, shown in Fig. 12.8, is naturally useful. Note especially that the passing of parameters by reference assures that the locations of the pointers are established before any values change.

Now we turn to the design of the nonrecursive marking procedure *nrdfs*. This procedure is one of these odd processes that is most easily understood when written with labels and gotos. Particularly, there are two "states" of the procedure, "advancing," represented by label 1, and "retreating," represented by label 2. We enter the first state initially, and also whenever we have moved to a new cell, either by an advance step or a switch step. In this state, we attempt another advance step, and only retreat or switch if we are blocked. We can be blocked for two reasons: (1) The cell just reached is already marked, or (2) there are no nonnull pointers in the cell. When blocked, we change to the second, or "retreating" state.

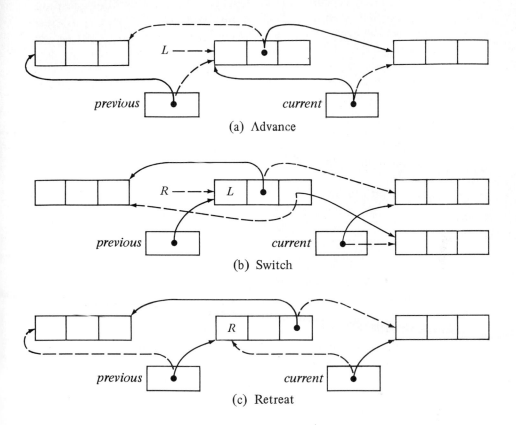

(a) Advance

(b) Switch

(c) Retreat

Fig. 12.7. Three basic steps.

```
procedure  rotate ( var p1, p2, p3: ↑ celltype );
    var
        temp: ↑ celltype;
    begin
        temp := p1;
        p1 := p2;
        p2 := p3;
        p3 := temp
    end; { rotate }
```

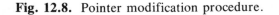

Fig. 12.8. Pointer modification procedure.

The second state is entered whenever we retreat, or when we cannot stay in the advancing state because we are blocked. In the retreating state we check whether we have retreated back to the dummy cell *source*1. As discussed before, we shall recognize this situation because *previous* = *current*, in which case we go to state 3. Otherwise, we decide whether to retreat and stay in the retreating state or switch and enter the advancing state. The code for *nrdfs* is shown in Fig. 12.9. It makes use of functions *blockleft*, *blockright*, and *block*, which test if the left or right fields of a cell, or both, have an atom or null pointer. *block* also checks for a marked cell.

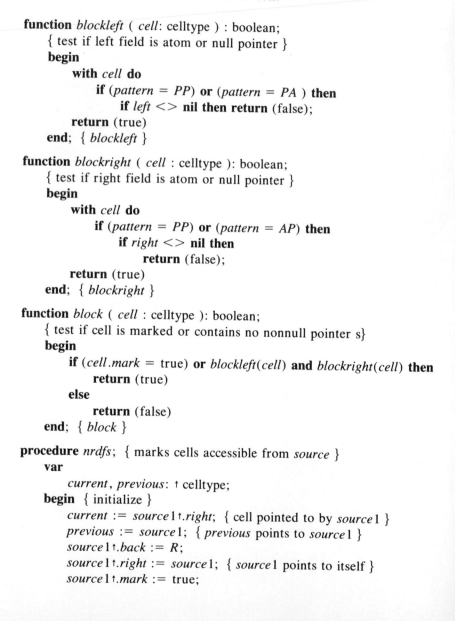

```
function blockleft ( cell: celltype ) : boolean;
    { test if left field is atom or null pointer }
    begin
        with cell do
            if (pattern = PP) or (pattern = PA ) then
                if left <> nil then return (false);
        return (true)
    end; { blockleft }

function blockright ( cell : celltype ): boolean;
    { test if right field is atom or null pointer }
    begin
        with cell do
            if (pattern = PP) or (pattern = AP) then
                if right <> nil then
                    return (false);
        return (true)
    end; { blockright }

function block ( cell : celltype ): boolean;
    { test if cell is marked or contains no nonnull pointer s}
    begin
        if (cell.mark = true) or blockleft(cell) and blockright(cell) then
            return (true)
        else
            return (false)
    end; { block }

procedure nrdfs; { marks cells accessible from source }
    var
        current, previous: ↑ celltype;
    begin  { initialize }
        current := source1↑.right; { cell pointed to by source1 }
        previous := source1; { previous points to source1 }
        source1↑.back := R;
        source1↑.right := source1; { source1 points to itself }
        source1↑.mark := true;
```

```
    state1:  { try to advance }
        if block(current↑) then begin  { prepare to retreat }
            current↑.mark := true;
            goto state2
        end
        else begin  { mark and advance }
            current↑.mark := true;
            if blockleft(current↑) then begin  { follow right pointer }
                current↑.back := R;
                rotate(previous, current, current↑.right);  { implements
                    changes of Fig. 12.7(a), but following right pointer }
                goto state1
            end
            else begin  { follow left pointer }
                current↑.back := L;
                rotate(previous, current, current↑.left);
                    { implements changes of Fig. 12.7(a) }
                goto state1
            end
        end;

    state2:  { finish, retreat or switch }
        if previous = current then  { finish }
            goto state3
        else if (previous↑.back = L) and
                not blockright(previous↑) then begin  { switch }
                    previous↑.back := R;
                    rotate(previous↑.left, current, previous↑.right);
                        { implements changes of Fig. 12.7(b) }
                    goto state1
                end
        else if previous↑.back = R then  { retreat }
            rotate(previous, previous↑.right, current)
                { implements changes of Fig. 12.7(c) }
        else  { previous↑.back = L }
            rotate(previous, previous↑.left, current);
                { implements changes of Fig. 12.7(c), but with
                    left field of previous cell involved in the path }
        goto state2
        end;

    state3:    { put code to link unmarked cells on available list here }
    end;  { nrdfs }
```

Fig. 12.9. Nonrecursive marking algorithm.

Deutsch-Schorr-Waite Algorithm Without an Extra Bit for the Field *Back*

It is possible, although unlikely, that the extra bit used in cells for the field *back* might cause cells to require an extra byte, or even an extra word. In such a case, it is comforting to know that we do not really need the extra bit, at least not if we are programming in a language that, unlike Pascal, allows the bits of the field *pattern* to be used for purposes other than those for which they were declared: designators of the variant record format. The "trick" is to observe that if we are using the field *back* at all, because its cell is on the path back to *source1*, then the possible values of field *pattern* are restricted. For example, if *back* = *L*, then we know the pattern must be *PP* or *PA*, since evidently the field *left* holds a pointer. A similar observation can be made when *back* = *R*. Thus, if we have two bits available to represent both *pattern* and (when needed) *back*, we *can* encode the necessary information as in Fig. 12.10, for example.

The reader should observe that in the program of Fig. 12.9, we always know whether *back* is in use, and thus can tell which of the interpretations in Fig. 12.10 is applicable. Simply, when *current* points to a record, the field *back* in that record is not in use; when *previous* points to it, it is. Of course, as these pointers move, we must adjust the codes. For example, if *current* points to a cell with bits 10, which we interpret according to Fig. 12.10 as *pattern = AP*, and we decide to advance, so *previous* will now point to this cell, we make *back=R*, as only the right field holds a pointer, and the appropriate bits are 11. Note that if the pattern were *AA*, which has no representation in the middle column of Fig. 12.10, we could not possibly want *previous* to point to the cell, as there are no pointers to follow in an advancing move.

Code	On path to *source1*	Not on path
00	*back=L, pattern=PP*	*pattern=PP*
01	*back=L, pattern=PA*	*pattern=PA*
10	*back=R, pattern=PP*	*pattern=AP*
11	*back=R, pattern=AP*	*pattern=AA*

Fig. 12.10. Interpreting two bits as *pattern* and *back*.

12.4 Storage Allocation for Objects with Mixed Sizes

Let us now consider the management of a heap, as typified by Fig. 12.1, where there is a collection of pointers to allocated blocks. The blocks hold data of some type. In Fig. 12.1, for example, the data are character strings. While the type of data stored in the heap need not be character strings, we assume the data contain no pointers to locations within the heap.

The problem of heap management has aspects that make it both easier and harder than the maintenance of list structures of equal-sized cells as discussed in the previous section. The principal factor making the problem easier is that marking used blocks is not a recursive process; one has only to follow the external pointers to the heap and mark the blocks pointed to. There is no need for a depth-first search of a linked structure or for anything like the Deutsch-Schorr-Waite algorithm.

On the other hand, managing the available space list is not as simple as in Section 12.3. We might imagine that the empty regions (there are three empty regions in Fig. 12.1(a), for example) are linked together as suggested in Fig. 12.11. There we see a heap of 3000 words divided into five blocks. Two blocks of 200 and 600 words, hold the values of X and Y. The remaining three blocks are empty, and are linked in a chain emanating from *avail*, the header for available space.

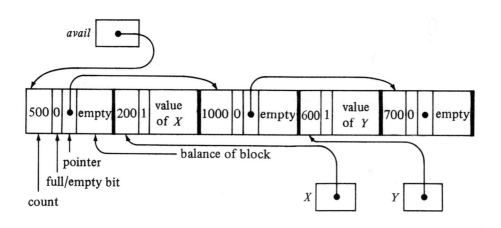

Fig. 12.11. A heap with available space list.

In order that empty blocks may be found when they are needed to hold new data, and blocks holding useless data may be made available, we shall throughout this section make the following assumptions about blocks.

1. Each block is sufficiently large to hold
   a) A *count* indicating the size (in bytes or words, as appropriate for the computer system) of the block,
   b) A pointer (to link the block in available space), plus
   c) A bit to indicate whether the block is empty or not. This bit is referred to as the *full/empty* or *used/unused* bit.

2. An empty block has, from the left (low address end), a count indicating its length, a full/empty bit with a value of 0, indicating emptiness of the block, a pointer to the next available block, and empty space.

3. A block holding data has, from the left, a count, a full/empty bit with value 1 indicating the block is in use, and the data itself.†

One interesting consequence of the above assumptions is that blocks must be capable of storing data sometimes (when in use) and pointers at other times (when unused) in precisely the same place. It is thus impossible or very inconvenient to write programs that manipulate blocks of the kind we propose in Pascal or any other strongly typed language. Thus, this section must of necessity be discursive; only pseudo-Pascal programs can be written, never real Pascal programs. However, there is no problem writing programs to do the things we describe in assembly languages or in most systems programming languages such as C.

Fragmentation and Compaction of Empty Blocks

To see one of the special problems that heap management presents, let us suppose that the variable Y of Fig. 12.11 changes, so the block representing Y must be returned to available space. We can most easily insert the new block at the beginning of the available list, as suggested by Fig. 12.12. In that figure, we see an instance of *fragmentation*, the tendency for large empty areas to be represented on the available space list by "fragments," that is, several small blocks making up the whole. In the case in point, the last 2300 bytes of the heap in Fig. 12.12 are empty, yet the space is divided into three blocks of 1000, 600, and 700 bytes, and these blocks are not even in consecutive order on the available list. Without some form of garbage collection, it would be impossible to satisfy a request for, say, a block of 2000 bytes.

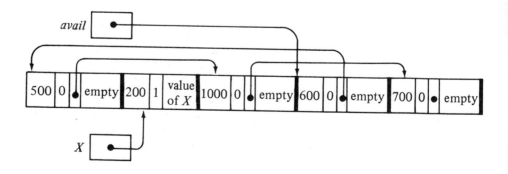

Fig. 12.12. After returning the block of Y.

Evidently, when returning a block to available space it would be desirable

---

† Note that in Fig. 12.1, instead of a count indicating the length of the block, we used the length of the data.

to look at the blocks to the immediate left and right of the block being made available. Finding the block to the right is easy. If the block being returned begins at position p and has count c, the block to the right begins at position $p+c$. If we know p (for example, the pointer Y in Fig. 12.11 holds the value p for the block made available in Fig. 12.12), we have only to read the bytes starting at position p, as many as are used for holding c, to obtain the value c. From byte $p+c$, we skip over the count field to find the bit that tells whether or not the block is empty. If empty, the blocks beginning at p and $p+c$ can be combined.

Example 12.4. Let us assume the heap of Fig. 12.11 begins at position 0. Then the block for Y being returned begins in byte 1700, so $p=1700$ and $c=600$. The block beginning at $p+c=2300$ is also empty so we could combine them into one block beginning at 1700 with a count of 1300, the sum of the counts in the two blocks. □

It is, however, not so easy to fix the available list after combining blocks. We can create the combined block by simply adding the count of the second block to c. However, the second block will still be linked in the available list and must be removed. To do so requires finding the pointer to that block from its predecessor on the available list. Several strategies present themselves; none is strongly recommended over the others.

1. Run down the available list until we find a pointer with value $p+c$. This pointer must be in the predecessor block of the block we have combined with its neighbor. Replace the pointer found by the pointer in the block at $p+c$. This effectively eliminates the block beginning at position $p+c$ from the available list. Its space is still available, of course; it is part of the block beginning at p. On the average, we shall have to scan half the available list, so the time taken is proportional to that length.

2. Use a doubly-linked available space list. Then the predecessor block can be found quickly and the block at $p+c$ eliminated from the list. This approach takes constant time, independent of the length of the available list, but it requires extra space for another pointer in each empty block, thus increasing the minimum size for a block to that required to hold a count, a full/empty bit, and two pointers.

3. Keep the available space list sorted by position. Then we know the block at position p is the predecessor on the available list of the block at $p+c$, and the pointer manipulation necessary to eliminate the second block can be done in constant time. However, insertion of a new available block requires a scan of half the available list on the average, and is thus no more efficient than method (1).

Of the three methods, all but the second require time proportional to the length of the available list to return a block to available space and combine it with its right neighbor if that neighbor is empty. This time may or may not be prohibitive, depending on how long the list gets and what fraction of the total program time is spent manipulating the heap. The second method –

doubly linking the available list – has only the penalty of an increased minimum size for blocks. Unfortunately, when we consider how to combine a returned block with its neighbor to the left, we see that double linking, like the other methods, is no help in finding left neighbors in less than the time it takes to scan the available list.

To find the block immediately to the left of the block in question is not so easy. The position p of a block and its count c, while they determine the position of the block to the right, give no clue as to where the block to the left begins. We need to find an empty block that begins in some position p_1, and has a count c_1, such that $p_1+c_1=p$. It appears we have three choices of strategy.

1. Scan the available list for a block at position p_1 and count c_1 where $p_1+c_1=p$. This operation takes time proportional to the length of the available list.

2. Keep a pointer in each block (used or unused) indicating the position of the block to the left. This approach allows us to find the block to the left in constant time; we can check whether it is empty and if so merge it with the block in question. We can find the block at position $p+c$ and make it point to the beginning of the new block, so these left-going pointers can be maintained.†

3. Keep the available list sorted by position. Then the empty block to the left is found when we insert the newly emptied block onto the list, and we have only to check, using the position and count of the previous empty block, that no nonempty blocks intervene.

As with the merger of a newly empty block with the block to its right, the first and third approaches to finding and merging with the block on the left require time proportional to the length of the available list. Method (2) again requires constant time, but it has a disadvantage beyond the problems involved with doubly linking the available list (which we suggested in connection with finding right neighbor blocks). While doubly linking the empty blocks raises the minimum block size, the approach cannot be said to waste space, since it is only blocks not used for storing data that get linked. However, pointing to left neighbors requires a pointer in used blocks as well as unused ones, and can justifiably be accused of wasting space. If the average block size is hundreds of bytes, the extra space for a pointer may be negligible. On the other hand, the extra space may be prohibitive if the typical block is only 10 bytes long.

To summarize the implications of our explorations regarding the question of how we might merge newly empty blocks with empty neighbors, we see three approaches to handling fragmentation.

---

† The reader should, as an exercise, discover how to maintain the pointers when a block is split into two; presumably one piece is used for a new data item, and the other remains empty.

1. Use one of several approaches, such as keeping the available list sorted, that requires time proportional to the length of the available list each time a block becomes unused, but enables us to find and merge empty neighbors.

2. Use a doubly linked available space list each time a ˋlock becomes unused, and also use left-neighbor pointers in all blocks, whether available or not, to merge empty neighbors in constant time.

3. Do nothing explicit to merge empty neighbors. When we cannot find a block large enough to hold a new data item, scan the blocks from left to right, merging empty neighbors and then creating a new available list. A sketched program to do this is shown in Fig. 12.13.

```
(1)     procedure merge;
        var
(2)         p, q: pointers to blocks;
            { p indicates left end of empty block being accumulated.
              q indicates a block to the right of p that we
              shall incorporate into block p if empty }
        begin
(3)         p:= leftmost block of heap;
(4)         make available list empty;
(5)         while p < right end of heap do
(6)             if p points to a full block with count c then
(7)                 p := p + c; { skip over full blocks }
(8)             else begin { p points to the beginning
                        of a sequence of empty blocks; merge them }
(9)                 q := p + c; { initialize q to the next block }
(10)                while q points to an empty block with some
                            count, say d, and q < right end of heap do begin
(11)                    add d to count of the block pointed to by p;
(12)                    q := q + d
                    end
(13)                insert block pointed to by p onto the available list;
(14)                p := q;
            end
        end; { merge }
```

Fig. 12.13. Merging adjacent empty blocks.

Example 12.5. As an example, consider the program of Fig. 12.13 applied to the heap of Fig. 12.12. Assume the leftmost byte of the heap is 0, so initially $p=0$. As $c=500$ for the first block, q is initialized to $p+c=500$. As the block beginning at 500 is full, the loop of lines (10)–(12) is not executed and the block consisting of bytes 0–499 is attached to the available list, by making *avail* point to byte 0 and putting a **nil** pointer in the designated place in that block (right after the count and full/empty bit). Then p is given the value 500 at line (14), and incremented to 700 at line (7). Pointer q is given value 1700 at line (9), then 2300 and 3000 at line (12), while at the same time, 600 and 700 are added to count 1000 in the block beginning at 700. As q exceeds the rightmost byte, 2999, the block beginning at 700, which now has count 2300, is inserted onto the available list. Then at line (14), p is set to 3000, and the outer loop ends at line (5). □

As the total number of blocks and the number of available blocks are likely not to be too dissimilar, and the frequency with which no sufficiently large empty block can be found is likely to be low, we believe that method (3), doing the merger of adjacent empty blocks only when we run out of adequate space, is superior to (1) in any realistic situation. Method (2) is a possible competitor, but if we consider the extra space requirement and the fact that extra time is needed each time a block is inserted or deleted from the available list, we believe that (2) is preferable to (3) in extremely rare circumstances, and can probably be forgotten.

Selecting Available Blocks

We have discussed in detail what should happen when a block is no longer needed and can be returned to available space. There is also the inverse process of providing blocks to hold new data items. Evidently we must select some available block and use some or all of it to hold the new data. There are two issues to address. First, which empty block do we select? Second, if we must use only part of the selected block, which part do we use?

The second issue can be dispensed with easily. If we are to use a block with count c, and we need $d<c$ bytes from that block, we choose the last d bytes. In this way, we need only to replace count c by $c-d$, and the remaining empty block can stay as it is in the available list.†

Example 12.6. Suppose we need 400 bytes for variable W in the situation represented by Fig. 12.12. We might choose to take the 400 bytes out of the 600 in the first block on the available list. The situation would then be as shown in Fig. 12.14.

Choosing a block in which to place the new data is not so easy, since there are conflicting goals for such strategies. We desire, on one hand, to be able to quickly pick an empty block in which to hold our data and, on the other hand, to make a selection of an empty block that will minimize the fragmentation. Two strategies that represent extremes in the spectrum are known as "first-fit" and "best-fit." They are described below.

1. *First-Fit.* To follow the *first-fit* strategy, when we need a block of size d, scan the available list from the beginning until we come to a block of size

---

† If $c-d$ is so small that a count and pointer cannot fit, we must use the whole block and delete it from the available list.

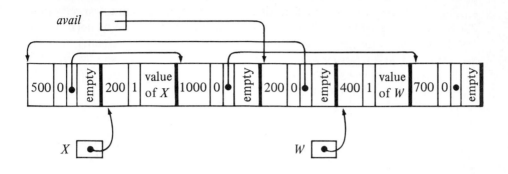

Fig. 12.14. Memory configuration.

$c \geq d$. Utilize the last d words of this block, as we have described above.

2. *Best-fit.* To follow the *best-fit* strategy, when we need a block of size d, examine the entire available list to find that block of size at least d, whose size is as little greater than d as is possible. Seize the last d words of this block.

Some observations about these strategies can be made. Best-fit is considerably slower than first-fit, since with the latter we can expect to find a fitting block fairly quickly on the average, while with best-fit, we are obliged to scan the entire available list. Best-fit can be speeded up if we keep separate available lists for blocks in various size ranges. For example, we could keep a list of available blocks between 1 and 16 bytes in length,† from 17-32, 33-64, and so on. This "improvement" does not speed up first-fit appreciably, and in fact may slow it down if the statistics of block sizes are bad. (Compare looking for the first block of size at least 32 on the full available list and on the list for blocks of size 17-32, e.g.) A last observation is that we can define a spectrum of strategies between first-fit and best-fit by looking for a best-fit among the first k available blocks for some fixed k.

The best-fit strategy seems to reduce fragmentation compared with first-fit, in the sense that best-fit tends to produce very small "fragments", i.e., left-over blocks. While the number of these fragments is about the same as for first-fit, they tend to take up a rather small area. However, best-fit will not tend to produce "medium size fragments." Rather, the available blocks will tend to be either very small fragments or will be blocks returned to available space. As a consequence, there are sequences of requests that first-fit can satisfy but not best-fit, as well as vice-versa.

---

† Actually, there is a minimum block size larger than 1, since blocks must hold a pointer, a count and a full/empty bit if they are to be chained to an available list.

Example 12.7. Suppose, as in Fig. 12.12, the available list consists of blocks of sizes 600, 500, 1000, and 700, in that order. If we are using the first-fit strategy, and a request for a block of size 400 is made, we shall carve it from the block of size 600, that being the first on the list in which a block of size 400 fits. The available list now has blocks of size 200, 500, 1000, 700. We are thus unable to satisfy immediately three requests for blocks of size 600 (although we might be able to do so after merging adjacent empty blocks and/or moving utilized blocks around in the heap).

However, if we were using the best-fit strategy with available list 600, 500, 1000, 700, and the request for 400 came in, we would place it where it fit best, that is, in the block of 500, leaving a list of available blocks 600, 100, 1000, 700. We would, in that event, be able to satisfy three requests for blocks of size 600 without any form of storage reorganization.

On the other hand, there are situations where, starting with the list 600, 500, 1000, 700 again, best-fit would fail, while first-fit would succeed without storage reorganization. Let the first request be for 400 bytes. Best-fit would, as before, leave the list 600, 100, 1000, 700, while first-fit leaves 200, 500, 1000, 700. Suppose the next two requests are for 1000 and 700, so either strategy would allocate the last two empty blocks completely, leaving 600, 100 in the case of best-fit, and 200, 500 in the case of first-fit. Now, first-fit can honor requests for blocks of size 200 and 500, while best-fit obviously cannot. □

12.5 Buddy Systems

There is a family of strategies for maintaining a heap that partially avoids the problems of fragmentation and awkward distribution of empty block sizes. These strategies, called "buddy systems," in practice spend very little time merging adjacent empty blocks. The disadvantage of buddy systems is that blocks come in a limited assortment of sizes, so we may waste some space by placing a data item in a bigger block than necessary.

The central idea behind all buddy systems is that blocks come only in certain sizes; let us say that $s_1 < s_2 < s_3 < \cdots < s_k$ are all the sizes in which blocks can be found. Common choices for the sequence $s_1, s_2, \ldots$ are 1, 2, 4, 8, . . . (the *exponential buddy system*) and 1, 2, 3, 5, 8, 13, . . . (the *Fibonacci buddy system*, where $s_{i+1} = s_i + s_{i-1}$). All the empty blocks of size s_i are linked in a list, and there is an array of available list headers, one for each size s_i allowed.† If we require a block of size d for a new datum, we choose an available block of that size s_i such that $s_i \geq d$, but $s_{i-1} < d$, that is, the smallest permitted size in which the new datum fits.

Difficulties arise when no empty blocks of the desired size s_i exist. In that case, we find a block of size s_{i+1} and split it into two, one of size s_i and

---

† Since empty blocks must hold pointers (and, as we shall see, other information as well) we do not really start the sequence of permitted sizes at 1, but rather at some suitably larger number in the sequence, say 8 bytes.

the other of size $s_{i+1}-s_i$.† The buddy system constrains us that $s_{i+1}-s_i$ be some s_j, for $j \leq i$. We now see the way in which the choices of values for the s_i's are constrained. If we let $j = i-k$, for some $k \geq 0$, then since $s_{i+1}-s_i=s_{i-k}$, it follows that

$$s_{i+1} = s_i + s_{i-k} \qquad (12.1)$$

Equation (12.1) applies when $i > k$, and together with values for $s_1, s_2, \ldots, s_k$, completely determines $s_{k+1}, s_{k+2}, \ldots$. For example, if $k=0$, (12.1) becomes

$$s_{i+1} = 2s_i \qquad (12.2)$$

Beginning with $s_1 = 1$ in (12.2), we get the exponential sequence 1, 2, 4, 8, ... Of course no matter what value of s_1 we start with, the s's grow exponentially in (12.2). As another example, if $k=1$, $s_1=1$, and $s_2=2$, (12.1) becomes

$$s_{i+1} = s_i + s_{i-1} \qquad (12.3)$$

(12.3) defines the *Fibonacci* sequence: 1, 2, 3, 5, 8, 13, ...

Whatever value of k we choose in (12.1) we get a kth *order buddy system*. For any k, the sequence of permitted sizes grows exponentially; that is, the ratio s_{i+1}/s_i approximates some constant greater than one. For example, for $k=0$, s_{i+1}/s_i is exactly 2. For $k=1$ the ratio approximates the "golden ratio" $(\sqrt{5}+1)/2 = 1.618$), and the ratio decreases as k increases, but never gets as low as 1.

Distribution of Blocks

In the kth order buddy system, each block of size s_{i+1} may be viewed as consisting of a block of size s_i and one of size s_{i-k}. For specificity, let us suppose that the block of size s_i is to the left (in lower numbered positions) of the block of size s_{i-k}.‡ If we view the heap as a single block of size s_n, for some large n, then the positions at which blocks of size s_i can begin are completely determined.

The positions in the exponential, or 0th order, buddy system are easily determined. Assuming positions in the heap are numbered starting at 0, a block of size s_i begins at any position beginning with a multiple of 2^i, that is, $0, 2^i, \ldots$. Moreover, each block of size 2^{i+1}, beginning at say, $j2^{i+1}$ is composed of two "buddies" of size 2^i, beginning at positions $(2j)2^i$, which is $j2^{i+1}$, and $(2j+1)2^i$. Thus it is easy to find the buddy of a block of size 2^i. If it begins at some even multiple of 2^i, say $(2j)2^i$, its buddy is to the right, at

---

† Of course, if there are no empty blocks of size s_{i+1}, we create one by splitting a block of size s_{i+2}, and so on. If no blocks of any larger size exist, we are effectively out of space and must reorganize the heap as in the next section.

‡ Incidentally, it is convenient to think of the blocks of sizes s_i and s_{i-k} making up a block of size s_{i+1} as "buddies," from whence comes the term "buddy system."

position $(2j+1)2^i$. If it begins at an odd multiple of 2^i, say $(2j+1)2^i$, its buddy is to the left, at $(2j)2^i$.

Example 12.8. Matters are not so simple for buddy systems of order higher than 0. Figure 12.15 shows the Fibonacci buddy system used in a heap of size 55, with blocks of sizes $s_1, s_2, \ldots, s_8 = 2, 3, 5, 8, 13, 21, 34,$ and 55. For example, the block of size 3 beginning at 26 is buddy to the block of size 5 beginning at 21; together they comprise the block of size 8 beginning at 21, which is buddy to the block of size 5 beginning at 29. Together, they comprise the block of size 13 starting at 21, and so on. □

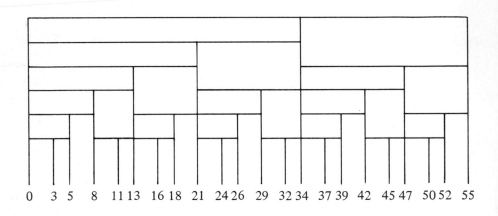

Fig. 12.15. Division of a heap according to the Fibonacci buddy system.

Allocating Blocks

If we require a block of size n, we choose any one from the available list of blocks of size s_i, where $s_i \geq n$ and either $i = 1$ or $s_{i-1} < n$; that is, we choose a best fitting block. In a kth order buddy system, if no blocks of size s_i exist, we may choose a block of size s_{i+1} or s_{i+k+1} to split, as one of the resulting blocks will be of size s_i in either case. If no blocks in either of these sizes exist, we may create one by applying this splitting strategy recursively for size s_{i+1}.

There is a small catch, however. In a kth order system, we may not split blocks of size $s_1, s_2, \ldots, s_k$, since these would result in a block whose size is smaller than s_1. Rather we must use the block whole, if no smaller block is available. This problem does not come up if $k=0$, i.e., in the exponential buddy system. It could be alleviated in the Fibonacci buddy system if we start with $s_1 = 1$, but that choice may not be acceptable since blocks of size 1 (byte or word, perhaps) could be too small to hold a pointer and a full/empty bit.

Returning Blocks to Available Storage

When a block becomes available for reuse, we can see one of the advantages of the buddy system. We can sometimes reduce fragmentation by combining the newly available block with its buddy, if the buddy is also available.† In fact, should that be the case, we can combine the resulting block with its buddy, if that buddy is empty, and so on. The combination of empty buddies takes only a constant amount of time, and thus is an attractive alternative to periodic mergers of adjacent empty blocks, suggested in the previous section, which takes time proportional to the number of empty blocks.

The exponential buddy system makes the locating of buddies especially easy. If we have just returned the block of size 2^i beginning at $p2^i$, its buddy is at $(p+1)2^i$ if p is even, and at $(p-1)2^i$ if p is odd.

For a buddy system of order $k \geq 1$, finding buddies is not that simple. To make it easier, we shall store certain pieces of information in each block.

1. A full/empty bit, as every block has.
2. The *size index*, which is that integer i such that the block is of size s_i.
3. The *left buddy count*, described below.

In each pair of buddies, one (the *left buddy*) is to the left of the other (the *right buddy*). Intuitively, the left buddy count of a block tells how many times consecutively it is all or part of a left buddy. Formally, the entire heap, treated as a block of size s_n has a left buddy count of 0. When we divide any block of size s_{i+1}, with left buddy count b, into blocks of size s_i and s_{i-k}, which are the left and right buddies respectively, the left buddy gets a left buddy count of $b+1$, while the right gets a left buddy count of 0, independent of b. For example, in Fig. 12.15, the block of size 3 beginning at 0 has a left buddy count of 6, and the block of size 3 beginning at 13 has a left buddy count of 2.

In addition to the above information, empty blocks, but not utilized ones, have forward and backward pointers for the available list of the appropriate size. The bidirectional pointers make mergers of buddies, which requires deletion from available lists, easy.

The way we use this information is as follows. Suppose k is the order of the buddy system. Any block beginning at position p with a left buddy count of 0 is a right buddy. Thus, if it has size index j, its left buddy is of size s_{j+k} and begins at position $p-s_{j+k}$. If the left buddy count is greater than 0, then the block is left buddy to a block of size s_{j-k}, which is located beginning at position $p+s_j$.

If we combine a left buddy of size s_i, having a left buddy count of b, with a right buddy of size s_{i-k}, the resulting block has size index $i+1$, begins at the same position as the block of size s_i, and has a left buddy count $b-1$. Thus, the necessary information can be maintained easily when we merge two empty

---

† As in the previous section, we must assume that one bit of each block is reserved to tell whether the block is in use or empty.

buddies. The reader may check that the information can be maintained when we split an empty block of size s_{i+1} into a used block of size s_i and an empty one of size s_{i-k}.

If we maintain all this information, and link the available lists in both directions, we spend only a constant amount of time on each split of a block into buddies or merger of buddies into a larger block. Since the number of mergers can never exceed the number of splits, the total amount of work is proportional to the number of splits. It is not hard to recognize that most requests for an allocated block require no splits at all, since a block of the correct size is already available. However, there are bizarre situations where each allocation requires many splits. The most extreme example is where we repeatedly request a block of the smallest size, then return it. If there are n different sizes, we require at least n/k splits in a kth order buddy system, which are then followed by n/k merges when the block is returned.

12.6 Storage Compaction

There are times when, even after merging all adjacent empty blocks, we cannot satisfy a request for a new block. It could be, of course, that there simply is not the space in the entire heap to provide a block of the desired size. But more typical is a situation like Fig. 12.11, where although there are 2200 bytes available, we cannot satisfy a request for a block of more than 1000. The problem is that the available space is divided among several noncontiguous blocks. There are two general approaches to this problem.

1. Arrange that the available space for a datum can be composed of several empty blocks. If we do so, we may as well require that all blocks are the same size and consist of space for a pointer and space for data. In a used block, the pointer indicates the next block used for the datum and is null in the last block. For example, if we were storing data whose size frequently was small, we might choose blocks of sixteen bytes, with four bytes used for a pointer and twelve for data. If data items were usually long, we might choose blocks of several hundred bytes, again allocating four for a pointer and the balance for data.

2. When merging adjacent empty blocks fails to provide a sufficiently large block, move the data around in the heap so all full blocks are at the left (low position) end, and there is one large available block at the right.

Method (1), using chains of blocks for a datum, tends to be wasteful of space. If we choose a small block size, we use a large fraction of space for "overhead," the pointers needed to maintain chains. If we use large blocks, we shall have little overhead, but many blocks will be almost wasted, storing a little datum. The only situation in which this sort of approach is to be preferred is when the typical data item is very large. For example, many file systems work this way, dividing the heap, which is typically a disk unit, into equal-sized blocks, of say 512 to 4096 bytes, depending on the system. As many files are much longer than these numbers, there is not too much wasted space, and pointers to the blocks composing a file take relatively little space. Allocation of space under this discipline is relatively straightforward, given what has been said in previous sections, and we shall not discuss the matter further here.

The Compaction Problem

The typical problem we face is to take a collection of blocks in use, as suggested by Fig. 12.16(a), each of which may be of a different size and pointed to by more than one pointer, and slide them left until all available space is at the right end of the heap, as shown in Fig. 12.16(b). The pointers must continue to point to the same data as before, naturally.

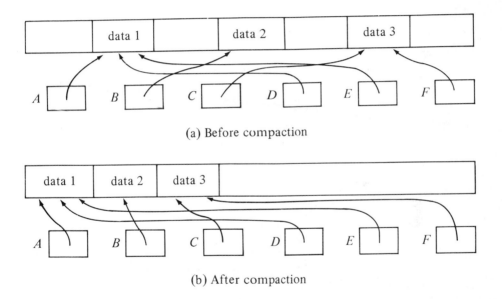

(a) Before compaction

(b) After compaction

Fig. 12.16. The storage compaction process.

There are some simple solutions to this problem if we allocate a little extra space in each block, and we shall discuss another, more complicated method that is efficient, yet requires no extra space in utilized blocks beyond what is required for any of the storage management schemes we have discussed, namely a full/empty bit and a count indicating the size of the block.

A simple scheme for compacting is first to scan all blocks from the left, whether full or empty, and compute a "forwarding address" for each full block. The *forwarding address* of a block is its present position minus the sum of all the empty space to its left, that is, the position to which the block should be moved eventually. It is easy to calculate forwarding addresses. As we scan blocks from the left, accumulate the amount of empty space we see and subtract this amount from the position of each block we see. The algorithm is sketched in Fig. 12.17.

```
(1)      var
                p: integer; { the position of the current block }
                gap: integer; { the total amount of empty space seen so far }
         begin
(2)             p := left end of heap;
(4)             gap := 0;
(5)             while p ≤ right end of heap do begin
                    { let p point to block B }
(6)                 if B is empty then
(7)                     gap := gap + count in block B
                    else { B is full }
(8)                     forwarding address of B := p−gap;
(9)                 p := p + count in block B
                end
         end;
```

Fig. 12.17. Computation of forwarding addresses.

Having computed forwarding addresses, we then look at all pointers to the heap.[†] We follow each pointer to some block B and replace the pointer by the forwarding address found in block B. Finally, we move all full blocks to their forwarding addresses. This process is similar to Fig. 12.17, with line (8) replaced by

$$\textbf{for } i := p \textbf{ to } p-1 + \text{count in } B \textbf{ do}$$
$$heap[i-gap] := heap[i];$$

to move block B left by an amount gap. Note that the movement of full blocks, which takes time proportional to the amount of the heap in use, will likely dominate the other costs of the compaction.

Morris' Algorithm

F. L. Morris discovered a method for compacting a heap without using space in blocks for forwarding addresses. It does, however, require an endmarker bit associated with each pointer and with each block to indicate the end of a chain of pointers. The essential idea is to create a chain of pointers emanating

---

† In all that follows we assume the collection of such pointers is available. For example, a typical Snobol implementation stores pairs consisting of a variable name and a pointer to the value for that name in a hash table, with the hash function computed from the name. Scanning the whole hash table allows us to visit all pointers.

from a fixed position in each full block and linking all the pointers to that block. For example, we see in Fig. 12.16(a) three pointers, A, D, and E, pointing to the leftmost full block. In Fig. 12.18, we see the desired chain of pointers. A chunk of the data of size equal to that of a pointer has been removed from the block and placed at the end of the chain, where pointer A used to be.

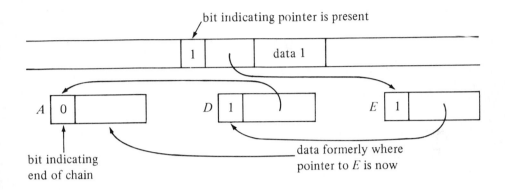

Fig. 12.18. Chaining pointers.

The method for creating such chains of pointers is as follows. We scan all the pointers in any convenient order. Suppose we consider a pointer p to block B. If the endmarker bit in block B is 0, then p is the first pointer found that points to B. We place in p the contents of those positions of B used for the pointer chain, and we make these positions of B point to p. Then we set the endmarker bit in B to 1, indicating it now has a pointer, and we set the endmarker bit in p to 0, indicating the end of the pointer chain and the presence of the displaced data.

Suppose now that when we first consider pointer p to block B the endmarker bit in B is 1. Then B already has the head of a chain of pointers. We copy the pointer in B into p, make B point to p, and set the endmarker bit in p to 1. Thus we effectively insert p at the head of the chain.

Once we have all the pointers to each block linked in a chain emanating from that block, we can move the full blocks far left as possible, just as in the simpler algorithm previously discussed. Lastly, we scan each block in its new position and run down its chain of pointers. Each pointer encountered is made to point to the block in its new position. When we encounter the end of the chain, we restore the data from B, held in the last pointer, to its rightful place in block B and set the endmarker bit in the block to 0.

Exercises

12.1 Consider the following heap of 1000 bytes, where blank blocks are in
use, and the labeled blocks are linked on a free list in alphabetical
order. The numbers indicate the first byte in each block.

0	100	200	400	500	575	700	850	900	999
a		b		c	d		e	f	

Suppose the following requests are made:

  i)   allocate a block of 120 bytes
  ii) allocate a block of 70 bytes
  iii) return to the front of the available list the block in bytes 700-849
  iv) allocate a block of 130 bytes.

Give the free list, in order, after executing the above sequence of
steps, assuming free blocks are selected by the strategy of

  a) first fit
  b) best fit.

12.2 Consider the following heap in which blank regions are in use and
labeled regions are empty.

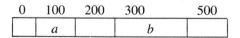

0	100	200	300	500
	a		b	

Give sequences of requests that can be satisfied if we use

  a) first fit but not best fit
  b) best fit but not first fit.

*12.3 Suppose we use an exponential buddy system with sizes 1, 2, 4, 8,
and 16 on a heap of size 16. If we request a block of size n, for
$1 \le n \le 16$, we must allocate a block of size 2^i, where
$2^{i-1} < n \le 2^i$. The unused portion of the block, if any, cannot be
used to satisfy any other request. If we need a block of size 2^i,
$i < 4$, and no such free block exists, then we first find a block of size
2^{i+1} and split it into two equal parts. If no block of size 2^{i+1} exists,
we first find and split a free block of size 2^{i+2}, and so on. If we find
ourselves looking for a free block of size 32, we fail and cannot
satisfy the request. For the purposes of this question, we never combine adjacent free blocks in the heap.

There are sequences of requests $a_1, a_2, \ldots, a_n$ whose sum is less than 16, such that the last request cannot be satisfied. For example, consider the sequence 5, 5, 5. The first request causes the initial block of size 16 to be split into two blocks of size 8 and one of them is used to satisfy the request. The remaining free block of size 8 satisfies the second request, and there is no free space to satisfy the third request.

Find a sequence a_1, a_2, . . . , a_n of (not necessarily identical) integers between 1 and 16, whose sum is as small as possible, such that, treated as a sequence of requests for blocks of size a_1, a_2, . . . , a_n, the last request cannot be satisfied. Explain why your sequence of requests cannot be satisfied, but any sequence whose sum is smaller can be satisfied.

12.4 Consider compacting memory while managing equal-sized blocks. Assume each block consists of a data field and a pointer field, and that we have marked every block currently in use. The blocks are currently located between memory locations a and b. We wish to relocate all active blocks so that they occupy contiguous memory starting at a. In relocating a block remember that the pointer field of any block pointing to the relocated block must be updated. Design an algorithm for compacting the blocks.

12.5 Consider an array of size n. Design an algorithm to shift all items in the array k places cyclically counterclockwise with only constant additional memory independent of k and n. *Hint.* Consider what happens if we reverse the first k elements, the last $n-k$ elements, and then finally the entire array.

12.6 Design an algorithm to replace a substring y of a string xyz by another substring y' using as little additional memory as possible. What is the time and space complexity of your algorithm?

12.7 Write a program to make a copy of a given list. What is the time and space complexity of your program?

12.8 Write a program to determine whether two lists are identical. What is the time and space complexity of your program?

12.9 Implement Morris' heap compaction algorithm of Section 12.6.

*12.10 Design a storage allocation scheme for a situation in which memory is allocated and freed in blocks of lengths one and two. Give bounds on how well your algorithm works.

Bibliographic Notes

Efficient storage management is a central concern in many programming languages, including Snobol [Farber, Griswold, and Polonsky (1964)], Lisp [McCarthy (1965)], APL [Iverson (1962)], and SETL [Schwartz (1973)]. Nicholls [1975] and Pratt [1975] discuss storage management techniques in the context of programming language compilation.

The buddy system of storage allocation was first published by Knowlton [1965]. Fibonacci buddy systems were studied by Hirschberg [1973].

The elegant marking algorithm for use in garbage collection was discovered by Peter Deutsch (Deutsch and Bobrow [1966]) and by Schorr and Waite [1965]. The heap compaction scheme in Section 12.6 is from Morris [1978].

Robson [1971] and Robson [1974] analyzes the amount of memory needed for dynamic storage allocation algorithms. Robson [1977] presents a bounded workspace algorithm for copying cyclic structures. Fletcher and Silver [1966] contains another solution to Exercise 12.5 that uses little additional memory.

Bibliography

Adel'son-Velskii, G. M., and Y. M. Landis [1962]. "An algorithm for the organization of information," *Dokl. Akad. Nauk SSSR* **146**, pp. 263-266. English translation in *Soviet Math. Dokl.* **3**, pp. 1259-1262.

Aho, A. V., M. R. Garey, and J. D. Ullman [1972]. "The transitive reduction of a directed graph," *SIAM J. Computing* **1**:2, pp. 131-137.

Aho, A. V., J. E. Hopcroft, and J. D. Ullman [1974]. *The Design and Analysis of Computer Algorithms,"* Addison-Wesley, Reading, Mass.

Aho, A. V., and N. J. A. Sloane [1973]. "Some doubly exponential sequences," *Fibonacci Quarterly* **11**:4, pp. 429-437.

Aho, A. V., and J. D. Ullman [1977]. *Principles of Compiler Design,* Addison-Wesley, Reading, Mass.

Bayer, R., and E. M. McCreight [1972]. "Organization and maintenance of large ordered indices," *Acta Informatica* **1**:3, pp. 173-189.

Bellman, R. E. [1957]. *Dynamic Programming,* Princeton University Press, Princeton, N. J.

Bentley, J. L. [1982]. *Writing Efficient Programs,* Prentice-Hall, Englewood Cliffs, N. J.

Bentley, J. L., D. Haken, and J. B. Saxe [1978]. "A general method for solving divide-and-conquer recurrences," CMU-CS-78-154, Dept. of CS, Carnegie-Mellon Univ., Pittsburg, Pa.

Berge, C. [1957]. "Two theorems in graph theory," *Proc. National Academy of Sciences* **43**, pp. 842-844.

Berge, C. [1958]. *The Theory of Graphs and its Applications,* Wiley, N. Y.

Birtwistle, G. M., O.-J. Dahl, B. Myhrhaug, and K. Nygaard [1973]. *SIMULA Begin,* Auerbach Press, Philadelphia, Pa.

Blum, M., R. W. Floyd, V. R. Pratt, R. L. Rivest, and R. E. Tarjan [1972]. "Time bounds for selection," *J. Computer and System Sciences* **7**:4, pp. 448-461.

Boruvka, O. [1926]. "On a minimal problem," *Práce Moraské Pridovedecké Spolecnosti* **3**:3, pp. 37-58.

Brooks, F. P. [1974]. *The Mythical Man Month*, Addison-Wesley, Reading, Mass.

Carter, J. L., and M. N. Wegman [1977]. "Universal classes of hash functions," *Proc. Ninth Annual ACM Symp. on Theory of Computing*, pp. 106-112.

Cheriton, D., and R. E. Tarjan [1976]. "Finding minimum spanning trees," *SIAM J. Computing* **5**:4, pp. 724-742.

Cocke, J., and F. E. Allen [1976]. "A program data flow analysis procedure," *Comm. ACM* **19**:3, pp. 137-147.

Coffman, E. G. (ed.) [1976]. *Computer and Job Shop Scheduling Theory*, John Wiley and Sons, New York.

Comer, D. [1979]. "The ubiquitous B-tree," *Computing Surveys* **11**, pp. 121-137.

Cooley, J. M., and J. W. Tukey [1965]. "An algorithm for the machine calculation of complex Fourier series," *Math. Comp.* **19**, pp. 297-301.

DBTG [1971]. *CODASYL Data Base Task Group April 1971 Report*, ACM, New York.

Demers, A., and J. Donahue [1979]. "Revised report on RUSSELL," TR79-389, Dept. of Computer Science, Cornell Univ., Ithaca, N. Y.

Deo, N. [1975]. *Graph Theory with Applications to Engineering and Computer Science*, Prentice-Hall, Englewood Cliffs, N. J.

Deutsch, L. P., and D. G. Bobrow [1966]. "An efficient incremental automatic garbage collector," *Comm. ACM* **9**:9, pp. 522-526.

Dijkstra, E. W. [1959]. "A note on two problems in connexion with graphs," *Numerische Mathematik* **1**, pp. 269-271.

Edmonds, J. [1965]. "Paths, trees, and flowers," *Canadian J. Math* **17**, pp. 449-467.

Even, S. [1980]. *Graph Algorithms*, Computer Science Press, Rockville, Md.

Even, S., and O. Kariv [1975]. "An $0(n^{2.5})$ algorithm for maximum matching in general graphs," *Proc. IEEE Sixteenth Annual Symp. on Foundations of Computer Science*, pp. 100-112.

Farber, D., R. E. Griswold, and I. Polonsky [1964]. "SNOBOL, a string manipulation language," *J. ACM* **11**:1, pp. 21-30.

Fischer, M. J. [1972]. "Efficiency of equivalence algorithms," in *Complexity of Computer Computations* (R. E. Miller and J. W. Thatcher, eds.) pp. 153-168.

Fletcher, W., and R. Silver [1966]. "Algorithm 284: interchange of two blocks of data," *Comm. ACM* **9**:5, p. 326.

Floyd, R. W. [1962]. "Algorithm 97: shortest path," *Comm. ACM* **5**:6, p. 345.

Floyd, R. W. [1964]. "Algorithm 245: treesort 3," *Comm. ACM* **7**:12, p. 701.

Floyd, R. W., and A. Smith [1973]. "A linear time two-tape merge," *Inf. Processing letters* **2**:5, pp. 123-126.

Ford, L. R., and S. M. Johnson [1959]. "A tournament problem," *Amer. Math. Monthly* **66**, pp. 387-389.

Frazer, W. D., and A. C. McKellar [1970]. "Samplesort: a sampling approach to minimal tree storage sorting," *J. ACM* **17**:3, pp. 496-507.

Fredkin, E. [1960]. "Trie memory," *Comm. ACM* **3**:9, pp. 490-499.

Friend, E. H. [1956]. "Sorting on electronic computer systems," *J. ACM* **3**:2, pp. 134-168.

Fuchs, H., Z. M. Kedem, and S. P. Uselton [1977]. "Optimal surface reconstruction using planar contours," *Comm. ACM* **20**:10, pp. 693-702.

Garey, M. R., and D. S. Johnson [1979]. *Computers and Intractability: a Guide to the Theory of NP-Completeness,* Freeman, San Francisco.

Geschke, C. M., J. H. Morris, Jr., and E. H. Satterthwaite [1977]. "Early expreience with MESA," *Comm. ACM* **20**:8, pp. 540-552.

Gotlieb, C. C., and L. R. Gotlieb [1978]. *Data Types and Data Structures,* Prentice-Hall, Englewood Cliffs, N. J.

Greene, D. H., and D. E. Knuth [1983]. *Mathematics for the Analysis of Algorithms,* Birkhauser, Boston, Mass.

Gudes, E., and S. Tsur [1980]. "Experiments with B-tree reorganization," *ACM SIGMOD Symposium on Management of Data,* pp. 200-206.

Hall, M. [1948]. "Distinct representatives of subsets," *Bull. AMS* **54**, pp. 922-926.

Harary, F. [1969]. *Graph Theory,* Addison-Wesley, Reading, Mass.

Hirschberg, D. S. [1973]. "A class of dynamic memory allocation algorithms," *Comm. ACM* **16**:10, pp. 615-618.

Hoare, C. A. R. [1962]. "Quicksort," *Computer J.* **5**:1, pp. 10-15.

Hoare, C. A. R., O.-J. Dahl, and E. W. Dijkstra [1972]. *Structured Programming,* Academic Press, N. Y.

Hopcroft, J. E., and R. M. Karp [1973]. "An $n^{5/2}$ algorithm for maximum matchings in bipartite graphs," *SIAM J. Computing* **2**:4, pp. 225-231.

Hopcroft, J. E., and R. E. Tarjan [1973]. "Efficient algorithms for graph manipulation," *Comm. ACM* **16**:6, pp. 372-378.

Hopcroft, J. E., and J. D. Ullman [1973]. "Set merging algorithms," *SIAM J. Computing* **2**:4, pp. 294-303.

Huffman, D. A. [1952]. "A method for the construction of minimum-redundancy codes," *Proc. IRE* **40**, pp. 1098-1101.

Hunt, J. W., and T. G. Szymanski [1977]. "A fast algorithm for computing longest common subsequences," *Comm. ACM* **20**:5, pp. 350-353.

Iverson, K. [1962]. *A Programming Language,* John Wiley and Sons, New York.

Johnson, D. B. [1975]. "Priority queues with update and finding minimum spanning trees," *Inf. Processing Letters* **4**:3, pp. 53-57.

Johnson, D. B. [1977]. "Efficient algorithms for shortest paths is sparse networks," *J. ACM* **24**:1, pp. 1-13.

Karatsuba, A., and Y. Ofman [1962]. "Multiplication of multidigit numbers on automata," *Dokl. Akad. Nauk SSSR* **145**, pp. 293-294.

Kernighan, B. W., and P. J. Plauger [1974]. *The Elements of Programming Style,* McGraw-Hill, N. Y.

Kernighan, B. W., and P. J. Plauger [1981]. *Software Tools in Pascal,* Addison-Wesley, Reading, Mass.

Knowlton, K. C. [1965]. "A fast storage allocator," *Comm. ACM* **8**:10, pp. 623-625.

Knuth, D. E. [1968]. *The Art of Computer Programming Vol. I: Fundamental Algorithms,* Addison-Wesley, Reading, Mass.

Knuth, D. E. [1971]. "Optimum binary search trees," *Acta Informatica* **1**:1, pp. 14-25.

Knuth, D. E. [1973]. *The Art of Computer Programming Vol. III: Sorting and Searching,* Addison-Wesley, Reading, Mass.

Knuth, D. E. [1981]. *TEX and Metafont, New Directions in Typesetting,* Digital Press, Bedford, Mass.

Kruskal, J. B. Jr. [1956]. "On the shortest spanning subtree of a graph and the traveling salesman problem," *Proc. AMS* **7**:1, pp. 48-50.

Lin, S., and B. W. Kernighan [1973]. "A heuristic algorithm for the traveling salesman problem," *Operations Research* **21**, pp. 498-516.

Liskov, B., A. Snyder, R. Atkinson, and C. Scaffert [1977]. "Abstraction mechanisms in CLU," *Comm. ACM* **20**:8, pp. 564-576.

Liu, C. L. [1968]. *Introduction to Combinatorial Mathematics,* McGraw-Hill, N. Y.

Lueker, G. S. [1980]. "Some techniques for solving recurrences," *Computing Surveys*, **12**:4, pp. 419-436.

Lum, V., and H. Ling [1970]. "Multi-attribute retrieval with combined indices," *Comm. ACM* **13**:11, pp. 660-665.

Maurer, W. D., and T. G. Lewis [1975]. "Hash table methods," *Computing Surveys* **7**:1, pp. 5-20.

McCarthy, J. et al. [1965]. *LISP 1.5 Programmers Manual*, MIT Press, Cambridge, Mass.

Micali, S., and V. V. Vazirani [1980]. "An $O(\sqrt{|V|} \cdot |E|)$ algorithm for finding maximum matching in general graphs," *Proc. IEEE Twenty-first Annual Symp. on Foundations of Computer Science*, pp. 17-27.

Moenck, R., and A. B. Borodin [1972]. "Fast modular transforms via division," *Proc. IEEE Thirteenth Annual Symp. on Switching and Automata Theory*, pp. 90-96.

Morris, F. L. [1978]. "A time- and space-efficient garbage compaction algorithm," *Comm. ACM* **21**:8, pp. 662-665.

Morris, R. [1968]. "Scatter storage techniques," *Comm. ACM* **11**:1, pp. 35-44.

Moyles, D. M., and G. L. Thompson [1969]. "Finding a minimum equivalent graph of a digraph," *J. ACM* **16**:3, pp. 455-460.

Nicholls, J. E. [1975]. *The Structure and Design of Programming Languages*, Addison-Wesley, Reading, Mass.

Nievergelt, J. [1974]. "Binary search trees and file organization," *Computer Surveys* **6**:3, pp. 195-207.

Papadimitriou, C. H., and K. Steiglitz [1982]. *Combinatorial Optimization: Algorithms and Complexity*, Prentice-Hall, Englewood Cliffs, N. J.

Parker, D. S. Jr. [1980]. "Conditions for optimality of the Huffman algorithm," *SIAM J. Computing* **9**:3, pp. 470-489.

Perl, Y., A. Itai, and H. Avni [1978]. "Interpolation search—a log log n search," *Comm. ACM* **21**:7, pp. 550-553.

Peterson, W. W. [1957]. "Addressing for random access storage," *IBM J. Res. and Devel.* **1**:2, pp. 130-146.

Pratt, T. W. [1975]. *Programming Languages: Design and Implementation*, Prentice-Hall, Englewood Cliffs, N. J.

Pratt, V. R. [1979]. *Shellsort and Sorting Networks*, Garland, New York.

Prim, R. C. [1957]. "Shortest connection networks and some generalizations," *Bell System Technical J.* **36**, pp. 1389-1401.

Reingold, E. M. [1972]. "On the optimality of some set algorithms," *J. ACM* **19**:4, pp. 649-659.

Robson, J. M. [1971]. "An estimate of the store size necessary for dynamic storage allocation," *J. ACM* **18**:3, pp. 416-423.

Robson, J. M. [1974]. "Bounds for some functions concerning dynamic storage allocation," *J. ACM* **21**:3, pp. 491-499.

Robson, J. M. [1977]. "A bounded storage algorithm for copying cyclic structures," *Comm. ACM* **20**:6, pp. 431-433.

Sahni, S. [1974]. "Computationally related problems," *SIAM J. Computing* **3**:3, pp. 262-279.

Sammet, J. E. [1968]. *Programming Languages: History and Fundamentals,* Prentice-Hall, Englewood Cliffs, N. J.

Schkolnick, M. [1975]. "The optimal selection of secondary indices for files," *Information Systems* **1**, pp. 141-146.

Schoenhage, A., and V. Strassen [1971]. "Schnelle multiplikation grosser zahlen," *Computing* **7**, pp. 281-292.

Schorr, H., and W. M. Waite [1967]. "An efficient machine-independent procedure for garbage collection in various list structures," *Comm. ACM* **10**:8, pp. 501-506.

Schwartz, J. T. [1973]. *On Programming: An Interrum Report on the SETL Project,* Courant Inst., New York.

Sharir, M. [1981]. "A strong-connectivity algorithm and its application in data flow analysis," *Computers and Mathematics with Applications* **7**:1, pp. 67-72.

Shaw, M., W. A. Wulf, and R. L. London [1977]. "Abstraction and verification in ALPHARD: defining and specifying iteration and generators," *Comm. ACM* **20**:8, pp. 553-563.

Shell, D. L. [1959]. "A high-speed sorting procedure," *Comm. ACM* **2**:7, pp. 30-32.

Shell, D. L. [1971]. "Optimizing the polyphase sort," *Comm. ACM* **14**:11, pp. 713-719.

Singleton, R. C. [1969]. "Algorithm 347: an algorithm for sorting with minimal storage," *Comm. ACM* **12**:3, pp. 185-187.

Strassen, V. [1969]. "Gaussian elimination is not optimal," *Numerische Mathematik* **13**, pp. 354-356.

Stroustrup, B. [1982]. "Classes: an abstract data type facility for the C language," *SIGPLAN Notices* **17**:1, pp. 354-356.

Suzuki, N. [1982]. "Analysis of pointer 'rotation'," *Comm. ACM* **25**:5, pp. 330-335.

Tarjan, R. E. [1972]. "Depth first search and linear graph algorithms," *SIAM J. Computing* **1**:2, pp. 146-160.

Tarjan, R. E. [1975]. "On the efficiency of a good but not linear set merging algorithm," *J. ACM* **22**:2, pp. 215-225.

Tarjan, R. E. [1981]. "Minimum spanning trees," unpublished memorandum, Bell Laboratories, Murray Hill, N. J.

Tarjan, R. E. [1983]. *Data Structures and Graph Algorithms,* unpublished manuscript, Bell Laboratories, Murray Hill, N. J.

Ullman, J. D. [1974]. "Fast algorithms for the elimination of common subexpressions," *Acta Informatica* **2**:3, pp. 191-213.

Ullman, J. D. [1982]. *Principles of Database Systems,* Computer Science Press, Rockville, Md.

van Emde Boas, P., R. Kaas, and E. Zijlstra [1977]. "Design and implementation of an efficient priority queue structure," *Math Syst. Theory,* **10**, pp. 99-127.

Warshall, S. [1962]. "A theorem on Boolean matrices," *J. ACM* **9**:1, pp. 11-12.

Weinberg, G. M. [1971]. *The Psychology of Computer Programming,* Van Nostrand, N. Y.

Weiner, P. [1973]. "Linear pattern matching algorithms," *Proc. IEEE Fourteenth Annual Symp. on Switching and Automata Theory,* pp. 1-11.

Wexelblat, R. L. (ed.) [1981]. *History of Programming Languages,* Academic Press, N. Y.

Wiederhold, G. [1982]. *Database Design,* McGraw-Hill, New York.

Williams, J. W. J. [1964]. "Algorithm 232: Heapsort," *Comm. ACM* **7**:6, pp. 347-348.

Wirth, N. [1973]. *Systematic Programming: An Introduction,* Prentice-Hall, Englewood Cliffs, N. J.

Wirth, N. [1976]. *Algorithms + Data Structures = Programs,* Prentice-Hall, Englewood Cliffs, N. J.

Wulf, W. A., M. Shaw, P. Hilfinger, and L. Flon [1981]. *Fundamental Structures of Computer Science,* Addison-Wesley, Reading, Mass.

Yao, A. C. [1975]. "An $O(|E| \log \log |V|)$ algorithm for finding minimum spanning trees," *Inf. Processing Letters* **4**:1, pp. 21-23.

Yao, A. C., and F. F. Yao [1976]. "The complexity of searching a random ordered table," *Proc. IEEE Seventeenth Annual Symp. on Foundations of Computer Science*, pp. 173-177.

Yourdon, E., and L. L. Constantine [1975]. *Structured Design*, Yourdon, New York.

Index